Soil Science
and Archaeology

Studies in Archaeological Science

Consulting editor G. W. DIMBLEBY

Soil Science and Archaeology

SUSAN LIMBREY

University of Birmingham, England

1975

ACADEMIC PRESS

London New York San Francisco

A Subsidiary of Harcourt Brace Jovanovich, Publishers

ACADEMIC PRESS INC. (LONDON) LTD.
24/28 Oval Road,
London NW1

United States Edition published by
ACADEMIC PRESS INC.
111 Fifth Avenue
New York, New York 10003

Library of Congress Catalog Card Number: 79 183466
ISBN: 0 12 7854 77 0

Text set in 11/12 pt. Scotch Roman, printed by letterpress,
and bound in Great Britain at The Pitman Press, Bath

Preface

Interest in and involvement with soils is inherent in the practice of archaeology. It is from the soil that people derive their nourishment and it is in, under and on it that their remains are found. The archaeologist needs to take the fertility and management of the soil into account in his consideration of the economy and way of life of the people he is studying, and he needs to understand it in order to interpret material remains whose distribution and condition may have been affected by soil processes. The soil scientist needs to draw upon the time-scales, the history of land use and the evidence of past soil conditions which can be provided by archaeology. Soil science and archaeology together contribute to the study of the landscape and its population.

Man's participation in the ecosystem has been changing in its mode and intensity ever since he began to use fire and to carry about quantities of materials in excess of his own body. People then began to redistribute the nutrient elements and to change the rates of and the energy involved in the cyclic processes by which the soil makes them available. By control of his food supply and the development of his social systems man has increased the separation between the sources of nutrients and their subsequent dispersal, until now the residues from food production often constitute a massive problem in waste disposal rather than a valuable resource. Archaeology can not only illuminate the history of this process, but by showing how and where problems of instability or irreversibility were encountered can contribute to the understanding of the soil/plant system which is necessary for the maintenance of the inflated and wobbly nutrient and energy cycle which has replaced the tight little local ones.

The first part of this book is an account of what soil is and how it works. Those archaeologists who feel that their knowledge of science is not enough for it to be comprehensible may turn to the matters

which interest them in the later parts and refer back to the early
chapters when their relevance becomes apparent, seeking help from
basic science books, or from friends, where necessary. In a subject
which involves almost every science one can name, and in some of
them touches the boundaries of present knowledge, simplification
is neither easy nor advisable. It would frustrate those who do
have the necessary background, and presentation of a drastically
over-simplified account can lead to fossilization in archaeological
thinking.

In part II the soils of the cool, wet areas are described, and the
history of some of them is discussed in detail, with particular reference
to those of the British Isles. This is a rapidly developing area of study,
almost every excavation can provide some detail of land use or soil
history, and there is enormous scope for increasing co-operation
between soil scientists and archaeologists. Part III is intended to do
no more than introduce the soils of regions in which I have little or
no experience, and indicate the possibilities for studies concerning
them. Archaeologists are strongly urged to get in touch with soil
scientists who have practical experience of the soils in their own
region.

Some of the more commonly encountered problems of survey and
excavation are dealt with in part IV. This is far from being an exhaus-
tive treatment since every site presents unfamiliar situations or
new combinations of familiar circumstances. I hope, however, that
excavators will become increasingly able to sort out for themselves
problems concerning the nature of the materials they encounter and
the processes that go on in them. There should always be someone on
site who has a basic understanding of physical, chemical and biological processes; it must be recognized that the education of the non-
scientist as an archaeologist is not entirely appropriate as a training
for the excavator.

Acknowledgements

I was introduced to the study of soils by Dr. I. W. Cornwall, whose
book *Soils for the Archaeologist* has provided many archaeologists with
a valuable insight into soils, and to whom I am grateful for passing
on his enthusiastic and pragmatic approach. Professor G. W. Dimbleby
read the text of this book at an early stage and made many very
useful suggestions, as well as providing constant encouragement, for
which I would like to thank him. I would like to thank the many soil
scientists with whom I have had valuable discussions, and also a

large number of excavators, those with whom I learnt to dig, and those who have welcomed me to their sites in recent years despite the fact that I sometimes spoiled their sections and disrupted the integrity of their open area excavation by digging holes.

Susan Limbrey
May 1975

Contents

15

Soils Associated with Archaeological Features   281

16

Reclaimed and Man-made Soils   335

PART I

Soil Materials and Processes

1

Mineral Components of Soil

The rocks of the earth's crust provide the *parent materials* from which the mineral components of soils are derived. Rocks and sediments are made up of crystalline and amorphous minerals which undergo alteration in the soil at rates which depend on their chemical composition and crystal structure and on the physical and chemical conditions in the soil. *Secondary minerals* are formed in the soil by re-arrangement and re-combination of the elements of the *primary minerals* as they undergo alteration. Water is the medium in which these changes occur, and in which life processes in the soil are carried on: without water, no soil forms.

The formation of secondary minerals does not involve all the elements released from the primary ones, and elements in solution in the water in the soil are available to plants and are taken in through their roots. Those elements whose abundance and solubility ensure a reliable supply in soils are the major mineral nutrients of plants, and are also essential to animals. Many other elements are needed by plants and animals in small quantities, and these are mainly ones whose mobility is low or whose abundance in rocks is either generally low or very variable. Table 1 shows what the plant and animal nutrients are and the approximate proportions in which they are needed.

Elements sufficiently mobile in the soil to be taken in by plant roots are also liable to be moved downwards through the soil, and eventually beyond reach of the roots and into the groundwater, by the percolation of water through it. The process is known as *leaching*. Elements leached from soils provide nutriment for living things in rivers, lakes and oceans, so soil processes support life in other than terrestrial situations, and far from the places in which they function.

Table 1. Elements which plants and animals require from or via the soil

	Plants	Animals
Nitrogen	M	M
Phosphorus	M	M
Potassium	M	M
Calcium	i–M	M
Magnesium	i	i
Sodium	(tr)	i
Chlorine		i
Iodine		tr
Fluorine		(tr)
Sulphur	i	tr
Iron	i	i
Manganese	i	tr
Copper	tr	tr
Zinc	tr	tr
Nickel		(tr)
Cobalt	tr	tr
Molybdenum	tr	tr
Boron	tr	
Vanadium	tr	
Selenium	(tr)	
Silicon	(tr)	(tr)

	Plants, amount taken from the soil each year, per hectare	Animals, amount needed per kilogramme of food
Macronutrients, M	tens to hundreds of kilogrammes	up to about 20 grammes
intermediate, i	up to a few kilogrammes	up to a few grammes
trace elements, tr	up to a few grammes	up to a few milligrammes

(tr) indicates elements which are only needed by some groups of plants or animals, or whose status as an essential nutrient is uncertain.

Weathering

The processes by which the parent material undergoes the alterations which produce the mineral component of soil come under the general heading of *weathering*, a term which implies that they take place under the influence of the atmosphere at the surface of the earth. It is difficult, and perhaps unneccessary, to define a boundary between weathering and soil formation, or to establish a point at which rock, undergoing physical fragmentation and the softening and disruption which are the result of the chemical processes of hydrolysis, hydration and oxidation, becomes soil.

Weathering can happen far from the point at which the sedentary soil forms on the resulting material, and it is part of the process of sorting and differentiation in a cycle of erosion and sedimentation. Physical weathering breaks solid rock into fragments by the forces of expansion and contraction on heating and cooling, by pressure when ice crystallizes in cracks, and by abrasion of the surface by finer particles carried in wind, water or ice. The sediments produced, which are called *clastic sediments*, may be scree from a rock face, moved only by gravity, or those aeolian, fluviatile or glacial deposits which form at greater distance from the source of the material. Chemical weathering is the result of interaction of the rock with rain water and the atmosphere. It can happen before or after physical weathering, and the two go hand in hand, each opening up further opportunity for the other.

Minerals interact with water by solution, hydrolysis, hydration and oxidation. It is by these reactions that the minerals are adapted to the cold, wet, acid and oxidized environment of the earth's surface: those that crystallize under conditions of high temperature and pressure in the formation of igneous and metamorphic rocks are unstable under normal surface conditions, and adjust, though mostly very slowly, by these reactions. Hydrolysis is chemical combination with the hydrogen, H^+, and hydroxyl, OH^-, ions which together form the water molecule, and when this happens at the surface of a crystal it modifies the ionic arrangements there, changing the bonding of the surface ions with those further in, so that they are less strongly bound and more available to further chemical reactions. Hydration involves entry of whole water molecules into the crystal; the re-organization needed to admit the molecules, in spacing and co-ordination, causes physical disruption, and the presence of water in situations where it can react further with the less strongly bound ions leads the way to further modification. Both hydration and hydrolysis prepare mineral surfaces and cleavage planes for oxidation; re-arrangements in co-ordination to accommodate the valency change which accompanies oxidation of iron and other multivalent elements causes further disruption. A weathered rock therefore tends to be less coherent and compact than a fresh one, more water and air can get in, and plant roots can penetrate.

Rainwater is always slightly acidified by the carbon dioxide it picks up in passing through the atmosphere. All the interactions of minerals with the water are greatly increased by this acidity. In a rock which is undergoing alteration beneath or within a soil, the solution reaching it is further modified by soil and plant substances, and by the consumption of some of the oxygen dissolved in it by soil organisms, and so the rates of, and the balances between, these reactions are changed.

Parent Materials

It has been said (Avery, 1972) that the parent material of a soil cannot be known, only inferred, since it no longer exists, An understanding of the development and history of a soil requires knowledge about those minerals no longer present, which have contributed their constituent elements to the growth of plants, to the formation of other minerals, and to the groundwater, but to assume that the rock or sediment forming the substratum to the profile is identical to that which formed its parent material can be misleading. All soils receive a continual or intermittent supply of dust, borne on the wind, or on the feet, fur and feathers of animals and birds. The dust contains mineral and organic substances, and the mineral component may be like or unlike that of the local soils. Some soils receive regular accretion to their surface of water borne materials; others receive materials moving downslope, perhaps from soils forming on different rocks cropping out further uphill. Where soils are forming on stratified materials, soil formation may so involve strata at the surface that they are no longer apparent, and those lower in the succession which remain identifiable below the soil appear to be its parent material.

The influence of parent material on soil is expressed in an interaction of physical and chemical factors. The parent material can only contribute to the soil those elements it contains, but the rate at which they become available, and the proportions in which they appear, depend on the mineralogy of the rock and the way in which different minerals yield to attack under the conditions of the soil. Minerals undergo attack by soil solution while still embedded in coherent rock and when loose in the soil. The texture of the rock and the nature of the binding mechanism which keeps its component particles stuck together controls the degree to which moisture has access to minerals in the interior, and to which the products of the interactions can get out: the equilibrium of a reaction depends on bringing new supplies of the reagents into contact and on removing the products.

Mineral grains in some rocks are entirely separate, and penetration of moisture along grain boundaries, dissolving a cementing substance or the surfaces of the grains themselves, will break up the rock; on the other hand, precipitation of an immobile product of the reaction may prevent access of more moisture. Thus, primary minerals may be released into the soil, ready to enter into soil processes, or may be trapped in a piece of rock, in the body of the soil or still attached to the underlying rock, and not released until alteration of the entire rock fabric leads to complete disaggregation.

The grain size of the crystals in the rock powerfully influences the rate of attack, since the ratio of surface area to volume is greater the smaller the particle: fine grained rocks present much more surface to attack than do coarse grained ones. The distribution of the various minerals relative to each other may also be important. Proximity of different minerals may result in release into immediate contact of materials which can then interact. An extreme case is the intergrowth of different minerals in composite crystals, but this can also result in some minerals being entirely enclosed within crystals of larger, perhaps more resistant, minerals, and so protected.

The crystal structure of individual minerals is the fundamental feature which decides the rate at which the elements they contain can be released into the soil or incorporated into secondary minerals. The arrangement of atoms building a crystal decides the strength of the crystal against disruption, and its resistance to chemical attack. Cleavage planes, along which a crystal will most easily break, are usually associated with the presence of divalent cations which are very reactive in soil conditions, so physical disruption and chemical susceptibility are closely related. Attack on the outside surface depends on the arrangements of ions, whether it is the more susceptible ions which are close to the surface or whether they are protected by more stable components, and the strengths of the bonds which hold them into the crystal.

Types of Rock

Igneous Rocks

The minerals of igneous rocks crystallize under high temperature or pressure or both. The chemical components are those of the magma, sorted by gravity into the more acid magmas rich in the lighter elements and the more basic magmas rich in the heavier elements. The

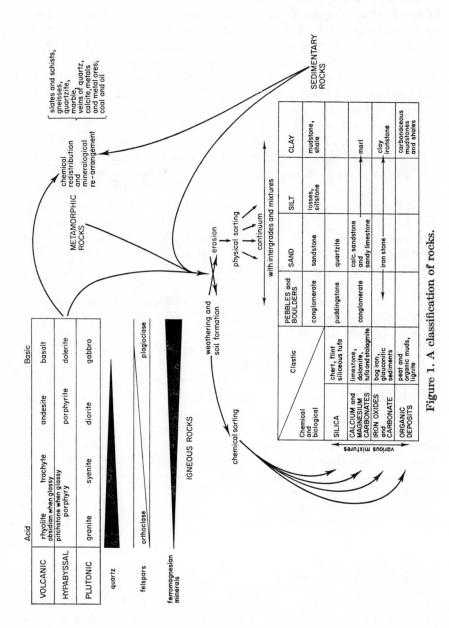

Figure 1. A classification of rocks.

minerals into which the elements present in the magma sort themselves during crystallization, and the crystal sizes, depend on the temperature and pressure at which they crystallize and on the rate of cooling.

Plutonic rocks crystallize under high pressure, and very slowly, deep beneath the earth's surface. They tend to crystallize completely, in a mosaic of large, uniform sized crystals which grow until they touch, having used up all the magma.

Volcanic rocks crystallize under atmospheric pressure and cool fast. They often contain "phenocrysts" of those minerals which crystallize early while the lava is still very hot, set in a fine grained matrix, a mass of small crystals. The crystals have boundaries representing the true crystal shape and between them is residual material in the form of glass, an unorganized solid with the atoms in random order, which remains after the magma has cooled too much for any further diffusion of ions towards growing crystals. Acid volcanic rocks may be entirely glassy or have only tiny proto-crystals representing the earliest stages of crystallization.

Intermediate in character between the plutonic and volcanic rocks are the *hypabyssal* ones. These are formed from volcanic magma which never quite gets out, being forced by erupting volcanoes into cracks and parting planes in the country rock or remaining in the neck of the volcano, and there solidifying under pressure rather above atmospheric and cooling more slowly than at the surface. In these rocks the porphyritic texture of large phenocrysts in a fine grained matrix reaches its fullest expression, and the groundmass is fully crystallized before solidification. Hypabyssal rocks are often modified by the effects which their own heat has on the surrounding rock. In the process of metamorphosing their neighbours they are themselves altered by the gaseous products of contact.

The igneous rocks contribute to the soil minerals which have crystallized under conditions very different from those in the soil and many of them are therefore relatively unstable and easily altered. Since they form from magma from deep in the earth's crust they can contain a full range of the elements of the crust and their potential for soil formation and plant nutrition can be high, particularly in the case of the more basic rocks.

Sedimentary Rocks

In this category we may include unconsolidated sediments as well as solid rocks. They have all undergone at least one cycle of weathering, and in the process have suffered a degree of sorting. Any sediment contains only part of the assemblage of elements which was present in the rock from which it is derived. Whatever the process of weathering or erosion, the more soluble components are taken, eventually, to the sea or to a lake. The most soluble remain in solution, contributing to the increasing salinity of the oceans and of the waters of closed basins, while some of the less soluble are precipitated in chemical and biological sediments.

The less soluble components of the original rock are left in clastic sediments, and physical sorting separates the coarse particles of more resistant minerals and the finer material containing more of the softer and more easily fragmented minerals. A chemical sorting follows from the physical sorting, since the resistance of the grains to fragmentation and abrasion is dependent on their crystal structure and this is related to composition: because of differences in density, small particles of heavy minerals, rich in the heavier elements, travel with larger particles of light ones.

From the same rock, therefore, may form in different places a chemical sediment and a range of clastic sediments from those containing small particles of minerals susceptible to weathering and containing abundant soil forming and plant nourishing elements, to those composed of large grains of resistant and sterile materials. Whereas the original rock would contribute the whole of its chemical riches to a soil, in the next cycle each of the sediments will lack some components, and the soils formed on them will be limited in expression of the possible range of development.

The cycle of weathering and erosion which produced a sedimentary rock may have included soil formation. In this case the range of sediments produced will include a suite of primary minerals depleted of those whose elements which were mobilized by soil formation and a suite of secondary minerals produced in the soil. Since the former will be in the larger size grades and the latter in the smaller, or in solution, they are easily separated in the sorting inherent in even the slightest degree of transport. The result will be a coarse sediment with a limited potential for further soil formation and a fine sediment already largely composed of secondary soil materials, well on the way to being a true soil. The floodloam and gravel of a river system is an example of this sorting.

Metamorphic Rocks

Some igneous rocks, especially volcanics, find themselves in situations where they undergo metamorphism, but mostly it is the sedimentary rocks which provide the raw material since it is the sedimentary basins which later suffer the uplift and folding which produce the high pressures and temperatures.

The process involves re-adaptation to high temperature and pressure of minerals which had already been partially altered at the surface so as to be more in equilibrium with atmospheric conditions. Recrystallization therefore takes rocks back towards the igneous rock's condition of potential instability, but retains the chemical sorting resulting from their surface history. The sorting is somewhat modified by the mixing up of different strata which goes on during folding and contortion, and afterwards, since different rocks which, unfolded, would not appear at the surface at the same time, are exposed in close proximity when erosion cuts across tilted and folded strata. Further modification results from the mobility within the metamorphosing rocks of liquid and gaseous materials derived from them under high temperature.

Recrystallization in response to directional pressure in solid material, rather than under hydrostatic pressure in a magma, results in minerals with strongly directional structures. It is important to soil formation on metamorphic rocks that these flat, plate-like crystals are closely related to the secondary minerals which form in soils.

Crystal Structure

In describing crystal structure it is convenient to talk about atoms as though they were solid spheres, and it helps a lot in visualizing the three dimensional structures to build them with, for example, apples: half an hour playing with a pound or two of fruit clarifies arrangements which few people can adequately construct in their minds from a written description and two dimensional drawings.

Though the solid spheres give the right spatial patterns, the properties of substances depend on the interactions and bonds between ions, and these can only be understood in the light of atomic structure. The atom is regarded as a small positively charged nucleus surrounded by a negatively charged cloud of electrons. An ion is an atom which has lost or gained electrons in the outer part of its cloud: if it has lost one or more it has an excess of positive charge and is called a *cation*; if it has gained electrons it has an excess of negative charge and is called an *anion*. The size of the charge on the nucleus decides how many electrons there should be to form a neutral atom. As the number increases, from hydrogen with only one, different numbers of electrons go to complete "shells" with different energy levels, successively farther out from the nucleus. The electron cloud is therefore not uniform but made up of different electron densities. When a shell is full another is started, though there are groups of elements which start new shells when there is still room in inner ones. It is the relative fullness of the outer shell which decides the *valency* of an element: how many electrons it can give away or accept to make an ion with a full outer shell.

Hydrogen and the metals most easily achieve complete outer shells by giving away electrons—in the case of hydrogen it is only one—to form cations. Some metals, and in rocks and soils the most important one is aluminium, can make a complete outer shell by either gaining or losing electrons, and which they do depends on whether the surrounding ions are more ready to give or receive them.

In the solid sphere concept it is the size of the electron cloud which is taken as the size of the ion, but it is the interactions between the outer parts of the clouds which decide the nature of the bonds which hold a crystal together. These bonds are in part ionic, that is, they are due to the electrostatic attraction between oppositely charged ions, and in part co-valent. In co-valent bonding, which is stronger than ionic, electrons do not pass completely from the sphere of influence of one nucleus to that of the other to form ions, but are shared between the two: the outer part of the electron cloud is common to the two ions. In crystals, the ions form bonds with all those around them to give a coherent structure. Overall valency balance is maintained but any one bond has ionic and co-valent components in proportions depending on the individual valencies, the size of the ions and how closely they are packed together.

The geometrical arrangements possible among ions of different sizes decides how many bonds ions of particular valencies can make with those around them. The resulting "co-ordination number" is conventionally expressed as the number of oxygen ions with which a given ion can surround itself. The arrangements taken up by different co-ordinations are named from the geometrical figure formed by the centres of the oxygen ions: four points equally disposed in space around a central point form a tetrahedron, so fourfold co-ordination is called tetrahedral; sixfold is octahedral (confusing, this), eight-fold is cubic and twelve-fold is dodecahedral.

The common rock-forming elements are listed in Table 2 with their ionic radii, valencies and co-ordination numbers, and their relative abundance. The multiple co-ordination numbers of the abundant elements aluminium and potassium are important in rock weathering and soil formation.

Silicate Structure

Silicate and alumino-silicate structures are based on the silica tetrahedron, in which four oxygens form bonds with one silicon. The silicon ion is small enough to fit into the space left in the middle when the oxygen ions lie in closest possible contact, all four touching in tetrahedral arrangement (Fig. 2a). The oxygens have a second valency and

Table 2. Valency, co-ordination number, ionic size and relative abundance of the predominant components of silicate minerals. (These elements make up over 95% of the earth's crust, no other element reaching 1% and most being less than 0·1%)

	Valency	Co-ordination number	Ionic radius, $\times 10^{-6}$ mm	Relative atomic abundance, %
Silicon, Si	4	4	0·41	20
Aluminium, Al	3	6, 4	0·50	6
Iron, Fe, ferric	3	6	0·64⎫	1·9
Iron, Fe, ferrous	2	6	0·76⎭	
Magnesium, Mg	2	6	0·65	1·8
Sodium, Na	1	8	0·95	2·5
Calcium, Ca	2	8	0·99	1·9
Potassium, K	1	8, 12, ?14	1·33	1·4
Oxygen, O	2		1·40	60

crystals are built up by using this in two ways. Tetrahedra can join together by sharing oxygens, so that each tetrahedron is no longer complete in itself, or they can join by using multivalent cations to form bridges between oxygens belonging to different tetrahedra. Silicate structures fall into groups according to the distribution of these two types of linkage, the stronger way building the tetrahedra into one another and the weaker making bridges between them.

Variety of composition is given in each structural group by the nature of the linking cations: calcium, magnesium and iron are the commonest. Further variety is introduced by the substitution of aluminium for silicon in some of the tetrahedra. The aluminium ion is bigger than that of silicon, and cannot quite fit into the closest packing arrangement; it pushes the oxygens further apart, reducing the strength of the bonds. The accommodation of aluminium, of valency three, leaves excess negative charge among the oxygens, and some monovalent cations such as sodium or potassium may be incorporated to take it up.

Aluminium in six-fold co-ordination is characteristic of the alumino-silicates, which are made up of layers of silica tetrahedra and layers of alumina, with aluminium in the centres of oxygen octahedra (Fig. 2b). These minerals are particularly important in soils, being formed in the soil as secondary minerals.

The structural groups of the silicates are: those with separate complete tetrahedra; those with tetrahedra joined together in pairs;

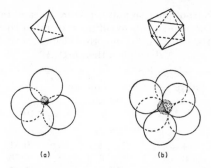

(a)   (b)

Figure 2. (a) Tetrahedral arrangement of four oxygen atoms around one silicon atom. (b) Octahedral arrangement of six oxygen atoms around one aluminium atom.

those with tetrahedra joined together in rings and in chains; and those in which there is a three-dimensional framework structure with the tetrahedra linked in all directions. The alumino-silicates, with their sheets of tetrahedra fit logically between the last two groups.

Separate Tetraheda

Olivine Group. In these minerals each tetrahedron is complete with four oxygens to each silicon, and they are linked together by the divalent cations of magnesium and ferrous iron. Not only do individual tetrahedra represent the most compact arrangement of four oxygens, but the tetrahedra pack together in a continuous structure, giving an overall closest packing of oxygens like apples in a box (Fig. 3a). In this arrangement the tetrahedral pores containing the silicon alternate with octahedral pores into which the iron or magnesium fit in six-fold co-ordination. The structure is very compact and physically strong; olivine has no good cleavage, only a tendency to break irregularly through the network of octahedral spaces. The high proportion of divalent ions to stable silica tetrahedra allows easy chemical attack at the surface, so it is highly susceptible to chemical weathering.

Variation among the olivines is only a matter of the relative proportions of iron and magnesium. When only iron is present the mineral is called fayalite, and when only magnesium, forsterite.

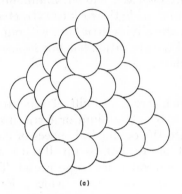

(a)

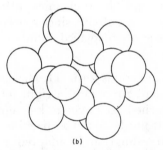

(b)

Figure 3. (a) Closest packing arrangement of oxygen atoms, giving an alternation of tetrahedral and octahedral pores. (b) Ring structure of six silica tetrahedra.

Usually there is some of each, and the name olivine covers all inter-mediates between these two end members of the "isomorphous series".

Olivine is formed early in crystallization of basic magmas, and so is common as phenocrysts in basalt. While the molten lava is seething and spurting in a crater the olivine phenocrysts which have already formed are exposed to the air, and the iron at the crystal surfaces is oxidized. At the oxidizing surface aluminium from the surrounding magma is incorporated into the crystal structure and a coating of iddingsite, a mixture of iron oxide and an alumino-silicate mineral, is formed on each olivine crystal. In old basalts the olivines are often found to be already altered within the apparently unweathered rock, and the iddingsite crystal outline is filled with a greenish alumino-silicate, chlorite. In this case, enough aluminium has been brought into the crystal, from outside the rock or extracted from neighbouring minerals, to modify the olivine structure without affecting visibly the rest of the rock, and the alteration has taken place within the confines of the crystal boundary.

Garnet Group. Garnets are also built of separate tetrahedra, linked together in this case by aluminium or ferric iron which, in six-fold co-ordination, fit into the octahedral spaces of closest packing. However, these trivalent ions cannot use up all the available pores as there are not enough oxygen valencies to go round. The balance is maintained by incorporation of some divalent ions in an eight-fold co-ordination which requires larger, cubic pores. The structure is therefore less compact. The divalent ions found in garnets are calcium, magnesium, manganese and ferrous iron. These, when exposed on a crystal surface, are susceptible to chemical attack, as in olivine, but garnet is more resistant than olivine since a porportion of surface sites is occupied by the less easily displaced aluminium and ferric iron.

Zircon. This resistant mineral is formed from separate silica tetrahedra linked by the four-valent zirconium in eight-fold co-ordination: each zirconium ion is attached to oxygens belonging to eight different tetrahedra by very strong bonds. The packing is near to the closest possible since the co-ordination is adapted to two different bond lengths, four to close oxygens and four to those further away, a compromise with the six-fold co-ordination which would best fit the packing.

Great physical strength is allied to a high degree of chemical resistance due in part to the low proportion of the metal and in part to the fact that this metal is very unreactive in both acid and alkaline conditions. The purest zircon is so stable that it is used as an index against which loss of other minerals is measured in assessing degree of weathering in a soil or sediment. It occurs in small quantities of small crystals in acid igneous rocks and in some metamorphic rocks, and its persistence leads to a wide distribution throughout the sedimentary realm.

Paired Tetrahedra

Two silica tetrahedra join together by sharing one oxygen between them, so that there are seven oxygens to two silicons. The pairs are linked together by divalent cation bridges, in the same way as the separate tetrahedra of the previous group. Minerals with this structure are not very common. Vesuvianite, a mineral which occurs in metamorphosed limestone, and has calcium, aluminium and iron as the linking cations, has a combination of separate and paired tetrahedra.

Ring Structures

A ring of three tetrahedra each sharing two oxygens with its neighbours, so that there are nine oxygens to three silicons, forms a unit which does occur, but not in any common minerals.

A six member ring (Fig. 3b), with eighteen oxygens to six silicons occurs in beryl and in tourmaline. The rings are stacked to form tubes, which are linked by cations. In beryl the linking ions are beryllium in tetrahedral spaces and aluminium in octahedral ones. In tourmaline, which is very variable in composition, the bridges between the rings in the stack are mostly magnesium, but aluminium, ferrous iron, and lithium sometimes take part, and a small proportion of boron is present; the tubes are linked together by aluminium and ferric iron in octahedral spaces; and inside the tubes are found sodium and calcium, and some fluorine and hydroxide ions can occur. The mineral, which is quite common in small quantities in many igneous and metamorphic rocks is both physically and chemically resistant and is therefore widespread in sediments and soils.

Chain Structures

If instead of closing to form rings tetrahedra continue to link by sharing oxygens with two neighbours chains of indefinite length are formed. The spare valencies of the two oxygens not taken up with chain forming are used to link the chains together through divalent cations. Chain structures are found in two groups of minerals, the single chain pyroxenes and the double chain amphiboles (Fig. 4).

Pyroxene Group. Calcium, magnesium and ferrous iron in different proportions in the bridging positions give a range of compositions in this group. Quite often the same crystal has concentric zones of different composition, formed as the composition of the residual magma changes during crystallization. Further variation is introduced by replacement of some of the silicon by aluminium, and the valency balance is maintained by incorporation of further aluminium in conjunction with ferrous iron among the bridging cations, and some sodium and lithium may occur.

The relatively weak links between chains allow ready cleavage in pyroxenes, in two planes which lie at 87° and 93° to the main crystal axis. On these planes the linking cations are exposed to chemical attack, so these minerals are susceptible to both physical and chemical

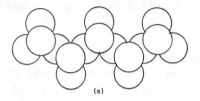

(a)

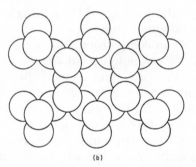

(b)

Figure 4. (a) Single chain arrangement of silica tetrahedra, as in minerals of the pyroxene group. (b) Double chain, as in minerals of the amphibole group.

weathering. This being so, their abundance in basic and intermediate igneous rocks gives them an important place in soil formation on igneous material.

The types of pyroxene are grouped first according to their crystal form and then according to composition, and in each group the members form an isomorphous series with a range of composition between the named end members. The orthorhombic pyroxenes are enstatite, which has only magnesium in bridging positions, and hypersthene, in which increasing amounts of ferrous iron occur. These have no aluminium. The monoclinic pyroxenes are: the diopside-hedenbergite series, with no aluminium and with calcium and magnesium in diopside and calcium and ferrous iron in hedenbergite; the augite series, the principal aluminous pyroxenes, with calcium, magnesium and iron in augite, and decreasing amounts of aluminium and ferric iron until pigeonite is reached; and the alkali pyroxenes aegerine and jadeite, which have some sodium or lithium and form series by increasing incorporation of aluminium to reach augite.

Amphibole Group. Chains like those of the pyroxenes are joined together to make double chains by further oxygen-sharing between alternate tetrahedra in each chain. The double chains are linked by bridging calcium, magnesium and ferrous iron, just as in the pyroxenes, but the use of some oxygens in doubling the chains leaves some excess valency among the cations, and some hydroxyl ions are incorporated to use it up; fluorine can appear in place of hydroxyl. Substitution of aluminium in tetrahedral positions gives a further valency imbalance, which is taken up by the large monovalent cations of sodium or potassium, which fit into the larger spaces between the chains. Where aluminium takes part in bridging, the valency balance is kept by incorporation of iron or manganese.

Like the pyroxenes, the amphiboles are readily cleaved between chains, the cleavages being at 56° and 124° to the main crystal axis, and the reactive cations are exposed. Their occurrence in igneous rocks and their behaviour under weathering and soil forming conditions is parallel to that of the pyroxenes, differences in composition in each group giving overlapping ranges of susceptibility. Within rocks, late changes during cooling of the magma can lead to resorption of an earlier crystallizing amphibole and its replacement within the same boundary by a pyroxene, with some left-over iron oxide crystallizing as granules of magnetite around the edge.

Isomorphous series analogous to those of the pyroxenes make up the subdivisions of the amphibole group. The hornblende series, analogous to the augite and augite-aegirine series is commonest. The other monoclinic amphiboles are the cummingtonite-grunerite series, with magnesium and iron, and the tremolite-actinolite series with calcium and magnesium and calcium, magnesium and iron. The alkali amphiboles which have sodium are arfvedsonite, glaucophane and riebeckite.

Sheet Structures

The chains of tetrahedra are joined together not just in twos as in the amphiboles but in a continuous sheet, in which three oxygens of each tetrahedron lie in the same plane and are shared with adjacent tetrahedra and the fourth sticks up, sitting above the pore in which the silicon lies, and can form attachments with other ions. In all the minerals of sheet structure the fourth oxygen is attached to another sheet, this time made of alumina octahedra, and the result is an alumino-silicate. These minerals will be described in the next section, because they include all the clay minerals, and if we leave them out of their logical position in the silicate scheme we can consider them later, when all the other silicates have been described, in the context of the way in which they form from the other silicates.

Framework Structures

When all four oxygens of each tetrahedron are shared with another a framework of indefinite extension in three dimensions is built up, consisting of an array of oxygen–silicon–oxygen links symmetrically disposed in space. Packing of the oxygens is no longer the closest possible in any part of the structure.

Felspar Group. These are the commonest rock-forming minerals, dominating the mineralogy of most igneous rocks. They are framework silicates in which a regular pattern of replacement of silicon by aluminium gives a valency excess which is taken up by incorporation of sodium, potassium and calcium into the lattice. They fall into two groups: the alkali felspars, in which a quarter of the silicons are replaced by aluminium and the compensating cations are potassium, in orthoclase, and potassium and sodium in the isomorphous series from microcline to anorthoclase; and the soda-lime, or plagioclase felspars in which the proportion of aluminium is between a quarter and a half of the tetrahedral positions. Plagioclases are isomorphous mixtures of two composition series: albite has a quarter of the silicons replaced by aluminium and contains sodium, and anorthite has half the silicons replaced by aluminium and contains calcium; oligoclase, andesine, labradorite, and bytownite are the intermediate members from the sodic to the calcic ends of the range. Near the sodic end, the structure is that of albite; increasing substitution of aluminium and incorporation of calcium leads towards anorthite, and towards that end the structure is that of anorthite; the middle members of the series have laminae, all with intermediate compositions but alternating in structure.

The felspar structures are not very compact, and big spaces occur into which the larger cations can fit. The spaces are regularly arranged, and cleavage occurs easily along planes passing through them, exposing the cations to chemical attack. Felspars are thus very susceptible to weathering and because of their abundance their behaviour under different conditions in the soil is very important in determining the formation of secondary minerals.

Silica. Silica itself has a number of different forms, all based on the tetrahedron, which are best described by considering the sequence of events when water is lost from a silica solution. In the solution the separate tetrahedra exist in fully hydrated form, that is, the second valency of each oxygen is satisfied by a bond with a hydrogen ion. There is a strong tendency for the tetrahedra to polymerize, joining up to share oxygens by eliminating one water molecule for each such link—losing both hydrogens and one of the oxygens.

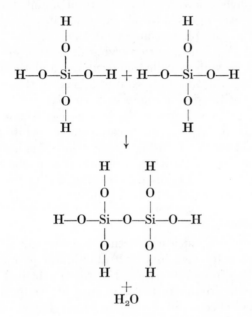

Completely anhydrous silica, that is, when polymerization is complete, can be a properly organized crystal, in which the three dimensional network is built up by spiral arrangements of the tetrahedra. This is quartz, which has different forms according to the temperature at which it crystallized. At ambient temperature crystallization is very slow, and most quartz is formed not by dehydration but by cooling from magma or by recrystallization in metamorphic rocks. When cooling is very rapid a disorganized arrangement occurs, which is silica glass, common in acid volcanic rocks. During dehydration from a silica solution, solidification occurs while a proportion of water is still present in the incompletely polymerized system, and a gel stage is reached which will readily absorb water and disperse again. Behaviour of such a system is highly dependent on the acidity, that is, the number of hydrogen ions present, and on other ions in the solution. Dehydration of the gel to form completely polymerized anhydrous silica generally leaves it in the form of a disorganized network of tetrahedra. Only very slow dehydration seems to allow organization into quartz: flint, which is formed by replacement of organic material by silica within a calcareous sediment, followed by dehydration on a geological time scale, has a mesh of tiny quartz crystals lying in solidified gel matrix. It is the rehydration of the gel and its subsequent leaching out which constitutes the patination of flint, the patina, or the cortext of flint as it occurs in the chalk, being the crystal mesh with air spaces where the amorphous component was.

Fully dehydrated silica, with varying degrees of crystallization of minute quartz crystals is called chalcedony, and the fully hydrated gel is opal. Apart from flint, combinations of these with varying amounts of impurities giving colour in decorative patterns are common as vesicle fillings in volcanic rocks and as replacement products in organic deposits. Amorphous silica can form as a white powder or as a continuous solid material during evaporation of siliceous water at a lake shore. The powder form has been observed in the lake sediments of the Basin of Mexico (Limbrey, 1975); the continuous solid provided raw material for palaeolithic tools at Olduvai Gorge (Hay, 1968, 1971).

Alumino-silicates

The basis of alumino-silicate structure is a sheet of silica tetrahedra and a sheet of octahedra in which aluminium or magnesium is in six-fold co-ordination with oxygen. Hydrogen attached to some of the oxygens gives a hydroxide rather than an oxide composition. Arrangements of the two sheets and the way in which they are linked together gives groups of minerals of which the micas and chlorites occur in igneous and metamorphic rocks and the clay minerals are the result of alteration and resynthesis in weathering and, in particular, in soil formation. Relationships between the groups are close, and transformations between some members require little structural reorganization. Substitution of aluminium for some of the silicon in the tetrahedral layer occurs in a regular pattern and monovalent and divalent cations are incorporated to maintain the valency balance. Figure 5 shows the arrangements of layers.

The simplest alumino-silicate minerals are talc and pyrophyllite. They have no substitution of aluminium in the tetrahedral layer, and the layers are arranged with one octahedral layer sandwiched between two tetrahedral and held together by the sharing of the fourth oxygens of the tetrahedra with the octahedra. Since there are no spare valencies

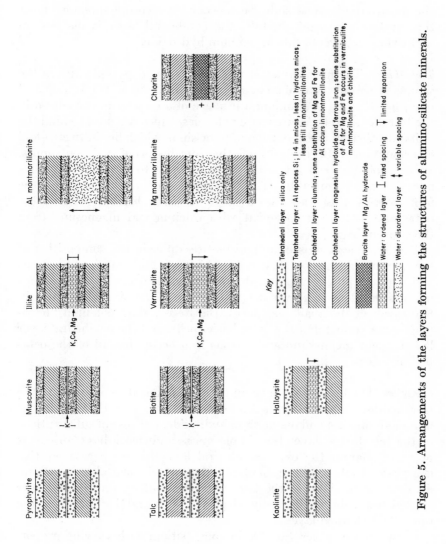

Figure 5. Arrangements of the layers forming the structures of alumino-silicate minerals.

arising from substitutions, and all the oxygens left over from tetrahedra linking point inwards to link with the middle of the sandwich, there are no bonds joining successive sandwiches as they are stacked in the crystal. The minerals are therefore soft and easily cleaved parallel to the sheets, and because sliding can occur between the unbonded units the crystals can bend. In talc, the octahedral layer is magnesium hydroxide, in pyrophyllite aluminium hydroxide.

The Micas. The mica minerals resemble talc and pyrophyllite in the sandwich arrangement of an octahedral layer between two tetrahedral ones, but they have aluminium in place of a quarter of the silicon in the tetrahedra so there are spare valencies, which are used to bind the units together by incorporation of potassium, which lies in twelve-fold co-ordination with oxygens of the tetrahedra of adjacent units. Water molecules can get in between the units to form the "hydrous micas" and vermiculite. The weathering of micas involves the formation of this water layer, together with leaching out of the potassium and its replacement with calcium and magnesium from the surroundings.

The two micas, muscovite and biotite, differ in the composition of their octahedral layers. Muscovite, which is a mineral of acid igneous and metamorphic rocks, has aluminium octahedra, and since aluminium is trivalent the layer is formed with an aluminium ion in only two-thirds of the available spaces. Biotite is a dark coloured "ferromagnesian" mineral of the more basic rocks, and the octahedral layer has divalent magnesium and ferrous iron occupying all of the octahedral spaces.

Chlorites. These minerals have an extra component in addition to the tetrahedra-octahedra-tetrahedra sandwich. This is another octahedral unit of aluminium and magnesium hydroxide. Because of substitutions in the tetrahedral layer the simple mica sandwiches have an excess negative charge; the extra octahedral layer has an excess positive charge due to its composition of a mixture of trivalent aluminium and divalent magnesium, so the two units are built up into multidecker sandwiches held together by electrostatic attraction between the oppositely charged layers.

Chlorites are rather variable in composition, with varying proportions of magnesium and aluminium and a ferrous iron component in the octahedral layers as well. They occur in metamorphic rocks, and in igneous rocks as an alteration product of olivine and hornblende which arises at a very early stage of weathering, when the other minerals are still fresh. They thus appear in soils as primary minerals, although they are often, strictly speaking, secondary.

Illite and Vermiculite. Related to both the micas and chlorite are the clay minerals in the illite, or hydrous mica group, and the vermiculites, the former being related to muscovite and the latter to biotite in the composition of the octahedral layer. The basic difference between the micas and these secondary minerals lies in the way in which the mica sandwiches are held together, the secondary minerals having a lower proportion of aluminium to silicon in the tetrahedral layer, and so less potassium between the units, and having a layer of water incorporated into the crystal structure in this position. In the vermiculites the water layer is double. The entry of water into the crystal is related to the removal of potassium by leaching and its replacement by the divalent cations of magnesium and calcium; in this way the micas could be converted into illite or vermiculite during weathering. Inter-stratifications of biotite and vermiculite and of chlorite and vermiculite occur in composite crystals.

The water in these crystals is associated both with the silica tetrahedra and with the divalent cations: it is not in the disordered state of free water but forms an ordered layer, one end of the polar molecules bonded to the oxygens of the tetrahedral layer and the other to the cations which lie in the middle of a double layer of water molecules. The relationship of vermiculite to chlorite is thus seen in the arrangement of mica sandwiches held together not by single ions of potassium as in the micas but by hydroxide sheets.

Montmorillonites. In these minerals the mica sandwich has a lower aluminium to silica ratio in the tetrahedra than it does in the vermiculites, and so has less surface charge to serve in holding them together. Instead of a structurally well defined double layer of water, which gives a fixed upper limit to the spacing between units, much more water can get in, pushing the units apart to several times the vermiculite spacing. The montmorillonites expand considerably on wetting and shrink on drying, and this habit is very important to their behaviour in soils, exerting a strong influence on the soil as a whole.

The montmorillonites have a variation in composition due to substitution of magnesium and ferrous iron for aluminium, the mica sandwich being of either muscovite or biotite type. Interstratifications with mica, chlorite and vermiculite are common.

Kaolin Group. The sheet structures described so far are all based on the mica sandwich, and their close relationship to each other is apparent. They differ mainly in the way the units are held together, and this is related to the number of substitutions in the tetrahedral layer, providing opportunity for different bonding mechanisms. These minerals are called 2:1 lattice minerals because they have two tetrahedral layers to one octahedral. Another class of minerals has only one of each forming a unit. These are the kaolin minerals, forming a group which is exclusively secondary in character, there being no primary minerals of related structure. They are based on a silica tetrahedral layer bound to an aluminium hydroxide octahedral layer by the fourth oxygens of the tetrahedral being also one of the octahedral oxygens. Instead of the octahedral layer being symmetrical as in the 2:1 minerals, so that it joins to tetrahedra on both sides, it has, on the side opposite to the tetrahedra, an array of hydroxyl ions sticking out. The crystal builds up by some degree of co-valent bonding between the hydrogens of the hydroxyl groups and the oxygens of the tetrahedra of the next unit; though all the tetrahedral valencies are satisfied, the hydrogens can share in them just enough to hold the units together.

Kaolinite is simply built up in the way described. The related mineral halloysite can incorporate a double layer of water between the units: it exists in three forms, dehydrated, partially hydrated and fully hydrated. Its crystals have a curious habit of curling into tubes, giving a fibrous form rather than the platy form of all other sheet structure minerals.

The kaolin minerals have little substitution of aluminium for silicon, and so lack the possibility of incorporation of other cations to give variability of composition. They are strictly alumino-silicate with no ferro-magnesian or alkali component.

Amorphous Alumino-silicate Minerals. The amorphous form of silica has its alumino-silicate equivalent in the mineral called allophane. Unlike free silica it occurs very commonly in soils and is abundant in some. This distinction is because alumina and silica are the least soluble of the soil components, and tend to be left behind as the more mobile ions are removed. In soils where such removal is very rapid, especially in soils on volcanic ashes where much of the primary material is in the form of the already amorphous and easily altered glass and the ash layers are very porous, the rate of release of leached alumino-silicate is too great for secondary mineral crystallization to keep up with it, so allophane accumulates. In all soils, the process of resynthesis which results in crystallization of clay minerals probably involves an intermediate gel or solution stage which would be comparable in its amorphous nature to allophane, and such an amorphous component is commonly found by mineral analysis techniques.

The nature of allophane appears to be that of silica tetrahedra into which are substituted aluminium ions up to a proportion of one in four of the silicons, beyond which a higher aluminium content is expressed as a separate octahedral phase, the two being bonded together by the negative excess charge due to the aluminium substitutions in the tetrahedra and the positive charges on the octahedra which are otherwise balanced by hydrogen ions.

Other Minerals in Soils

Alumina

To complete the silicate–alumino-silicate range there is alumina itself. As a primary mineral alumina occurs in the form of corundum, less hard only than diamond and in its coloured forms being the gemstones ruby, sapphire and some of the emeralds, amethysts and topazes. In corundum alumina octahedra are packed together in a continuous octahedral framework analogous to the tetrahedral framework of silica.

As a secondary form of alumina we have gibbsite, a hydrous form with the aluminium enclosed in octahedral pores between two layers of hydroxyl ions. Since alumina is less mobile only than silica under soil conditions, gibbsite occurs only where silica has been removed— otherwise an alumino-silicate would form. Alumina accumulates only under humid tropical conditions by leaching in an acid environment under high temperatures. There is usually some iron oxide left behind or redeposited in the gibbsite horizon, which is characteristic of lateritic soils. The bauxite ores which are the source of aluminium for industrial purposes are found in regions where intense tropical leaching occurred for long periods during the Tertiary period. The release of the earth's most abundant element from the silica with which it is usually combined requires too much energy for its industrial extraction to be a feasible proposition, so it has to be taken from situations where the energy of the atmosphere has acted long enough and intensely enough to make the separation.

Iron Minerals

Iron is dispersed throughout the earth's crust as a component of the ferromagnesian minerals, and it also occurs abundantly in the more basic igneous rocks as magnetite, Fe_3O_4. Haematite, Fe_2O_3, is not original in igneous rocks but appears by oxidation in an early stage of alteration of ferromagnesian minerals, even within the still liquid magma as it makes contact with the atmosphere. Haematite is common in metamorphic rocks, and forms veins and seams which are exploited for iron ore. Hydrated oxides of iron are very abundant as products of weathering and soil formation, since the immobility of iron under the conditions usually prevailing in weathering situations results in its being left behind when other components are leached away, or excluded from the crystal lattice during the rearrangements involved in secondary mineral formation.

There has been much investigation of the forms in which iron oxide in soil occurs. On release from the crystal lattice of primary minerals the iron would be in disorganized form, interacting with the soil solution to form an oxidized and hydrated gel; it would seem that crystallization occurs as goethite or lepidocrocite, both of which are hydrated ferric oxide, $FeO.OH$. The crystals tend to be very fine and are either dispersed over clay mineral surfaces or form a powdery mass; more water is incorporated on crystal surfaces, giving the light brown substance known as limonite. If this becomes dehydrated, concretions of goethite and eventually haematite occur.

Other iron minerals which can be important under circumstances which arise in archaeological contexts are pyrites, iron sulphide FeS, and vivianite, a hydrated iron phosphate $Fe_3P_2O_8.8H_2O$. They both form in wet and poorly aerated situations where an abundance of organic matter is subject to microbial attack until oxygen is depleted. Iron rendered mobile under reducing conditions can combine with sulphur or with phosphate derived from the organic matter.

Pyrites forms in lacustrine muds; it appears as small metallic looking crystals which are often perfect cubes or octahedra. Its strong tendency to oxidation when the sediment in which it formed dries out or receives more oxygen in the water means that it often presents a rusted surface, or total alteration to a nodule of limonite, and it is in this form that it appears in sub-aerial soils subsequently formed on the sediment. Its presence is a useful indication of previously waterlogged conditions.

Vivianite forms in waterlogged floodloams and in some buried soils. It can appear as small nodules of white powdery material, and quickly turns blue on exposure to the air.

Further discussion of minerals important in soil formation can be found in Marshall (1964).

2

Organic Components of Soil

Soil Animals

The soil animals are those which live entirely within the soil or the litter layer. They range in size from the nematodes, which with a thickness of a few microns overlap in dimension the larger micro-organisms, to the burrowing rodents. The smaller members of the fauna, which can move through the pore spaces of the soil, are often called the mesofauna or meiofauna, the macrofauna being those which disturb the soil in passing through it, enlarging pores, pushing particles aside or eating their way through.

Nematodes, or eel worms, are thread-like worms, some 0·5 to 1·5 mm long, which live in water-filled pores and water films in the soil. They feed by piercing plant, other nematode, and micro-organism cells to get at the liquid contents. Some are entirely parasitic in habit, and live within plant roots. There are normally a few million beneath each square metre of land surface, representing a biomass of a few grams.

Enchytraeid worms are rather bigger, several millimetres in length, and they feed by ingesting plant remains and fungi. They are common in acid soils, where there may be up to hundreds of thousands per square metre, giving a biomass of up to 50 grams.

Earthworms vary in size, up to 25 cm in temperate regions and much more in the tropics. They ingest and comminute plant remains and usually mineral soil as well, so that the organic and mineral material is mixed in the gut. Earthworms are rarely found in soils with a low calcium content, since they need a constant supply and excrete it as calcium carbonate in the digestive secretions. They are sensitive to acidity, and also apparently dislike the taste of certain substances in the leaves of plants growing on very acid soils. Their numbers decrease with increasing acidity, and only a few acid-tolerant species are found where the pH is less than 4·5. In neutral to alkaline soils with an abundance of plant residues there may be up to 500 earthworms per square metre, giving a biomass of up to 250 grams. It has been estimated that the weight of earthworms beneath pasture may equal the weight of animals grazing upon it.

The total number of arthropods in the soil can reach a quarter of a million per square metre of surface. Of these, there would be some hundred thousand each of mites and collembola, and the rest would be insects and insect larvae, spiders, centipedes, millipedes and woodlice. Their total biomass might be tens of grams per square metre.

Figure 6 shows the forms and relative sizes of some of the soil animals.

Most of the soil animals live by eating dead and decaying plant remains. They have very efficient chewing apparatus and break tissue into fragments which may be as small as a few cells only. Their actual source of nutriment is largely the micro-organisms which live on the decaying material, since they do not themselves possess the enzymes necessary for digestion of cellulose and so can not get at the cell contents. Some micro-organisms live in the gut of soil animals, digesting the food and sharing with their host the products of digestion.

The effect of soil animals on the litter is therefore mostly only comminution, but this is of enormous importance in exposing more material to microbial attack. The initial colonization of plant remains by micro-organisms before comminution is followed by a second, massive invasion after comminution.

The soil animals also carry out important mixing operations, both within the litter layer and in the mineral soil. Earthworms take leaves from the surface down into their burrows, where they eat both leaf and mineral soil. Most worms leave the mixed mineral and organic excreta in voids in the soil, but some species come to the surface to cast. The activity of surface casters in taking soil from below and depositing it on the surface is familiar to archaeologists as an effective means by which artifacts, from potsherds to floors and the remains of walls, are buried. The process forms the "worm sorted layer" of a soil; all material larger than about 2 mm is too big to be ingested by the worms and is left behind, and the finer material forms a layer whose thickness depends on the time which has elapsed since it began to form. Anything dropped on the surface moves down through the worm sorted layer until it rests on the coarse material below.

The complement of the worm sorted layer is the stony layer left behind. An additional feature is the "pea grit" or "split pea" layer familiar to excavators, which is the result of certain species of worms taking down little stones, shell fragments, and even seeds to line a chamber in which they curl up in dry weather. The lining helps to

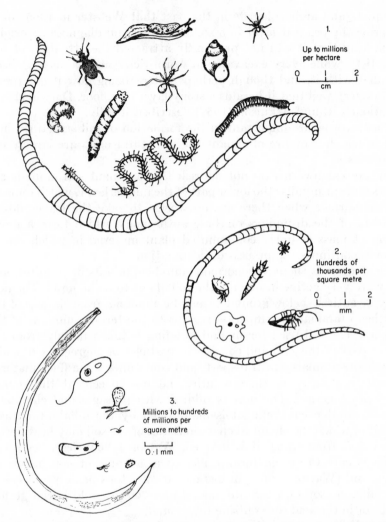

Figure 6. A selection of soil animals. These are sketches only, to give an idea of the ranges of forms, sizes and numbers. 1. Earthworms, slugs and snails, spiders, centipedes and millipedes, insects, and larvae. 2. Enchytraeid worms, collembola, protura, mites, and the larger kinds of amoebae. 3. Nematodes and protozoa; sizes of the latter range from those indicated down to very much smaller forms.

maintain a suitable atmosphere in the chamber. The worms go down until they come to a layer which they can't easily get into, such as a gravel or the buried floor of a building, and a layer of the little stones accumulates as the chambers formed in successive seasons collapse. It is symptomatic of the small amount of communication between soil scientists and archaeologists in the past that Webster in 1965 could describe the pea grit layer in a soil on gravel and diagnose its origin, remarking that it had not previously attracted the attention of soil scientists, while every excavator in worm-rich country has long been familiar with it, and though perhaps not realizing how it got there, knows very well that it heralds a stony layer or a floor. Darwin (1881) mentioned it, and Atkinson (1957) described it fully.

Earthworms are important in the formation of soil structure: in a soil in which they are abundant the structural units are largely old worm casts.

Where earthworms do not flourish the dominant soil animals are mites and springtails; the lower part of the humus layer, which remains on the surface when there are no worms to take the leaves down, consists of the droppings of these anthropods, which form a moist compact mass of finely comminuted plant material in which fungal growth continues the process of degradation.

In some soils there are enormous numbers of ants or termites, and they are very active in moving the soil about to make nests. The nest may be entirely below ground or may be a mound, from the size of the ant heap familiar in temperate regions to the termite mounds of the tropics. The soil used for mound building is taken mainly from the lower part of the soil, and consists of particles or aggregates of a size which the animals can transport, and sometimes some finer material which they carry in their mouths and use to cement the coarser particles together. The nest is riddled with passages and chambers, and a suitable environment is maintained by the inhabitants, as a result of which the chemical characteristics of the soil may be different from that from which it is derived. Watson (1962) describes a calcareous core to an old termite mound in an area of non-calcareous soils, and (Watson, 1967) an instance of skeletons being preserved in the alkaline soil of a termite mound in a 700 year old burial ground but not in the acid soils outside the mound.

Erosion and destruction of the mound distributes its material over the surrounding soil, forming a layer which, in areas where termites are abundant, may be a continuous "termitogenic" horizon. It is customary in some tropical areas to spread termite soil deliberately to prepare ground for agriculture; since it is derived from the subsoil and is therefore poor in organic matter it will tend to be infertile, as ant soil in temperate regions is, but it may be that this practice is confined to areas of lateritic soils, in which a greater clay content of the sub-soil will give better water and nutrient retention. Some termites incorporate organic material into their nests by their practice of

6

Soil Groups and Profile Types

Incipient and Immature Soils

The earliest stages of soil formation are somewhat different on solid rocks and incoherent materials. In the latter, plant roots can grow and nutrients are available for them to absorb as soon as dissolution of mineral materials begins, or, in many alluvial situations, as soon as nutrient-bearing waters subside to leave the soil moist. The limiting factor in soil development is likely to be shortage of nitrogen; some micro-organism activity is necessary to fix atmospheric nitrogen and make it available to plants. Colonization of sediments by bacteria starts with their introduction in dust or flowing water and can proceed as soon as moisture is available and there is a small amount of organic matter present. Incipient or "raw" soils on incoherent material in which alteration of minerals can begin throughout the rooting zone and organic matter begin to accumulate within the upper part, but in which as yet no horizon differentiation is apparent are known as *regosols*, or (Avery, 1973) as *raw sands*, *raw alluvial soils*, or *raw earths* depending on the nature and origin of the parent material. Where soil formation is beginning in a shallow accumulation of hard rock fragments, the soil is a *skeletal raw soil*.

Hard rock surfaces provide no root hold for plants and initially no moist and porous environment for soil fauna and micro-organisms. Soil formation begins with colonization by algae or by lichens. Lichens are symbiotic associations of algae, which photosynthesize to provide energy, and fungi, which exude substances which attack mineral surfaces and make available nutrient elements. The macronutrients are released from the rock by simple solution; the micronutrients are taken from the minerals by chelating substances in the exudate. Nitrogen is fixed by blue-green algae as components of lichen. The rough surface of the lichen-encrusted rock traps airborne dust, contributing to accumulation of weatherable minerals in finely divided

form and organic materials, which include the spores and eggs of the soil population. Contraction of the gummy exudates when the lichens dry plucks fragments of rock from the surface, and with the accumulation of these and the lichen residues a primitive soil develops which can hold enough moisture for mosses to grow. The cushions of moss contribute to creation of a moist environment in which the soil fauna can live and higher plants can take root. The solid rock begins to yield to hydration and oxidation as acid moisture percolates from the organic layer at the surface into cracks and along crystal boundaries, softening the rock and providing channels for penetration by plant roots. Cycles of freezing and thawing, and of heating and cooling, contribute to fragmentation in appropriate climatic situations.

Ranker

With the establishment of a distinct **A** horizon, either, in the case of the hard rock, predominantly organic, or, in the incoherent material,

Figure 12. A ranker profile.

a crude mineral and organic mixture, the soil is known as a *ranker* (Fig. 12). Rankers and ranker-like alluvial raw soils are the immature, **A/C**, profile type in freely draining situations on all parent materials except highly calcareous ones. On non-calcareous rocks, and on very hard calcareous ones, the **A** horizon develops initially by the activity of arthropods, which deposit their faecal pellets of finely comminuted plant residue among rock fragments initially present or loosened from the surface, or within the upper part of incoherent sediments. Earthworms cannot participate until there is enough calcium in the organic cycle to satisfy their requirements and until there is sufficient depth of soil to provide them with a moist refuge during dry periods.

Rendzina

On calcareous rocks the dissolution of calcium carbonate under the influence of rainwater and the acids of organic materials proceeds

much more rapidly than does attack on other rock forming minerals. Solution widens cracks sufficiently for plant roots to penetrate and contribute further to development of channels reaching deeply into the C horizon. In the organic horizon calcium is available for earthworms from the beginning, and dark coloured stable associations of humus substances and minerals are formed; earthworms can retreat down cracks and root channels when necessary. A *rendzina* (Fig. 13) is a soil consisting of an almost black calcareous mull humus formed entirely of worm casts lying on relatively unaltered calcareous rock.

Progress from the ranker or rendzina stage depends on the establishment of a **B** horizon. On steep slopes, where erosion decapitates the

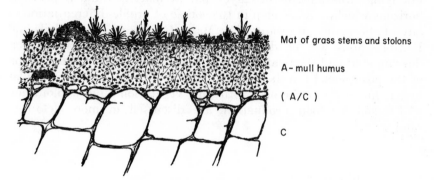

Mat of grass stems and stolons

A - mull humus

(A/C)

C

Figure 13. A rendzina profile.

profile from time to time, or where continuous and rapid soil creep occurs, the immature stage may be maintained indefinitely. Rendzinas can persist on any surface of pure limestone, such as chalk, provided there has been no contribution to the soil from superficial deposits, or after soils developed partly within such deposits have been eroded away. Much of the chalk contains so little mineral material other than calcite that there is nothing to form an appreciable **B** horizon. Solution of the chalk lowers the land surface without contributing more than a very small percentage of its volume to the soil, and development of a **B/C** horizon cannot proceed beyond oxidation of any iron present in the chalk at the surface of structural planes and in root channels. As the calcite goes, the mineral residue is immediately incorporated into the **A** horizon by worm activity. On the purer chalks it would take many thousands of years of soil formation without any loss from the surface by erosion to build up a depth of soil greater than that normally mixed by worms. Loss by erosion, even under stable

vegetation, can easily exceed the rate of growth of the **A** horizon from below.

From a ranker or a rendzina, various paths of profile development are possible, and the direction taken depends on a complex interaction between the characteristics of the parent material and environmental factors. The soils formed under the deciduous forests of the temperate zone are the Major Group of *Brown Soils*, of which the *brown forest soil* is the central character. The Brown Soils merge into the *Podzolic Soils* which are characteristic of the coniferous forests, and into the *Chernozems* of the steppes. Associated with all the soils are the immature soils and the gleyed soils. Gleyed forms are characteristic of low lying situations, or of clayey parent materials and impervious horizons which may develop in the course of profile differentiation in any of the main soil groups. The Major Groups of Brown Soils and Podzolic Soils will be described here, before the history of their development and the relationship between them, under the formation of the post-glacial forests, under changing climate, and in particular under human interference with the vegetation and exploitation of the soil for food production, are discussed in more detail in Chapter 7.

Brown Soils

The Major Group of Brown Soils as defined by Avery (1973) contains soils which have been called *brown earths* and *brown forest soils* under other classification systems. *Brown earth* has previously been used in the British system to cover the greater part of the Major Group, but is now introduced at the first level of subdivision, as a group. *Brown forest soil* is equivalent to brown earth at this level and corresponds to the usage of the F.A.O. Soil Map of Europe. It will be used here because of the way it draws attention to the forest history of these soils, and thus to their ancestral status with respect to soils in this and in other Major Groups, which is argued here.

Brown Forest Soil

Under a deciduous forest growing on a brown forest soil the vegetation is well nourished, and the leaves are rich in bases and poor in phenolic compounds. Rainwash from the leaves and litter and a high rate of microbial attack on plant residues return bases to the upper part of the soil and tend to keep the exchange complex saturated. Abundance of divalent cations prevents dispersion of the colloidal component of

cultivating fungus for food on plant debris carried in for the purpose. Some of this organic matter takes up extremely stable forms when incorporated into the soil, and might improve the physical characteristics of the soil though decaying too slowly to provide nutriments for the growing plants. The chemical modifications brought about to maintain the environment of the nest might also benefit subsequent plant growth.

Where ants are active beneath and within a stone structure the accumulation of soil between and above the stones, and among plants growing in cracks in the structure, is an important factor in its burial. The accumulation occurs by the dumping of soil brought out from the construction of passages and chambers of an underground nest, or by the nest itself being built above the original ground surface. In a stone mound or ruined building which has developed a soil and plant cover over the top, the ants may be operating from this new surface, but to get material from their nest they go down to the buried surface, and in taking soil from beneath the stones they lower them into the buried soil just as do worms.

A timber structure provides great opportunity for ant activity; as the timbers decay the ants feed on the fungi which grow on the decaying wood, and may make their nests below or within the structure, close to the food supply. The wood is then replaced by soil as the nest is developed. This may be the main source of the "post-hole fill" whose uniform texture and fine, soft structure is so characteristic.

Various creatures live in ant and termite nests, some having developed a symbiotic relationship with the primary inhabitants. Abandoned nests house spiders, woodlice and other soil animals which may enlarge and modify the passages. Earthworms may rework the ant soil or cast in the passages and chambers soil taken from elsewhere. The typical appearance of ant soil will thus be modified and wormcast structure will supersede that of the nest material. Only the position of the soil may give a clue to its origin as an ant's nest.

There are many carnivorous beasts in the soil, preying on other soil animals or living on their remains and those of larger animals which die on the surface or in burrows. Some ants and termites, all centipedes and many beetles, are carnivores. Many insects lay eggs in the soil or in dung or decaying organic material on the surface; the carnivorous larvae hatch out and crawl through the soil eating as they go until they metamorphose. Some beetles bury the corpses of even very large animals by taking soil out from under them. They then lay their eggs in the carcase and by the time they hatch the bacteria have rotted it to a suitable state for the grubs to devour.

Larger animals and birds contribute to the soil processes by participation in the cycling of organic materials and mineral salts. Their activity in disturbing and mixing soil is also significant: birds turn over the litter in hunting for insects, and probe the ground with their beaks to get at worms and larvae; burrowing animals shift the soil in foraging for worms or plant roots and in making living places. Burrows can be very confusing in an archaeological context since they disturb stratification and allow material from the surface to get into lower levels, or that from below to be distributed on the surface. Field workers are accustomed to kick open mole heaps to look for flints and potsherds brought up from within or below the worm-sorted layer; they perhaps don't realize that if a soil in which moles have been active is buried the artifacts found on the buried surface do not necessarily just pre-date burial but may have been deposited at any earlier time and brought out by moles. Where there are worms there are usually moles feeding on them, and the whole of the worm-sorted layer is being continually tunnelled by them.

Gophers, in the New World, have a similar effect in tunnelling after roots, but don't confine their activities to the topsoil. In a stratified deposit in which roots have grown particularly well in some moist layer, gophers will follow this layer and may completely alter its internal stratification while leaving layers above and below relatively undisturbed.

Micro-organisms

The micro-organisms are bacteria, fungi, actinomycetes, algae and protozoa. The bacteria, fungi, actinomycetes and algae have generally been regarded as belonging to the plant kingdom and the protozoa to the animal, but the distinction becomes blurred, some protozoa having chlorophyl and being able to photosynthesize like plants, some bacteria and fungi having habits of predation, and some algae and fungi having protozoa-like stages in their life cycles. They are therefore all put together as Protista, more primitive than either plants or animals and having features of both. There are also the even smaller beings, the viruses, and the virus-like particles called phages which attack bacteria and actinomycetes. These consist of genetic material in a protein sheath, and they invade cells and pervert their functions to producing materials for their own reproduction.

The micro-organisms have in common the basic organization of the living cell. A semi-liquid cytoplasm is enclosed in a membrane which forms a barrier through which can pass certain substances but not others. Similar membranes form a network within the cell and enclose various bodies in which are sited the enzymes responsible for the cell's respiration and for synthesis of its components. The genetic information of the cell is in the nuclear material. The micro-organisms are divided into the more primitive prokaryotes in which the nuclear material is free in the cytoplasm and the eukaryotes in which, as in higher organisms, it forms a discrete body surrounded by a membrane. The two groups differ in other ways consistent with the more primitive nature of the former: the degree of differentiation of functional materials into separate bodies is less, the reproductive process is almost entirely asexual, whereas in the eukaryotes some interchange of genetic material can occur.

The familar names of the groups are misleading since they were classified before details of their structure and chemistry were fully elucidated. The actinomycetes were thought, as the name implies, to be related to the fungi, since their macroscopic manifestation, in producing mycelium and propagating by spores, is similar. Their structural materials and internal organization, however, show them to be prokaryotes and related to the bacteria. Similarly the blue-green algae, in spite of their greater size and the possession of chlorophyl and other pigments which enable them to photosynthesize, are much more primitive than the other algae and are again found to be prokaryotes.

Bacteria. Bacteria are single cells which have a cell wall which gives them shape as rods or spheres. Some have flagellae which impart some motility, and many are surrounded by slimy or sticky substances which they exude. They multiply by division into identical cells, which either separate or remain in contact to form strings or clusters. Many can survive adversity, such as nutrient deficiency, desiccation, heat or cold, by forming spores: the genetic material and essential enzymes are enclosed in a protective coating and the rest of the cell can succumb to destructive processes leaving the spore to survive and germinate when conditions again become favourable. Some form rather less well protected cysts by enclosing the whole cell in a protective coat.

Fungi. Most forms of fungi grow vegetatively as mycelium, branched, filamented hyphae, which in simpler forms consists of a single multi-nucleate cell, and in the more complex are septate. The yeasts are simple unicellular fungi which do not grow as mycelium but are globular in shape and reproduce by budding. The mycelial fungi reproduce by means of spores, which develop in spore bodies growing on a form of hyphae different from the vegetative type. In many forms the spores grow on hyphae which extend out into air spaces from the moist substrate on which the fungus is growing vegetatively. Some fungi can encyst, enabling vegetative cells to survive adverse conditions. Large complex structures and thick strands of mycelium produce a resemblance to the tissue of higher plants.

Actinomycetes. In many of their external characteristics the actinomycetes resemble fungi. They grow vegetatively to produce mycelium and reproduce by spores which develop singly or in spore bodies. However, they are prokaryotic, and have a cell wall of the same substances as those of bacteria, and produce external slimes like many bacteria.

Algae. Among the algae are micro-organic forms which live in soil. They are more advanced than the other micro-organisms, having larger cells, a cell wall which in many cases is cellulose, and in the organization of their cells they come closer to the higher plants. Many produce mucilages and some lay down inorganic deposits within the mucilage but outside the cell wall; diatoms produce silica tests, and other forms are covered with silica or calcium carbonate scales. Many algae are motile, and if they become buried in the soil can find their way up to the light. They possess chlorophyl and other pigments and need to live in the upper part of the soil so as to photosynthesize.

Protozoa. The protozoa are single cell creatures which lack a cell wall and are motile. The amoebae move by modification of their shape and others by undulating motion of flagella or cilia. They multiply by division, and either separate or remain in interdependent colonies which take up a characteristic form. Some have tests made of organic materials or of silica.

Many protozoa engulf their food in particle form, and in the soil the bacteria are the usual prey.

Micro-organism Population

Numbers of micro-organisms in the soil are difficult to estimate: it involves "counting" microscopic particles which number in millions and vary in concentration from single individuals to exponentially growing colonies. Since the population fluctuates greatly over short periods, as temperature and moisture conditions change, and over longer periods with seasonal change in nutrient supply, absolute numbers have little meaning.

Estimates of bacteria are in the region of several thousand million per gram of soil, giving a biomass of some hundreds of grams beneath each square metre of the land surface. Actinomycetes and fungi, because of their mycelial habit of growth, do not allow of realistic figures for numbers of individuals; they would generally be fewer than bacteria, and actinomycetes more than fungi. Their biomass is of the same order as that of bacteria but generally rather less, except that in acid soils fungi predominate over both bacteria and actinomycetes. Algae give counts of tens to hundreds of thousands per gram, with a biomass of tens to hundreds of grams per square metre. Protozoa, whose numbers are related to those of the bacteria on which they mostly feed, give counts of thousands to hundreds of thousands per gram, and a biomass which is normally only a few grams per square metre but can go up to a hundred.

Micro-organisms are dependent on the soil water for their existence. They live in or on moist organic residues, in water filled pore spaces and in films of water on mineral and organic particles. They spread through the soil so as to utilize all possible food sources. Colonization starts from a single cell or fragment of mycelium, or from a protective spore or cyst or a reproductive spore. Transport of these units may be by soil animals in the course of their movement through the soil and in the mixing of material in their gut, in dust blown in the wind over the surface, or in percolating water. Protozoa, the motile stages of some fungi and algae, and those bacteria which have flagellae, can move through the soil. Fungi and actinomycetes spread by growth of mycelium and can cross barren areas in this way as long as contact with one source of nutriment is maintained until another is reached. Once growth begins, the colony spreads over or through the food source and by doing so can reach and utilize all parts of it.

Growth of micro-organisms implies multiplication: it is the colony, or the fungal mycelium, which grows rather than the individual cell. The cell divides, buds, or introduces septa instead of growing bigger than about twice the normal size.

Death of micro-organisms is not an inevitable stage in the life cycle but occurs in response to hostile conditions. Many forms produce endospores or cysts, which can survive adverse conditions, but others will die under temperatures of either extreme, desiccation, shortage or excess of oxygen depending on their mode of respiration, or starvation. Endospore formation allows death of the rest of the cell. Invasion by phages is usually followed by death when the newly formed phages burst out through the cell wall. Toxic substances exuded by some micro-organisms will kill others, and some produce enzymes which break down the structural and protective materials of others, either externally or after ingestion.

Death is follows by lysis. During life, the substances of the cell are kept separate to some extent, partly by membranes and partly by chemical mechanisms. This prevents digestion of the cell by its own enzymes. On death, the separation ceases to function, membranes break down under enzyme attack and the substances of the cell can interact with each other and can escape through the cell wall. Autolysis is digestion by the cell's own enzymes, lysis is digestion by those of another.

In the soil, most of the time a very high proportion of the micro-organism population is inactive, as resting cells, spores and cysts. A limiting supply of one or more nutrients, usually nitrogen or phosphorus, appears to be the normal condition. On addition of a food source there is an immediate exponential increase in growth as colonies multiply, and a succession of micro-organisms follows, as each form exhausts those components which it can use, rendering the substrate less suitable for itself but providing, in the material it leaves unused and in its own cells and waste products, a substrate which favours another group. What appears macroscopically as a process of decay is, microscopically, a succession of life, but involves a progressive exhaustion of nutrients until in the end only spores and other resting cells remain.

The physical and chemical conditions in the soil have a considerable effect on the level of activity, and so on the rate of transformation of organic residues. Moisture content is critical: no micro-organism can grow in a Relative Humidity of less than about 63%, and most need much more water than is available at this percentage. Low temperatures reduce activity; few micro-organisms can grow below 10°C, and many not below 15°C. The rates of chemical reactions increase with temperature, so growth rates increase, up to the temperature at which proteins begin to coagulate, so optimum temperatures are towards the upper end of the range in which growth occurs, usually lying between 25 and 40°C. Since soil temperatures rarely exceed the temperature at which optimum growth of a high proportion of the population occurs, it is rarely a limiting factor. High temperatures are only reached in bare soils in tropical regions, and desiccation would limit growth long before heat. Many organisms can survive both high temperatures and freezing by forming spores.

Chemical conditions in the soil affect the balance of activity between the different groups. The main effect of pH is in its control of the supply of nutrients, and many organisms have a wide range of tolerance of acidity or alkalinity as such. Bacteria, actinomycetes and algae grow best in slightly alkaline conditions, fungi in slightly acid. With increasing acidity therefore, fungi increasingly dominate, and below pH 4 are the only active forms; they too cease to grow at about pH 2·5. With increasing alkalinity, bacteria and actinomycetes dominate, but at the highest pH values reached in soils salinity becomes the limiting factor, since only a few species normally found in soils can tolerate high salt concentration.

Respiration

Some micro-organisms respire aerobically, some anaerobically, and some can change their mode of operation to suit prevailing conditions. The obligate aerobes cannot grow without oxygen, but there is a range of degree of oxygen depletion and of carbon dioxide concentration which different species can tolerate. Similarly, oxygen is toxic to some anaerobes but others can live in well-aerated conditions.

Algae are almost all aerobic, though there are some which can function anaerobically if they have to. Fungi are mostly aerobes, but the yeasts have an anaerobic mode of operation to which they switch when the supply of oxygen runs out. Bacteria and protozoa cover a wide range of oxygen requirement and include obligate aerobes and obligate anaerobes as well as versatile types.

Since the soil has a complex system of water and air filled pores of various sizes, and undergoes continual change of its water content, the rate of supply of oxygen varies greatly from point to point and with time. Oxygen to replace that used by micro-organisms and soil animals must be brought in with the water moving through the pores or must diffuse from an air–water interface. In the interiors of aggregates to which only fine pores give access oxygen depletion is liable to occur and anaerobic activity takes over. Similarly, a wet soil which is poorly drained will suffer oxygen depletion, particularly at sites of high organic activity centred on fragments of decaying tissue.

Nutrition

Micro-organisms can be either autotrophic or heterotrophic. Autotrophs are able to use the energy of the sun by photosynthesis and to synthesize their materials starting from the inorganic carbon of carbon dioxide, or to obtain energy by oxidizing inorganic substances. Heterotrophs need organic materials as sources of both energy and carbon, and vary greatly in the range of organic compounds that they require ready made and those that they can synthesize for themselves.

The photosynthesizing autotrophs are the algae, the blue-green algae, some of the bacteria, and some of the protozoa. The chemo-autotrophs are various bacteria which live by oxidizing inorganic nitrogen, iron or sulphur compounds. All the fungi and all the actino-mycetes are heterotrophs, as are most of the protozoa and many bacteria. Some algae can change to heterotrophic function in the absence of light.

Feeding among the heterotrophs is saprophytic, that is, feeding on dead organic matter, which includes the predacious habit of protozoa and some fungi, or parasitic, in which the organism lives on or in the tissue of a living plant, animal or other micro-organism and derives part of its nutriment from the metabolic processes of its host.

Saprophytic organisms take in their food either by engulfing other organisms or particles, or by absorbing molecules and ions from solution in the moisture of the substrate. Small organic molecules and ions, as well as the inorganic nutrients, are taken in through pores in the cell wall and membrane; larger molecules are carried across the membrane in combination with transfer substances. Within the cell, enzymes carry out the oxidations necessary to release energy, and break down complex substances into units which can be re-assembled to form the structural and functional materials needed for cell growth and multiplication. Some enzymes are excreted, to break down outside the cell large molecules which cannot be ingested into units which can.

In the soil, photosynthesizing organisms can only function in the upper levels where light has access, and so play a relatively small part in soil processes, except in the early stages of soil formation. When there is not as yet an organic carbon source on a bare rock surface, and no chance for plants to take root, only autotrophs can grow. Blue-green algae may form a crust on, or in cracks in, a rock, and colonization by lichens is an essential step in incipient soil formation. A lichen is the manifestation of a symbiotic relationship between a fungus and an alga: the alga synthesizes organic materials which the fungus can utilize, while the fungus exudes substances which release mineral nutrients from the rock and make them available to the algae as well as to itself. The effect of contraction of a drying lichen, adhering strongly to the rock by its fungal mucilage, in detaching flakes and particles is important in forming the mixture of mineral particles and organic residues which constitutes the primitive soil. Other micro-organisms can then contribute to degradation of the residues, and plants can take root.

The inhabitants of the soil and their life processes are vividly described in Russell, E. J. (1957). Burges and Raw (1967) has chapters on the various groups, in more detail.

Soil Organic Matter

The organic component of the soil consists of the following groups of substances, whose interelationships are shown in Fig. 7.

1. Plant and animal residues.

2. Living micro-organisms, soil animals and plant roots.

3. Specific organic substances. Some are derived directly from living plants by rain washings from leaves and as exudates from roots, and by leaching of dead tissue. Some are excreted by living animals. Some are released by micro-organisms as exudates and extra-cellular enzymes, or, in conditions of nutrient deficiency, excreted as waste products. Utilization by micro-organisms of some components of residues releases other substances into the soil, and lysis of micro-organism cells releases their materials, unaltered or after some degree of interaction with each other.

4. Humus substances. These are very large molecules derived from some of the substances of the previous group by condensation reactions. They can be called "hetero-poly-condensates", that is, they are composed of large numbers of different compounds joined together by condensation. They have an acid nature imparted by peripheral reactive groups, and a strong reddish- or yellowish-brown colour. They make up 85 to 90% of the soil organic matter and constitute a reserve of plant nutrients which is mobilized by the activity of some micro-organisms which can break up the large molecules. As a component of the soil colloid complex they contribute to the texture and take part in forming the structure of the soil and in controlling the mobilities of other mineral and organic substances. As acids they take part in the alteration of primary minerals and in controlling the environment in which secondary minerals crystallize.

The specific organic substances are present free in the soil in quantities which depend on the rate at which they are utilized by micro-organisms. The simple sugars, amino-acids and soluble proteins, and other soluble compounds of low molecular weight can be absorbed by most micro-organisms and are efficiently metabolized; they are found in the soil as traces only. Insoluble substances such as fats and

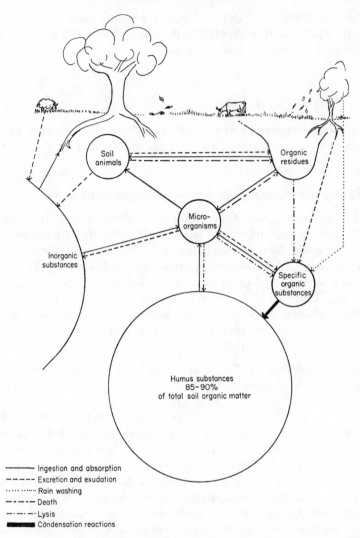

Figure 7. An indication of the pathways in transformations between different forms of organic matter in the soil.

fatty acids, waxes and resins, disappear more slowly. Large molecules of polymerized substances have a longer life in the soil and are therefore found in larger amounts. The number of organisms which possess enzymes capable of breaking the polymers into the smaller units which can be absorbed is relatively restricted, and the amount of energy needed to carry out the degradation, that is, to synthesize the necessary enzymes, is high in relation to the energy and nutrient content of the components, so growth of the organisms is slow.

These polymers are the cellulose from plant and alga cell walls, other polysaccharides from plant and micro-organism tissue and exudates, chitin and other nitrogenous polymers from micro-organisms and from the exoskeletons of arthropods.

Plants lay down in the cell walls as they age the substance called lignin. Pearl (1967) gives an account of the chemistry of lignin. It is a condensed aromatic substance based on phenyl-propane units, that is, a benzene ring with a chain of three carbons attached. The structural units vary between different plants: three phenolic aldehydes, characterized by different reactive groups on the benzene ring, occur in the plant groups in different combinations: vanillin and p-hydroxy-benzaldehyde in conifers, vanillin and syringaldehyde in woody dicotyledons, and all three in monocotyledons.

Substitutions on the 3-carbon chain attached to these units characterize the different types of lignin. Units are built up into larger units by condensation reactions between the OH group opposite to the side chain on the benzene ring and the carbons of the side chains of other units, and further condensation occurs between the benzene rings themselves to produce a polycyclic structure. Details of the lignin structure and the degree and nature of its variability have yet to be worked out. It forms very large molecules, whose molecular weight is in thousands, and it is only slowly degraded by micro-organisms, which can split off individual structural units using external enzymes and after absorbing the simple aromatic compounds can break the benzene ring via quinoid intermediates and complete metabolization of the resulting aliphatic substances.

Other phenolic substances found in plant tissue are the catechin and flavonoid compounds in which the phenyl-propane unit has another benzene ring attached directly to the end of the 3-carbon chain and often turned back to form further linkages with the first ring, and combinations of these with the anthocyanin form, which has only one benzene ring.

The humus substances, whose molecular weight reaches tens of thousands, are built up from structural units which are in part the same as those of lignin, in part those of the flavonoid compounds, an in part could be derived from either.

Quinoid units also take part; these represent an intermediate stage in either forming or breaking the benzene ring, and they are found in both plant tissue and micro-organisms. Micro-organisms can synthesize aromatic compounds, and some of the structural units of the humus substances are synthesized in their cells and not necessarily derived from either lignin and its components or the other plant polyphenols.

The condensation reactions joining up the structural units may take place under the influence of enzymes either free in the soil or within micro-organism cells. The high concentration of ingredients within a cell undergoing lysis, and their protection from attack by other micro-organisms until after the more stable, condensed molecule has formed, suggest that the latter is more probable, and the formation of high molecular weight, dark coloured substances has been observed within the dark coloured mycelium of certain fungi as it undergoes lysis, and within the cells of bacteria which have multiplied inside decaying plant tissue.

There is no certainty yet as to whether the humus substance molecule is a relatively open structure or whether there is a core of more closely condensed, polycyclic structure with more loosely attached units at the outside.

The study of soil organic matter is a large and prolific field of research. The account given here represents a simplified synthesis of published views, based on the following sources in particular: Kononova (1964); Hallsworth and Crawford (1965); McLaren and Petersen (1967), particularly the chapter by Hurst and Burges; Burges and Raw (1967); papers in Soil Science Vol. 11 no. 1, 1971, special issue on Soil Organic Matter; McLaren and Skujiņš (1971); Stephenson and Goh (1971); Kononova and Alexandrova (1973); Filip et al. (1974).

Many phenolic compounds are tannins. Coagulation of protein by a tannin depends on formation of a number of linkages between the OH group and the protein molecule to build up a stable, cross-linked structure. The molecule has to be large enough and have enough OH groups to reach different sites on the protein: polyphenols have more than one OH group to the benzene ring; those having large enough molecules are tannins. They fall into two groups: the hydrolysable tannins, based on gallic acid and its derivatives combined with glucose, and the condensed tannins, based on the double ring structure of the catechin and flavone molecules and combinations of these. Plants produce polyphenols in types and in quantities which vary with plant species, rate of growth and nutritional status (Brown et al., 1963; Feeney and Bostock, 1968). Some produce no tannins at all, many produce hydrolysable tannins, and a more restricted number also produce condensed tannins. *Calluna* produces condensed tannin alone. Under conditions of nutrient shortage, possibly when phosphorus is not supplied fast enough, there is an increased tendency to produce condensed tannins, and the variation with season, hydrolysable tannins being produced in the spring and early summer and condensed ones in middle and late summer, may reflect the rates at which nutrients are used at different stages of growth.

On death of a leaf, lysis of the cell allows coagulation of proteins by the tannins. The hydrolysable tannins, having a glucose component which is readily attacked by micro-organisms, are not long lasting, but the condensed ones are only slowly degraded. Litter derived from plants which have produced condensed tannins is therefore resistant to attack; not only are the proteins themselves protected, but tanning of proteins in the matrix of the cell wall protects the cellulose fibres embedded in it (Handley, 1954).

The distribution of OH groups on a large molecule, which gives polyphenols their tanning ability, also makes them chelating agents: bonds are formed between one polyvalent metal ion and two or more OH groups on the same organic molecule. Whereas only the larger polyphenol molecules can tan proteins, smaller ones and other organic substances having the required reactive groups suitably disposed can take a metal ion in their grip, and the resulting chelate is soluble.

Rain washes from the surface of the growing leaf substances exuded by the leaf, liberated by mechanical and microbial damage, and produced by metabolism of the micro-organism population on the leaf. After leaf fall, further soluble compounds are washed into the soil; most of the chelating agents, the polyphenols and organic acids, probably enter the soil by these means rather than during the later stages of litter decomposition, particularly since it is those plants producing abundant polyphenols whose litter will be protected from rapid attack by the tanning effect of some of them.

The importance of polyphenols in soil lies in their contribution to the formation of humus substances, their influence on the rate at which organic residues are degraded, and their function as chelating agents in mineral alteration and soil profile development. Much work has been done on their origin, function, and fate in soils. In addition to the papers cited above, the following papers indicate the progress of work in this field: Bloomfield (1953, 1954, 1957); Coulson et al. (1960); Davies et al. (1964); Brown and Love (1961); Bocks et al. (1963, 1964); Burges et al. (1964); Lewis and Starkey (1969); Martin and Haider (1969); Haider and Martin (1970).

In the archaeological context, an understanding of the methodology and the progress of research in soil organic matter studies must form the basis of any work which is done on characterization of amorphous residues of organic materials encountered during excavation. It is important too in the establishment of criteria for the selection of samples and for the development of methods of pre-treatment to remove contaminating substances in radiocarbon dating.

Dating of soil organic matter from active surface soils yields apparent dates which represent the average life of the humus substances, some just formed, some which have lain in the soil for thousands of years, giving the concept of *mean residence time*. Fractionation of the organic matter by different methods of chemical extraction yields fractions of different apparent ages: the more soluble forms being less stable in the soil and thus predominantly younger, and the more stable fractions having more of the more ancient humus substances (Campbell *et al.*, 1967; Scharpenseel *et al.*, 1968). Attempts to date buried soils by dating their humic horizons give results which are too high by the amount of the mean residence time. Contamination of dating samples of simple age may be by younger humus substances derived from the soils which have formed later over the area or by older ones from the soils which existed there at the time the sample was deposited. Pretreatment by conventional methods does not necessarily remove all of it, since the more stable forms are not soluble in the warm, dilute alkali used. Where the sample has undergone humification, rather than being a charred material, there is no sure way of separating sample and contaminant.

3

The Soil System as a Whole

The Components

We now have assembled the components of the soil. To summarize:

Primary Minerals. These are derived from rock and are usually almost all to be found in the coarser particle size grades, gravel, sands and silt, as coherent rock fragments containing several different minerals and as individual crystals. Since they are altered and destroyed by processes which act principally at their surfaces the rate of disappearance increases, relative to volume, as the particle size decreases, and only the most resistant minerals are found in appreciable quantities in the smallest size grade, the clay.

Secondary Minerals. These are formed by interaction of primary minerals and the soil solution. They increase in size by accretion of ions and groups of ions at their surfaces, so their proportional growth rate decreases as their size increases, and since growth depends on the balance between accretion and loss of ions in a continually changing environment the chance of their reaching a large size is low. Having thin platy shapes they are easily broken during the slight movements always going on in the soil as moisture content changes and as roots and animals disturb it. Secondary minerals are therefore almost confined to the clay grade, though stable aggregates of particles may persist in the soil and bring some of them into the silt and even the sand grade.

Living Components. These are plant roots, soil animals, and microorganisms. They range in size from the large roots of trees and the burrowing mammals through the smaller roots and the mesofauna which are comparable in size to the mineral particles of sand and coarse silt size, to the micro-organisms and finest root hairs which are comparable in size to the fine silt and the larger clay particles.

Non-living Organic Components. These are the plant and animal residues in all stages of decay and re-utilization, and range in size again from dead trees and carcases of large animals through the increasingly comminuted residues and fragments released by degradation of more susceptible components, to individual molecules of organic substances. The largest organic molecules are comparable in size to small clay minerals, and from here they grade down into simple molecules in solution in the soil moisture.

The Soil Solution. This is water containing molecules and ions which are derived from mineral materials by solution and hydrolysis, from organic materials by lysis of cells, from living organisms within the soil by exudation, excretion and respiration and from animals and plants above the soil surface by excretion and by rain washing of plant tissues, and from the atmosphere.

The Soil Atmosphere. This is composed of the gases of the above-ground atmosphere in proportions modified by the respiration of soil organisms. There is continual interchange with the air above, by diffusion and by mixing as the wind plays over the surface of the soil, and with the gases in solution in the soil moisture. In general, the soil atmosphere contains more carbon dioxide than the air above, and its humidity is higher.

Soil Structure

The components of the soil are in intimate and dynamic association. Physical and chemical conditions are continually changing, with the daily and annual changes of temperature and moisture conditions and the associated cycles of organic activity. There are spatial variations on the scale of local topographic and geological inequalities, and related to the positions of individual plants, and on a smaller scale related to the mutual disposition of the mineral and organic components and the pores filled with soil solution or atmosphere. The soil is a mosaic of continually changing micro-environments.

The arrangement of the components forms the soil structure. Units of structure are aggregates of solid material known as *peds*; the soil solution may occupy pores in the peds and spaces between them and form films on the surfaces of peds and individual particles. Part of the water content of the soil is represented by the soil solution and part by bound water in association with the crystalline particles, the amorphous mineral materials and the complex organic molecules which go to make up the peds. The soil atmosphere occupies those voids which are not occupied by the soil solution. Micro-organisms live in the soil solution, with extension of mycelium into atmosphere-filled spaces, and the soil animals live in pores containing atmosphere or solution according to the needs of their life style. Plant roots, being unable to move, cannot retreat from a dry pore in which they cannot obtain nourishment to a wet one, and so are found in both, but by growing when moisture is available to them they appear to seek those parts of the soil which are more continuously moist.

The formation of structure by the aggregation of soil into peds is achieved by a number of interrelated mechanisms. The hyphae of fungi and the gummy substances exuded by other micro-organisms bind materials together. In the gut of soil animals, particularly earthworms, mucilages are produced and micro-organisms multiply, and many peds in that part of the soil which is rich in organic materials are faecal pellets and casts of these animals. However, most of these primary binding materials are readily attacked by micro-organisms and so have a short life in the soil. A succession of factors must be postulated to account for the long term stability observed for peds in some soils, each developing in turn as the previous binding material is used up and the population of micro-organisms is modified to make use of new substrates.

Plant roots contribute to formation of soil structure in several ways. As they grow and thicken they compress the adjacent soil, and as they withdraw moisture from it they cause it to contract; fine roots bind particles together and also provide sites of intense micro-organism activity producing binding substances, and as the roots decay they leave open pores. The structure of soils under grass, which has an extensive array of fine roots of which a high proportion dies each year, is particularly well developed.

The importance of soil structure for the growth of plants lies not in the structural aggregates themselves but in the voids within them, created by the passage of animals and left by the decay of roots, and between them, created by the shrinking process or left by the lack of fit between the faecal deposits of the soil animals. These voids, together provide the pore space of the soil, an interconnecting system of tubular, planar and irregular shaped spaces through which the soil solution and atmosphere move and into which roots grow.

The stability of the peds which define the structure is important for the maintenance of pores, since if they collapse and the individual soil particles fall into uniform close packing there is less space for movement of water and for access of air, and in wet conditions the soil may become waterlogged. Coarse textured soils may allow adequate percolation and provide enough spaces for plant roots even in the absence of structural pores, but fine textured soils are almost entirely dependent on structural pores for maintenance of suitable conditions for plant growth. The mechanisms by which structure is formed and maintained are fully discussed by Russell (1971). The relationship between good structure and high crop yields is shown in a study by Low (1973).

The Soil Colloid

The mutual arrangements of the soil components are in themselves sources of structural stability. The fine particles and amorphous gels of secondary minerals and the complex molecules of humus substances behave in the presence of water according to the dictates of surface chemistry: they comprise the soil colloid. An explanation of the nature and behaviour of colloidal materials is given in Childs (1969).

Colloids can be of many types, but they have in common the presence of two phases, which in soil are the liquid phase of the soil solution and the solid phase of mineral and organic substances. Interactions between the phases are dominated by the behaviour of ions at the surfaces of the solid materials.

Mineral particles have electrically charged surfaces because, as will be clear from the account of crystal structure in Chapter 1, although charges are balanced within a crystal there must be ions at or close to the surface whose valency is not fully satisfied. Silica tetrahedra have a very stable structure, so the silicon ion itself is rarely exposed at a surface; the oxygens are firmly held into the structure but those close to the surface have an excess of negative charge. Alkali metal and alkaline earth cations are less strongly held. If they stay in place they may provide an excess of positive charge at the surface, but they are more likely, under soil conditions, to be detached and leave behind a further excess of negative charge. Iron in minerals is normally present in the reduced, divalent condition; when it is near to a surface in an environment which contains ions which will accept electrons, it becomes oxidized to the trivalent condition and the consequent re-arrangement of bonds can produce an excess of positive charge which tends to be rapidly taken up by hydroxyl ions from the surroundings. Aluminium behaves in different ways according to the ionic environment at the surface and is a special case which will be considered later.

In a liquid, molecules and ions are in rapid motion. During collisions between them, and between the molecules and ions and the surfaces of particles in contact with the liquid, ions are knocked off surfaces and molecules are split into their component ions; other ions are trapped into association with each other and with surfaces: there is a continual interchange between free ions and those to a greater or lesser degree associated with each other or with a surface charge. Small particles have a high ratio of surface area to mass, and interactions between their surface charges and the ions in the environment becomes an increasingly important factor in controlling their physical as well as their chemical behaviour as size decreases. Particles of the size of clay minerals, two microns and less, can be regarded as large bodies moving slowly among the very much larger numbers of darting and colliding ions and molecules, and buffetted by collisions with them. The smaller the particle the more it is affected by such collisions: it is this factor which accounts for the different rates of settling of particles in a liquid, the smaller particles falling more slowly because their downward movement under gravity is more easily slowed up and deflected by bombardment by ions and molecules.

Water molecules are polar, that is, though the molecule as a whole is neutral the positive and negative charges of the ions of which it is composed are asymmetrically disposed, so that one "end" is more positively charged and the other more negatively charged. These charged ends are attracted towards the charges on particle surfaces and towards ions in the solution, and the molecules arrange themselves as layers of bound or adsorbed water adjacent to the surface or around the ion. The thickness of the ordered water layer depends on the concentration of charges on the surface or on the size and valency of the ion.

In company with the water molecules, ions of opposite charge to that on a surface are attracted to it, and form a layer over the surface, and ions of like charge to the surface are repelled by the partially masked surface charge but attracted to the ions forming this layer, so they form a more diffuse cloud or outer layer. The double layer of ions may be entirely within the ordered water layer or partly outside it. When two particles carrying their ordered water and ionic atmosphere approach one another, the way in which they interact depends on the extent to which the surface charge and those of the successive layers of ions are masked by the ordered water layer: if they are almost completely masked the much weaker forces of attraction between any particles, dependent on their mass, can be expressed, and the particles are drawn into contact. If the ionic double layer lies to any great extent outside the ordered water layer, the like charges of the outer layers repel each other and the particles move apart. Which happens depends on the concentration of charges on the surfaces, the concentration of ions in the solution, and the nature of the ions in the solution. Since clay minerals have predominantly negatively charged surfaces, it is the concentration and nature of the cations forming the inner layer which is usually the controlling factor, and in soil these are mostly ions of hydrogen, sodium, potassium, calcium, and magnesium, and the ammonium ion in much smaller quantity. The relative effectiveness of these ions in causing particles to come together, that is, to flocculate, or to move apart or disperse, tends to vary somewhat with other environmental factors, but in general the monovalent cations sodium and potassium tend to cause dispersion, and calcium causes flocculation. The hydrogen ion, being very small and monovalent, might be expected to be even more effective in causing dispersion than sodium, but perhaps because it is usually in the form of the much larger hydronium ion it behaves more like a divalent ion, lying somewhere between potassium and calcium. Magnesium is also anomalous, generally being either slightly more or slightly less effective in causing flocculation than calcium, but under some conditions it behaves like a monovalent ion and causes dispersion.

In practice the larger clay particles are too big to be much involved in mobility of the colloid. They tend to be clustered together, since the high ionic concentration in which the minerals can grow is also conducive to flocculation, and once flocculated they remain so. It is particles of the order of size of 0·5–1 micron and less which are easily moved by the forces involved; the combination of numerous such particles in a floc forms a body too heavy to be moved, but when dispersed as single particles they move in the soil solution as a sol. Thus, peds contain flocculated clay whose cohesion contributes to the aggregation of the ped; as moisture moves in the soil pores, if its ionic concentration is such that the finer of the clay particles close to the walls of the pores disperse, they float off into the liquid and move with it until they reach an ionic environment which induces flocculation, when they cluster against the walls of the pore. Conditions can occur in which the whole of the clay component of the peds disperses, and then the peds disintegrate.

The clay minerals form the greater part of the soil colloid but the other components can be very important in deciding the behaviour of the system as a whole. They are the oxides and hydrous oxides of iron, aluminium and silicon and the larger molecules of humus substances, and they exist as particles generally smaller than the clay minerals and as amorphous gels which when dispersed form sols which merge with the true solution phase.

Iron oxide, already partially hydrated by attachment of hydroxyl ions as soon as it becomes oxidized, is released from minerals as other, more soluble ions are removed and exists initially probably as an amorphous gel which slowly ages into crystalline form, either goethite or lepidocrocite. It has a strong tendency to further hydration by adsorbing water molecules. The fully hydrated crystalline form and the hydrous gel lie on the surfaces of the primary minerals from which the iron is released, on the surfaces of secondary minerals, and also can be present as flocs separate from the other minerals. Iron oxide in these forms gives yellowish-brown colours to soils, and together is known as limonite.

Dissociation of hydroxyl ions from the hydrous iron oxides leaves them with positively charged surfaces. This engenders close association with the negative charges on mineral surfaces, and partially masks those charges, so that a clay/iron colloid is less likely to disperse than the same clay would be alone. The association also generally prevents dispersion of the iron oxides alone under conditions which would otherwise allow it. Iron goes into solution only under reducing conditions or under more acid conditions than those which normally occur in soils. It can be mobilized by chelating agents, and the amorphous gel form can disperse into a sol if it is associated with silica, but otherwise ferric iron in soil only moves if the clay particles on whose surface it is carried move.

Aluminium is also released into the soil in the form of an insoluble hydrous oxide, as groups of alumina octahedra are hydrated and detached from primary minerals. It is slightly more soluble than the equivalent forms of iron and small quantities do go into solution at pH values reached in acid soils. Since it is more abundant than iron in most minerals and since it lacks the possibility of solution in reduced form, it accumulates relative to other soil components. However, whereas little iron can be accommodated in crystals of secondary minerals, and none at all in the case of the kaolinitic ones, most of the alumina released goes into combination with silica, first, probably, as an amorphous gel and then in crystalline form. Residual alumina, which may accumulate when the secondary minerals being formed are of kaolinitic kind, with only one alumina layer to one silica layer, forms a hydrated layer on mineral surfaces. In acid conditions it loses hydroxyl ions and its effect on the behaviour of the colloid is similar to that of iron oxides. In increasingly alkaline conditions the possibility arises of losing hydrogen ions to leave negative charges. As this happens the octahedral groups join together by way of the linking of positively and negatively charged sites, with the elimination of one water molecule from each such linkage, that is, the hydroxide polymerizes, and the surplus of surface charges tends to disappear. Crystallization of gibbsite from the polymerized alumina occurs eventually in tropical soils where the accumulation of alumina is particularly marked.

Silica is rather more soluble than the hydroxides of both iron and aluminium, and its solubility increases markedly with temperature, so that considerable quantities go into solution in soils in warm regions, leaving an excess of alumina. In cold climates silica is appreciably soluble only in acid soils, where, as tetrahedra are released from primary minerals they take up hydrogen ions to form a completely hydrated system, as described on p. 21. Dissociation of the hydrogen ions leaves negative charges, and silica behaves similarly to the clay minerals as a negatively charged colloid. The strong tendency to polymerization, enhanced by the availability of hydroxyl ions as pH increases, rapidly reduces solubility, and silica is precipitated in alkaline conditions, as the amorphous silica gel. Crystallization from this stage does occur in some soil conditions, but is a very slow process. At intermediate pH, the negatively charged silica can form complexes with the positively charged iron or aluminium hydroxides; their surface charges being taken up in forming the association, they are much less susceptible to flocculation than either would be alone, and in this form some mobility of all three hydroxides occurs in many soils in which they are individually insoluble.

Humus substances occur in the form of molecules which range in size up to a few thousandths of a micron. From their peripheral groups of predominantly acid nature dissociation of hydrogen ions leaves negative charges. Basic peripheral groups of nitrogenous nature occur, giving the possibility of positive surface charges but this does not seem to be expressed to any extent in their behaviour: like clay minerals and silica they form positively charged colloids. The surface charges are weaker than those of the mineral colloids but on the other hand they have a stronger tendency to adsorb water.

The differentiation of humus into fractions, which have been given different names suggesting that they are different substances, is based upon its behaviour when subjected to different solvent solutions. It therefore reflects the variations in colloidal properties. The fractions form a continuum among the molecules of varying size and chemical composition, from more condensed molecules with relatively fewer reactive groups at their surfaces to less condensed molecules with an abundance of reactive groups, and involving different degrees of association with the mineral components of the soil. The less condensed forms, *fulvic acids*, readily disperse even in acid conditions to form sols which merge with the true solution phase. The more condensed forms, *humic acids*, are flocculated in acid conditions but disperse in alkaline, especially if sodium ions, and sometimes magnesium, dominate the cations present. *Humin*, the fraction which remains flocculated in alkaline conditions, probably represents humus molecules in association with calcium or with the mineral components of the colloid.

Association of humus substances with iron and aluminium oxides occurs, and via these positively charged components humus molecules can be linked to clay minerals or to silica so as to form stable complexes. Direct association with montmorillonite can occur and is extremely resistant to dispersion and to micro-organism attack. It is thought that the molecules of humus get in between the layers of the crystal structure, in the interlayer water, and cannot be reached by even the smallest bacteria. The montmorillonite/humus association takes on a very dark colour.

Cation Exchange and Soil Acidity

Water molecules dissociate into their constituent hydrogen and hydroxyl ions to the extent of one molecule in 10^7. The hydrogen ions tend to associate with another whole molecule to form the hydronium ion, H_3O^+, but it is conventional to talk of the hydrogen ion except where its behaviour is affected by the different geometrical relations of the larger ion. Carbon dioxide from the atmosphere and derived from respiration of plant roots, soil animals and micro-organisms is dissolved in the soil water and forms carbonic acid, which may dissociate into the hydrogen and carbonate ions. Organic acids, washed into the soil from the vegetation or the litter layer, released in the soil during decomposition and excreted by micro-organisms and by animals in and above the soil, also dissociate to a greater or lesser degree to contribute hydrogen ions to the system. The number of hydrogen ions present is a measure of acidity: the more there are the more acid is the solution, and acidity is conventionally expressed by pH, the negative logarithm of the hydrogen ion concentration. Since in pure water the concentration of hydrogen ions is equal to that of hydroxyl ions, this concentration is an indication of neutrality; one hydrogen ion for 10^7 water molecules gives pH 7. A greater concentration of hydrogen ions gives a more acid solution and a lower pH, for example, one hydrogen ion for every 10^4 water molecules gives an acid solution of pH 4, and conversely fewer hydrogen ions, say one in 10^9, gives an alkaline solution of pH 9.

The other cations in the soil solution are predominantly those of sodium, potassium, calcium and magnesium, most abundant of the soluble components of rock and predominant among the plant nutrients. Initially detached from surfaces of minerals a proportion of them is thereafter kept in circulation by uptake from the soil solution by plant roots, incorporation into plant tissue, maybe a stage or two of incorporation in animal tissue, and return to the soil solution via excretion or microbial decay. Ammonium ions are released into the soil by excretion by animals and micro-organisms, and participate in the nitrogen cycle. In the soil all these cations are in continual interchange between free existence in the soil solution and varying degrees of association with anions and with the charged surfaces of the soil colloid. The total amount of cations which can be held by the colloid is the "*cation exchange capacity*" of the soil. Most of it arises from the clay minerals, and it varies with the degree to which their surface charges are taken up in stable associations with the positively charged components of the colloid, and with the type of mineral. Kaolinite, having no dissociable cations has a low surface charge, only that representing the incompletely satisfied oxygen valencies, and a low exchange capacity. The 2:1 minerals have large numbers of substitutions in their crystal lattices and so large numbers of dissociable cations and a high exchange capacity, montmorillonite the highest of all. Some cation exchange occurs on humus substances too.

The proportions in which the cations present in the system occur in the exchange complex depends on their concentrations in the solution and the strengths with which they can be held. Some are more easily displaced than others: in general sodium is more easily displaced than potassium and both more easily than calcium and magnesium, and the difference is enhanced in dilute solutions. Clearly, since the tendency of the colloid to disperse is dependent on the relative numbers of different ions present and since they compete with each other with different results at different concentrations, the system is rather complex, but in general the greater the concentration of a particular cation in the system the greater its concentration in association with the colloid and the greater its predominance in deciding the behaviour of the colloid. The behaviour of potassium is anomalous. The ion is close in size to that of oxygen and there are many spaces in crystal lattices into which it can fit. It tends to move from the exchange complex into the lattice of secondary minerals and so becomes fixed. There is much less potassium in soil solutions than its mobility and its abundance in rocks would suggest, and since it is one of the major plant nutrients its behaviour is critical.

The pH of the soil as measured by standard techniques is the pH of the soil solution. Since the concentration of all cations is greater in the inner layer of the ionic atmosphere around charged surfaces, forming the exchange complex, than it is in the free solution, hydrogen ion concentration of the solution is lower than that of the soil as a whole. That is, the measured pH is generally higher than the true pH of the soil.

However, the concept of soil acidity is further complicated by the reactions of the aluminium ion. In acid conditions, as cations are lost from the exchange complex and increasingly displaced from positions in the mineral surface, aluminium ions close to the surface are detached from their octahedral units and take up residence in the exchange complex. They are very strongly held there, and though they are exchangeable, they occupy the exchange complex in preference to hydrogen ions. The hydrogen ion concentration of the soil solution is therefore increased, and hydrolysis of the aluminium, taking OH ions out of solution, increases acidity still further. In very acid soils the measured pH is lower than that of the soil as a whole.

Movement of Water in the Soil

The water in the soil is present within the colloid as adsorbed water on particle surfaces and in gels and between the layers of clay minerals, and in the soil solution forming films on surfaces of peds and particles and filling some of the pores. Movement of the solution through the pores is limited by surface tension, and the finer the pore the greater the pressure difference needed to move the solution through it. Drainage of a soil under gravity leaves moisture in the finer pores and as surface films, the quantity of water so held being known as the *field capacity* of the soil. More water can be removed by plant roots, exerting suction which can take water from finer pores than will drain, but roots cannot move water through the finest pores, and the water content at which they can no longer draw supplies is known as *permanent wilting percentage*. Beyond this water is lost only by evaporation and it is lost not only from pores but from the colloid too. Those clay minerals which expand by taking water into interlayer positions shrink, and in soils rich in such clays the soil as a whole shrinks markedly, cracks opening as it does so. In such soils repeated drying is the predominant factor in forming soil structure. Clay minerals normally flocculate in "edge to face" arrangements, which leaves micropores in the floccs. As water is lost by evaporation from these pores, particularly if it happens repeatedly, they may collapse and the flat crystals fall into face to face packing which is much more difficult to disperse. Gels of humus substances and hydroxides lose water as desiccation proceeds. Such dehydration can be irreversible, and in soils subject to intense desiccation concretions form of iron oxides and silica, humus and silica, humus and iron oxides, or all three, with some alumina too.

As rainwater soaks into a soil it fills up the pores and a boundary of saturation moves down the profile. If the soil is initially drier than field capacity, the water may take considerable time to move into the finer pores and to swell the clay minerals and the gels. As the soil swells the pores close up and the rate of percolation decreases. As the fresh water enters the soil it takes soluble substances into solution, or dilutes the solution already there, and the process of ion exchange proceeds. As the solution moves down through the soil, soluble components are carried from the upper levels to the lower parts of the soil, and in a freely draining soil a long period of rain may carry them out of the profile and into surface waters or the regional groundwater. The ionic concentration of the soil solution increases downwards and may reach a point at which substances taken into solution further up are precipitated. In the upper part of the soil dilution of the soil solution and exchange of cations in the exchange complex for those of hydrogen in the fresh water may reach the point at which the colloid begins to disperse. The dispersed material travels down through the pores until it reaches a point at which the concentration of the cations leached ahead of it is sufficient to cause flocculation and is then deposited on the walls of the pores. The alignment of clay minerals in these circumstances is parallel to the direction of flow, and can be seen to be so under the polarizing microscope.

The processes by which soluble substances are released from mineral and organic materials continue all the time the soil is moist. Leaching is intermittent, occurring during each period of rain and long enough afterwards for surface run-off to be complete and the soil to drain to field capacity. The chemical conditions in the soil therefore change in cyclic fashion, concentration of the solution building up as water is

taken by plants and lost by evaporation and at the same time alteration processes go on, and falling again when next it rains. The depth to
which the dilution of the soil solution occurs depends on the amount of
rain which falls and the amount of water the upper parts of the soil
absorb before any drains into the soil below: short showers may do no
more than wet part of the soil. The overall effect of the cyclic changes
depends on the amount of weatherable minerals in the soil and the rate
of supply of organic residues, on the temperature, which has a very
strong effect on the rates of chemical and biological processes, and on
the frequency and amount of rain. In cool rainy areas in soils on rocks
poor in bases leaching predominates strongly over rate of release of
bases, in the upper part of the soil and sometimes throughout the
profile, and the exchange complex is saturated by hydrogen, and
aluminium, ions which replace basic cations as fast as they are released.
At the other end of the scale, in a climate in which temperatures are
high and the soil usually moist but the rainfall not excessive, soils on
basic rocks have their exchange capacity saturated by bases. Variations in climate and parent material can provide combination of rates
of release of bases and rates of leaching which gives a complete range
of soil conditions between these extremes.

Mineral Alteration

Hydrolysis and oxidation of mineral surfaces proceeds by the replacement of basic cations by hydrogen ions and the association of hydroxyl
ions with aluminium and iron. In acid conditions the movement of
aluminium from lattice positions to the exchange complex and its
hydration there releases more hydrogen ions to displace basic cations.
If leaching is active the cations do not build up in the soil solution so as
to slow down the reaction, and hydrolysis proceeds rapidly. The surface
of the mineral becomes unstable and silica tetrahedra and alumina
octahedra are detached. Aluminium in tetrahedral co-ordination,
where it is in substitution for silica in the primary mineral, reverts to
its preferred octahedral state once free of tetrahedral geometry, with
further hydration. Formation of secondary minerals depends on the
behaviour of silica and alumina once released and on the concentration
of other ions in the system. Kaolinite is formed by polymerization of
silica and alumina into sheets and the arrangement of the sheets in
pairs; it requires no other constituents and has the lowest silica
requirement of any of the secondary minerals, and so is most likely to

form when bases are leached away and some of the silica too has gone into solution. It is characteristic of strongly leached soils.

The other clay minerals need varying amounts of bases and iron, and twice as much silica as alumina. Some of the aluminium is in tetrahedral co-ordination and it may be that it has to be incorporated into the secondary minerals before the alumina tetrahedra released from primary minerals have a chance to revert to octahedral arrangement, since the low pressure and temperature at the earth's surface probably preclude the formation of alumina tetrahedra from solution. To enable 2:1 secondary minerals to form, the products of surface hydrolysis have to remain in close association and their formation is characteristic of soils not being actively leached. Which mineral it is which crystallizes depends on concentrations of the various cations present and to a certain extent on the structure of the primary mineral. Micas have a very strong tendency to alter to micaceous clay minerals, by loss of potassium and increased hydration.

The mixed mineralogy of most parent materials and the variation in conditions from point to point and from moment to moment in the soil lead to the formation of mixtures of secondary minerals, except in the two extreme cases of only kaolinite forming in strongly leached soils and only montmorillonite in very unleached basic conditions, where once it has begun to form its swelling nature further inhibits leaching and enhances conditions for its own formation. In most soils, the clay mineral suite changes through the profile, with usually a higher percentage of kaolinite in the surface and more 2:1 minerals lower down, reflecting the degree of leaching. Interstratifications of different minerals among the 2:1 forms is common and may arise from changing soil conditions. Chlorite layers in montmorillonite and vermiculite may form by the incorporation of hydrated alumina into interlayer positions instead of on external surfaces.

Change in the clay mineral suite with time occurs, due to changes in the conditions within the soil as mineral alteration proceeds and as the vegetation changes in response to the changing availability of nutrients. External changes such as climatic variation, change in ground water level or interference by man can produce further changes. In the course of these changes secondary minerals themselves become subject to alteration. The surface attack suffered by primary minerals may be suffered by secondary ones if leaching is increased so far as to remove the materials necessary for their continued formation. Kaolinitic minerals can only suffer total degradation if silica goes into solution, and alumina left behind crystallizes as gibbsite. This has happened in very old soils in tropical regions where leaching is active:

rate of alteration of primary minerals is high and if their products continue to be removed, the supply of bases and silica will eventually slow down to the point at which kaolinite begins to degrade. In less extreme conditions, 2:1 minerals are stripped of their basic cations, and surface degradation leads to loss of silica layers, leaving kaolinitic minerals. This probably happens in the course of development of most soils, except those on rocks too acid to allow 2:1 minerals to form in the first place. Starting at the surface, the 2:1 minerals formed in the earlier stages of soil formation are gradually degraded as leaching proceeds. As the particles get smaller they become more susceptible to dispersion and once dispersion begins to occur and part of the colloid migrates down the profile, leaching can become more effective still in a soil left more porous by the loss of fine material. The larger particles may then begin to degrade and in their turn suffer dispersion. The process is helped by the presence of chelating agents which take iron and aluminium from the hydrated layers on surfaces of minerals, increasing their susceptibility to dispersion and exposing their surfaces to attack. With the mobilization of aluminium by this means the kaolinitic minerals are readily degraded.

Loughnan (1969) discusses processes of mineral alteration in soils.

Cycles in the Soil

The cycle of growth and decay of which all living things are part is made up of the individual cycles of each constituent element. In the land based system the soil feeds into the cycle those elements derived from the mineral world and the nitrogen derived from the atmosphere. Hydrogen and oxygen are supplied from the atmosphere via the soil in the form of water, which is not only essential in itself but also provides the medium in which the chemical and biological processes which make nutrient elements available take place and in which these nutrients are taken into the plants. The soil acts as a reservoir for nutrients. They are present in available form in solution and in the exchange complex, and in less available form in the organic residues undergoing decomposition, and in the mineral particles undergoing alteration, from which fresh nutrients are fed into the cycle. The amount of material participating in the cycle can build up to the point at which losses from leaching, erosion and export of organic material are just replaced by nutrients released from minerals, or in the case of nitrogen incorporated from the atmosphere. The level of activity at which the whole cycle is maintained is limited by the rate at which the nutrient in shortest supply can be made available.

The Carbon Cycle

Carbon is taken in by green plants from the air as carbon dioxide and is combined with hydrogen and oxygen from water to form carbohydrates in the process of photosynthesis. Carbohydrates provide much of the structural materials of plant tissue and are the raw materials from which all other biochemical synthesis starts. Photosynthesis is also the means by which the sun's energy is utilized by plants, and the carbohydrates represent a store of energy for later use by the plant itself and which can be used by the animals which feed upon the plant.

Animals and aerobic micro-organisms use the carbohydrate component of plant and animal tissue to provide their requirements of both carbon and energy. Since the breaking down of carbohydrates to obtain energy yields more carbon than is needed the excess is released as carbon dioxide in respiration. In the soil, as residues are utilized by soil animals and micro-organisms, carbon dioxide released by their respiration is returned to the atmosphere by diffusion between the soil atmosphere and the free atmosphere at the soil surface. The carbon dioxide content of the soil atmosphere is higher than that of the free atmosphere, and it is in constant interchange with carbon dioxide in solution in the soil moisture. In some soils excess carbon dioxide is immobilized by formation of calcium carbonate and is thus isolated from the carbon cycle until environmental conditions change and it is either re-incorporated within the soil or released elsewhere after erosion or solution in ground or surface waters. In soils forming on carbonate-containing parent materials carbon which was immobilized and taken out of the carbon cycle in earlier geological periods is released. In these soils some of the carbon dioxide in the soil moisture and the soil atmosphere is therefore made from "old" carbon. Many micro-organisms seem to need small quantities of carbon dioxide even when functioning aerobically, and under the anaerobic conditions which can occur locally and transiently even in apparently well aerated soils and are a constant feature of poorly drained soils, some micro-organisms switch to anaerobic modes of operation in which they utilize carbon dioxide. They therefore incorporate some old carbon into their tissues, and pass it on into the humus substances which form from their residues. Results of radiocarbon dating of soil organic matter, and particularly of fen peats and lake deposits, in areas of carbonate rocks may therefore be higher than they should be.

The possibility also exists that plants growing in soils on carbonate rocks may assimilate a higher proportion of old carbon than those growing elsewhere. The quantities of carbohydrates taken in by plant roots are probably insignificant, though those plants supporting mycorhyzal fungi may assimilate organic compounds whose carbon is derived in part from soil CO_2. The atmosphere immediately above the soil contains a concentration of carbon dioxide released from the soil, and though mixing with the free atmosphere probably rapidly reduces the proportion of old carbon to levels at which the error in a radiocarbon date would be negligible, the risks may be significant in situations where dense vegetation prevents air movement close to the ground and the rate of soil processes is high, such as under dense forest, or a luxuriant crop, in a warm, humid climate.

The fall to the ground of carbohydrate-containing plant and animal residues is the means by which the sun's energy is transferred to the life processes in the soil environment. Release of the plant nutrients from the residues is dependent on these life processes, and the release of fresh supplies of nutrients from minerals during development of a soil/plant system, and for replacement of losses, is increased by the presence of organic substances. The carbon cycle thus drives all the other nutrient cycles, and the soil is the gear box where the power is transferred from the engine to all the other components of the system.

The Nitrogen Cycle

The nitrogenous compounds in all organisms are based on the amino-acids and the purine and pyramidine bases. More than twenty amino-acids are known, and they and the compounds formed from them make up much the greater part of nitrogenous material. Amino-acids are formed in plant leaves by interation between the ammonium ion and carbohydrates. Some micro-organisms can synthesize some or all of the amino-acids they need in the same way, but others have to take them in ready made, as do all animals. Proteins are formed by polymerization of amino-acids and the linking together of the polymer chains. All enzymes are proteins, and all cytoplasm contains proteins as a reserve food awaiting utilization in the cell. Proteins participate in the formation of the structural and functional membranes in all cells, and in the cell walls of some micro-organisms they act as structural materials. In animals the structural components of skin, sinew and bone include the protein collagen, and the non-living structures of hair, feather, hoof and horn are the protein keratin. In arthropods the exoskeleton is chitin, the amino-sugar polymer which also forms the cell walls of some fungi. Amino compounds are present in the exudates of some micro-organisms.

The five purine and pyramidine bases are organic compounds in which carbon and nitrogen form five member ring structures. Their sequence in nucleic acid molecules contains the genetic information in all types of organism. One of these bases is also a component of the sugar–phosphate compounds which carry out all the energy transfers involved in cell processes.

The nitrogen cycle is illustrated in Figure 8. Nitrogen is taken in by plants in the form of the nitrate or ammonium ion from solution in the

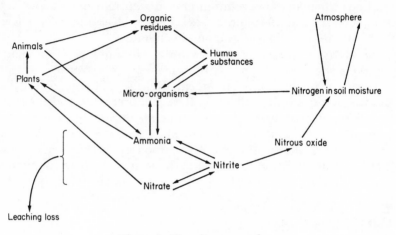

Figure 8. The nitrogen cycle.

soil moisture. Nitrate is reduced to ammonium within the plant before it can be used in biosynthesis. The animals which feed upon the plants and upon each other, and those micro-organisms which need carbohydrate to provide energy, concentrate nitrogen. The herbivores at the start of the food chains consume plant material with a high ratio of carbon to nitrogen, but since most of the carbohydrate has to be used to provide energy what remains for synthesis of structural and functional materials is so rich in nitrogen that not only is protein used for structural purposes which in plants are performed by carbohydrates, but excess nitrogen is excreted. Mammals excrete urea and insects uric acid. Soil animals in general excrete either or both, and micro-organisms excrete ammonia, and sometimes more complex nitrogen compounds. Urea and uric acid are rapidly converted to ammonia in the soil by inorganic processes.

In the course of decomposition of the plant and animal residues which fall to the soil the nitrogenous compounds are utilized by a succession of animals and micro-organisms and at each stage some of the nitrogen is mineralized, that is, converted to the inorganic form of ammonia which yields the ammonium ion on solution in the soil moisture. All the nitrogen taken in by plants from the soil is thus eventually returned to the soil in the form in which it can be taken in again. If there were no losses, the nitrogen cycle could continue indefinitely in this simple form, though it would not explain how nitrogen, derived from the atmosphere in the elemental form of nitrogen gas, came to be in the soil in the first place. Losses do occur, gaseous ammonia returning to the atmosphere and the ammonium ion becoming strongly adsorbed in the exchange complex and trapped between the layers of clay minerals. In further loops of the nitrogen cycle, elemental nitrogen returns to the atmosphere and the nitrate ion is lost by leaching. Particular groups of micro-organisms are responsible for these parts of the cycle.

Nitrogen Fixation. Fixation, that is, incorporation of elemental nitrogen from the atmosphere into the soil/plant system, is clearly essential in the early stages of soil formation and is also necessary to replace losses by leaching, adsorption, erosion, by cropping and grazing and as a result of fire. It is carried out by two groups of organisms, those which live free in the soil and those which live in symbiotic relationships with plants.

The free living nitrogen fixers are some groups of autotrophic and heterotrophic bacteria, of both aerobic and anaerobic habit, and the blue-green algae, which are aerobic autotrophs. Nitrogen in solution in the soil moisture is assimilated and incorporated into amino compounds, probably after reduction to ammonia in the cell. The organisms exude soluble nitrogen compounds during or after active growth and these and their cell materials can be used by other micro-organisms, with the production of excreted ammonia. Fixation is inhibited by the presence of ammonia, and so is performed most actively in soils which are poor in nitrogenous residues, though there must be sufficient carbohydrate-containing residues present to nourish the heterotrophic nitrogen fixers. Even those nitrogen fixers which are of aerobic habit are more active in conditions of oxygen depletion and many are entirely anaerobic in habit, so fixation is a feature of soils which are waterlogged and of localized areas of oxygen shortage in otherwise well aerated soils. It is therefore commoner in fine textured soils which drain badly and in which diffusion of oxygen to centres of organic activity is slower than in coarse textured soils.

Symbiotic nitrogen fixers are certain groups of aerobic, hetero-trophic bacteria which live in nodules in the roots of certain plants. The leguminous plants are the largest group of plants which can participate in this symbiotic relationship, and it is thought that the ability was developed in the tropical legumes, from which those now found in temperate regions are derived, in response to nitrogen deficiency in strongly leached tropical soils. Some plants which live in wet habitats, notably alder, bog myrtle and sea buckthorn, also have root nodules containing nitrogen fixing bacteria. The nitrogen taken in from soil moisture is used to make amino compounds for the use of the bacteria and for supply to the plant, while the plant supplies carbo-hydrates to the bacteria. The respiration rate of these bacteria is high since they need energy to carry out the fixation reactions, and so in contrast to fixation by free living organisms they require well aerated soil to supply plenty of oxygen and allow carbon dioxide to get away. The nitrogen fixed is not available in the soil for use by other plants until the nitrogen-rich residues of the nodulated plants are decomposed in the soil or broken down by animals. Exploitation of leguminous plants for their high protein content, in particular the use of their seeds, takes away all the nitrogen fixed. Increase in the nitrogen status of a soil for the benefit of non-leguminous crops is achieved by "green manuring", that is, cutting or ploughing-in the legumes before the nitrogen has been concentrated in the seeds. The value of legu-minous plants in pasture is also dependent on the high nitrogen content of the green foliage, and excretion by the grazing animals returns part of the nitrogen to the soil in available form.

Nitrification. The process of nitrification forms a loop of the nitrogen cycle which accounts for a large part of the nitrogen loss from soils. Certain bacteria derive their energy from oxidation of ammonium ions to nitrite and of nitrite to nitrate. The nitrate ion can be assimilated by plants but it is also susceptible to leaching since soils have little power to retain soluble anions. The nitrite ion generally has only a transient existence in the soil since the bacteria which oxidize it to nitrate work faster than those which produce it. Nitrification proceeds in all well aerated soils provided the pH is not too low for growth of the bacteria and provided there is sufficient organic matter to satisfy their carbo-hydrate requirements. In poorly aerated soils there is some risk of accumulation of nitrite, which is toxic to plants, because the bacteria which produce it are more tolerant of oxygen shortage than those which oxidize it. The process of denitrification generally prevents nitrite building up, however.

Denitrification. With fixation putting nitrogen into the soil and nitrification exposing it to leaching, without a mechanism to return it to the atmosphere all the nitrogen would end up in the oceans. Denitrification is carried out by further groups of bacteria which derive the oxygen they need for oxidation of carbohydrates by reducing nitrite and nitrate to nitrogen and nitrous oxide, which can escape from the soil; nitrous oxide is quickly reduced to nitrogen once released. Since the bacteria which can perform this operation have a source of oxygen denied to those which cannot, they can compete successfully for foodstuffs in poorly aerated soils. Denitrification is characteristic of wet situations, particularly marshes and estuarine mud flats into which nitrate bearing waters drain, and it is here that most of the nitrogen fixed in the soil is returned to the atmosphere. Paddy fields are also places of active denitrification, and the nitrogen released in the soil is utilized by algae which form a bloom on the water standing in the fields.

Sulphur

Sulphur is a component of all proteins: two of the amino-acids have sulphur on a side chain, and in those proteins consisting of more than one polypeptide chain the chains are held together by bonds between the sulphur atoms in each. Protein molecules are folded into configurations which are braced by further sulphur bonds between adjacent loops.

Plants take in sulphate ions from the soil solution and use the sulphur in the synthesis of amino-acids. In amino-acids and in the intermediate compounds formed during their synthesis the sulphur is in a more reduced state than in the sulphate ion. Plant and animal residues bring these reduced organic sulphur compounds into the soil. Many micro-organisms satisfy their sulphur requirements by taking in the amino compounds ready made. In the course of mineralization of organic residues the sulphur not incorporated into the materials of successive generations of micro-organisms or into humus substances is released into the soil still in reduced form. Under anaerobic conditions hydrogen sulphide and the evil smelling organic sulphur compounds called mercaptans are released. Under aerobic conditions hydrogen sulphide is released but is immediately oxidized in the soil moisture with the formation of the sulphate ion.

A cycle involving only assimilation of sulphate by plants and release of hydrogen sulphide by micro-organisms followed by inorganic oxidation would be adequate to maintain the system. Unlike nitrogen, sulphur is derived from the mineral world and its incorporation into the soil does not require a microbial pathway. Any losses from the soil to the atmosphere in the gaseous form of hydrogen sulphide are returned in the rainfall after oxidation to sulphate, and losses of sulphate in the soil solution by leaching can be replaced from the parent material by mineral alteration. Nevertheless, the sulphur cycle does have loops through microbial pathways which are analogous to those involving nitrogen, the common factor being the multiplicity of oxidation states in which nitrogen and sulphur can exist. There are micro-organisms which can obtain oxidizing power by reduction of the higher oxidation states, enabling them to obtain energy in the usual way by oxidation of carbohydrates, and which can derive their energy directly by oxidation of the more reduced sulphur or nitrogen compounds.

In aerated soils many micro-organisms reduce sulphate, taken from the soil moisture, and partly oxidized sulphur compounds which are released during decomposition of organic matter, releasing hydrogen sulphide which is then inorganically oxidized to sulphate. Under anaerobic conditions a few forms of bacteria become very successful in the competition for nutriment by virtue of their ability to derive oxygen by reduction of sulphate, just as do those which reduce nitrate and nitrite. Hydrogen sulphide is produced in quantity in these situations, and is responsible for much of the characteristic smell of the mud and decaying plant remains beneath stagnant water. In unaerated organic muds accumulation of hydrogen sulphide begins before reduction of iron oxide, so that when ferrous iron is released the hydrogen sulphide is available for its immediate conversion to ferrous sulphide, which imparts the black colour to such deposits. Accumulation of hydrogen sulphide when there is insufficient iron to use it up can reach levels toxic to plants.

Several groups or organisms derive energy by oxidation of the more reduced forms of sulphur instead of or as well as oxidizing carbohydrates. They take in elemental sulphur, sulphide ions or some partly oxidized inorganic sulphur compounds and release sulphate. One group of bacteria, the *Thiobacillus* genus, is almost entirely chemoautotrophic, obtaining all their energy in this way, and many bacteria, fungi and actinomycetes carry out partial or complete oxidation less efficiently and apparently only incidentally to a more normal mode of existence. The activity of sulphur oxidizers is beneficial in using up accumulations of hydrogen sulphide as soon as waterlogged soils become re-aerated as they drain. During reclamation of drained estuarine marshes the oxidation of sulphides is essential to prevent toxicity to plants, but the great abundance of sulphide in these soils, derived from the sulphates in sea water and accumulated over a long period of anaerobic decomposition, leads to formation of sulphuric acid as it is oxidized, and the resulting extremely low pH has to be counteracted by addition of lime. The lime also displaces the undesirable high quantities of sodium from the exchange complex, assisting the flocculation of the colloid and the development of soil structure.

In arid regions production of sulphate can lead to accumulation of gypsum in the soil, forming a gypsic horizon at depth or an encrustation at the surface. However, in saline soils a beneficial reduction in pH from extremely alkaline levels results from the formation of sulphuric acid by certain of the sulphur oxidizers.

There are some photosynthetic bacteria which use a sulphur-containing pigment instead of chlorophyl. They are not important in terrestrial soils since they require light, but are active at the soil/water surface of aquatic soils and can be an important part of the micro-organism population there.

Phosphorus

Phosphorus in the soil/plant system is almost entirely in the form of the orthophosphate radical, in inorganic combination with hydrogen and metal cations and in organic combination with various organic substances. Adenosine triphosphate, a compound of the nitrogenous base adenine, the sugar ribose and three phosphate radicals contains two "high energy" bonds between phosphates. The formation and breaking of these bonds is the means by which energy in all living cells is incorporated in sugars by photosynthesis and transferred from them to all the energy-using reactions involved in life processes. Phosphate is part of the structure of nucleic acids and is also present in phospholipids, fatty substances in structural materials and food reserves, and is a component of some proteins. In seeds, phosphorus supplies for the seedling are stored as phytin, a calcium/magnesium phosphate compound of the aromatic alcohol inositol. The actual quantity of phosphorus present in plant and micro-organism materials is low in relation to its fundamental importance in life processes. In the vertebrate animals phosphorus is used in much greater quantity in the formation of bone, whose mineral component is the calcium phosphate hydroxyapatite. Animals concentrate phosphate just as they do nitrogen, the great quantities of plant food which herbivores need to eat to provide energy supply sufficient phosphate for structural purposes, and a surplus which is excreted.

Plants take in the phosphate ion. Phosphorus differs from nitrogen and sulphur in that the form which can be taken in from solution is an ion of low solubility in most normal soil conditions. Phosphate is released into the soil from primary minerals, bone and organic combinations in a number of valency states, PO_4^{---}, HPO_4^{--} and $H_2PO_4^{-}$, which form salts with calcium, iron and aluminium whose solubility varies greatly with conditions of pH and state of oxidation in the soil. In alkaline conditions insoluble calcium phosphates form, and if calcium carbonate is present a combined carbonate–phosphate forms by substitution of phosphate for carbonate in the crystal lattice. In acid conditions calcium phosphates become more soluble and phosphate associates instead with iron and aluminium hydroxides and clay minerals, displacing hydroxyl ions. The reaction can be regarded as adsorbtion by anion exchange or as formation of iron and aluminium phosphates as thin layers on the hydroxide surfaces. To begin with, when the layer is newly formed and very thin and the strength of the adsorbtion low, the phosphate can easily be displaced by hydroxyl ions and is in equilibrium with the phosphate ions in solution. With accumulation of more phosphate, strongly bound crystalline iron and aluminium phosphates, closely associated with the hydroxides, become inaccessible to interchange with the ions in solution and are unavailable in the short term to uptake by plants. The phosphate in equilibrium with the ions in solution, and therefore available to replace that taken by plants, is known as the "labile" phosphate of a soil.

In strongly leached and very acid soils, phosphate ions which displace hydroxyl ions associated with aluminium in the surfaces of clay minerals can cause degradation of the mineral, the aluminium phosphate breaking away as a group, disrupting the crystal structure. Some phosphorus is found in the crystal lattice of clay minerals, and here it appears to replace silicon in tetrahedral positions.

Phosphate can associate with organic molecules such as humus substances by displacement of hydroxyl ions on peripheral groups. Complexes involving organic molecules and iron and aluminium phosphates can form, and are important in the protection of iron from corrosion in situations where phosphate is available in the soil and a substance such as a proteinaceous part of an artifact is in close contact with the iron.

Since neutral and alkaline soils usually contain enough calcium, and almost all soils enough iron and aluminium hydroxides, to immobilize large quantities of phosphate, most of the inorganic phosphate in soils is in one or other of the insoluble forms. In soils near to neutral all forms are present. When pH rises there are more hydroxyl ions available to displace phosphate from the iron, and to a lesser extent from the aluminium phosphates, but the phosphate displaced immediately associates with calcium ions in the exchange complex. Similarly, when pH falls some calcium phosphate goes into solution but the phosphate ions become adsorbed onto the iron and aluminium hydroxides. Frequent slight changes in pH in a soil near to neutral do keep a high proportion of the phosphate in the labile pool, however, by preventing ageing of the insoluble forms. In strongly acid and strongly alkaline soils, the mobilization of phosphate in sufficient quantities to maintain the supply to plants is probably in part due to the activity of some of the organic acids formed by micro-organisms in chelating the metals with which the phosphate is associated and so releasing it into solution, as well as by the release of phosphate itself in the inorganic form by micro-organisms degrading organic residues.

In anaerobic situations where there is an abundance of organic matter undergoing degradation and releasing phosphate and where some of the iron is in reduced form, the ferrosoferric phosphate vivianite can form. It is found in marsh clays and in floodplain deposits which do not drain adequately, and has been found in the humic horizons of buried soils from which air is excluded, in the form of a white powder which can aggregate into nodules or pellets, and which turns bright blue on exposure to air. It is sometimes seen as a blue bloom on iron artifacts from such situations, its formation protecting the iron from further corrosion.

The primary form of phosphate is the calcium phosphate mineral apatite, normally fluorapatite, which is widely distributed in igneous and metamorphic rocks but often in very small quantities and in tiny needle-like crystals which can be totally enclosed in some other mineral and so protected from dissolution. Because of its low solubility apatite is found in many sedimentary rocks, together with iron and aluminium phosphates, but again may be in small quantities. Phosphate is thus present in all soil parent materials: without it no true soil could form since plant growth and micro-organism activity would be impossible. Phosphorus is not released from soils or rocks in any gaseous forms which can be brought down in the rainfall, unlike sulphur and nitrogen, and so losses from soils can only be replaced from the parent material. In spite of its low solubility, some phosphate is lost from all soils, since during the transformations within the soil it is briefly exposed to leaching. The cycle of plant growth and decay concentrates phosphates in the upper horizons of soils to a greater extent than other nutrients, since any taken in by roots at depth is returned to the soil surface in residues and is likely to be immobilized close to its point of release from organic combination. Erosion therefore causes serious loss. Many strongly leached soils are very low in phosphate and bear vegetation limited to a sparse cover of a restricted number of undemanding species. Any export of phosphatic material such as that resulting from use of the soils for cropping or grazing can completely destroy the vegetation. This applies particularly to some of the soils on the ancient land surfaces of southern Africa and Australia which have undergone very long periods of tropical or subtropical soil formation and have been continually losing material from their surfaces at the same time as losing material in solution by leaching. It has been suggested (Walker, T. W., 1965) that many soils on phosphate-poor rocks are limited in their development from the start, and are kept in a permanently immature state, bearing a vegetation limited in abundance and variety, and that the pattern of soil types in some landscapes is the pattern of phosphate content of the parent material rather than of any other variable.

The redistribution of phosphate brought about by movements of animals from grazing ground to fold, by concentration by people of their rubbish and excreta close to their dwellings, and by deposition of highly phosphatic corpses in burial places tends to become fossilized by the immobilization of phosphate in the soil.

All archaeologists are aware of the potential value of phosphate studies in survey and in interpretation of archaeological deposits. However, in any normal soil phosphate content varies considerably from point to point (Ball and Williams, 1968), and before a phosphate survey can be interpreted archaeologically the range of lateral variation and the distribution of phosphate through the normal soil profiles of the area must be investigated. Success of a phosphate survey depends on the use of a very large number of samples, and in some apparently successful studies much of the information yielded by the phosphate survey is either just what one would expect anyway, or such as would be derived from a proper study of soils in relation to field evidence without the need for analytical work (e.g., Sieveking et al., 1973).

Iron and Manganese

Iron and manganese, like nitrogen and sulphur, provide micro-organisms with the possibility of using changes in oxidation state to obtain energy, either directly by oxidation of the metal, or indirectly, using reduction of the metal to obtain oxidizing power which is used to get energy from carbohydrates.

Many bacteria can turn to this latter mode of operation in soils when they have used up the available oxygen faster than fresh supplies can reach them. Other bacteria in aerated soils can oxidize the ferrous and manganous forms of the metals, though not as an exclusive energy source, and in most soils, where conditions of oxygen shortage are short lived or localized, iron or manganese reduced by the first group is likely to be rapidly re-oxidized by the second. Immobilization of manganese as the dioxide by this means can be a source of deficiency in its availability to plants.

Iron bacteria which do use oxidation of the ferrous iron as their sole source of energy are found in streams draining from waterlogged soils and deposits. Marsh soils forming on or in iron rich materials provide an abundance of organic residues and of iron, so the iron reducing activity in the soil is high. As water carrying the ferrous ions drains from the marsh, a florescence of bacteria develops at the point where it begins to flow freely in the open air, and any vegetation or plant residues in the aerated water become thickly coated with a deposit of ferric hydroxide.

In plants and animals iron is present in the pigmented substances involved in respiration, such as haemoglobin, and in chlorophyl. Manganese is a component of the enzymes of respiration and protein synthesis. The quantities needed are very small, and these metals can be considered as micro-nutrients. While iron is abundant in the soil, its deficiency for plant growth is of common occurrence, particularly in certain crop plants, because of interference by other metals with its uptake and its mobilization within the plant. A high quantity of calcium in the soil often causes iron deficiency. Manganese may be fairly abundant in the soil, enough to be visible as black deposits of manganese dioxide where concentration and immobilization in this form has taken place, or it may be present in trace quantities only. Deficiency as a plant nutrient occurs as a result of the immobilization process as well as because of low initial quantities in the soil parent material.

Potassium and Sodium

All living cells contain large quantities of potassium in ionic form in the cytoplasm. It does not form part of structural materials but is essential in the processes of energy transfer within the cell, including the first stage of photosynthesis, and it takes part in the enzymic reactions involved in protein synthesis.

Potassium is abundant in most soil parent materials, much of it in the easily weathered potassium-bearing felspars and micas, and it is incorporated into the micaceous secondary minerals. In the soil it is one of the main cations of the exchange complex and soil solution, and in many soils its availability to plants and micro-organisms is likely to be limited more by relative abundance of other cations, especially calcium, than by absolute shortage of potassium.

The potassium in the micaceous clay minerals forms a reserve which is released only when the concentration in the soil solution drops to low levels. In soils of low potassium content, withdrawal by plants from the exchange complex and soil solution results in some release from clay minerals. When demand by plants ceases an increase of potassium by release from plant residues and primary minerals can result in fixation again in the clay minerals.

Soils are rarely totally lacking in reserve supplies of minerals containing potassium, but since it is highly susceptible to leaching, and since cropping or grazing without replacement of organic residues results in continuing depletion, old, heavily leached soils and those exhausted by usage can show sufficient shortage of potassium for photosynthesis to be inefficient and for protein formation to be reduced in spite of an adequate nitrogen supply. In the old, leached soils of tropical regions, which have a clay mineralogy dominated by koalinitic minerals which do not fix potassium and which have a low exchange capacity, much of the potassium in the soil plant system is in the vegetation and likely to be lost in forest clearance. That which remains in the soil or is returned to it in ash or residues of the forest trees is highly susceptible to loss by cropping or grazing.

Where soils are deficient in nitrogen or phosphorous, potassium deficiency may not matter since growth is in any case limited. When nitrogen or phosphorous are supplied in fertilizers, potassium may then have to be supplied too, since the demand in the soil is increased.

In strong contrast to potassium, sodium does not appear to play an essential role in any plant, though some grow better with than without it. In conditions of potassium shortage certain groups of plants can use sodium instead of potassium for some of their cell functions. The relative importance of sodium and potassium in the organic world is illustrated by the fact that whereas in igneous rocks they occur in approximately equal amounts, in the oceans there is about ten times as much sodium as potassium. The rest of the potassium is held in the soils, mainly in the clay minerals, and in the organic world. Without the participation of potassium in clay mineral formation, land plants so dependent on it would never have evolved, since it would have been just as susceptible to loss as sodium, on which no plant can depend since there are no sodium salts sufficiently insoluble under soil conditions to ensure retention against leaching.

Calcium and Magnesium

Calcium is a component of the middle lamella of plant cell walls, and it is essential to the proper growth and function of roots. Calcium accumulates in leaves to an extent which varies greatly between species; for some plants a high calcium content is a necessity and they cannot grow unless it can be achieved; for others demand for calcium is tailored to supply. Variations in calcium content of litter are an important factor in differentiation of soils on parent materials of different calcium content or of soils subject to different amounts of loss by cropping and grazing.

Magnesium is a constituent of chlorophyll, and it also plays a part in the transport of phosphate through the plant to the cells in which it is used.

Calcium and magnesium are abundant in intermediate and basic igneous rocks, most of the calcium being in the felspars and the magnesium in the ferromagnesian minerals; these minerals are mostly fairly easily weathered under soil conditions. Earlier cycles of weathering and erosion have produced the enormous quantities of limestone and other sedimentary rocks rich in calcium and magnesium carbonates which now form soil parent materials over large areas of the earth's surface.

In neutral and alkaline soils calcium and magnesium are normally abundant, but in soils on acid rocks one or other may be in short supply and in the case of calcium the species composition of the vegetation may be modified. The low calcium content of the litter may dominate soil processes either from the beginning, or, probably more usually, after loss of calcium in cropping and grazing has reached a certain point.

In soils dominated by sodium, such as those on reclaimed coastal and estuarine sediments and those in arid regions in areas of internal drainage and where irrigation is practised, calcium and magnesium deficiency in plants occurs in spite of high contents in the soil. Under the alkaline conditions associated with sodium dominance, calcium and magnesium are immobilized in insoluble form, and the readiness with which sodium is displaced into the soil solution from the exchange complex, and the ease with which it is absorbed by plant roots tends to exclude the less mobile divalent cations from uptake by plants.

Other Micro-nutrients

In addition to the macro-nutrients, nitrogen, phosphorus, potassium and calcium, those of intermediate status, sulphur and magnesium, and the micro-nutrients, iron, manganese and sodium, which have been discussed already because of their importance in soil processes and their relationship to the other nutrients considered, plants and micro-organisms need small quantities of boron, cobalt, copper, molybdenum, zinc, and possibly chlorine and vanadium. Plants take in selenium and iodine, which may not be essential to them but which are to the animals which feed upon them, as are all the other plant nutrients except perhaps boron. Plants also take in elements which do not seem to be essential either to them or to animals, such as silica, which is accumulated in the cells of grasses and some other plants as silica bodies or phytoliths.

Most of the metallic micro-nutrients as well as other trace elements in soil are toxic either to plants or to animals if their presence exceeds trace amounts. Where toxicity to plants sets in at low level, as in the case of arsenic, danger to animals in grazing is minimal, but a risk lies in the toxic element mobilized from rock or from industrial residues by soil processes entering drinking water by drainage from the soil. Some elements toxic to animals but not to plants may be accumulated by plants even though their need is only for trace quantities, as in the case of molybdenum and boron, or for none at all, such as lead and selenium, and then the risk to grazing animals is present.

For many of the micro-nutrients the margin between deficiency and toxicity is not great. Mobilization and availability to plants is strongly related to soil conditions and to mutual interference, in the soil and in the plant, affecting uptake and use of the nutrient elements. The balance in the soil may be critical; disturbance of the vegetation, which has developed in concert with the establishment of balance, and losses arising from agricultural practices, can so upset it that deficiencies or toxic excesses set in. Application of mineral fertilizers carries a particularly high risk of interference with supplies of nutrients other than those which the fertilizer is intended to replace; use of organic manures is much less likely to introduce such risks because the process of degradation of the organic residues provides a buffer against imbalance.

Soil Horizons

The processes of mineral alteration, incorporation of organic matter, leaching of soluble substances and mobility of the soil colloid lead to the development of a number of zones, more or less parallel to the surface, which differ from one another in physical and chemical characteristics. These zones are called *soil horizons* and together they make up the *soil profile*. Profiles differ, that is, the nature and degree of development of their horizons differ, according to a complex interaction of mineralogy and texture of parent material, the yearly patterns of temperature and rainfall, vegetation and land use, slope and position on the slope, and time. The presence or absence of various horizons and their degree of development form the basis of most systems of soil classification.

There are a number of systems of horizon nomenclature in current use, the most widely accepted one being that of the International Society of Soil Science (1967), which will be used here. Some of the systems in use in different countries are based on this, with local modifications introduced to take account of differing importance accorded to diagnostic horizons under different classifications and varying views as to the genetic relationships among soil groups. Horizons are given letter designations, capital letters indicating their predominant characteristics and lower case subscripts denoting mode of formation, modifying processes or particular materials.

Organic Horizons—**L, F, H, O**

The loose litter which lies on the surface of the soil and usually represents a single year's leaf fall, not yet fully involved in soil processes but washed of soluble exudates and products of microorganism activity is denoted by the letter **L**.

F represents the organic residues undergoing comminution by the soil fauna and decomposition by micro-organisms. Plant fragments are still recognizable as such but are darkened by the formation of humus substances and are bound into a moist, compact mass by fungal mycelium.

More completely humified material in which cell structure is no longer detectable forms the almost black horizon, much of it in the form of faecal pellets of arthropods, of amorphous humus, designated **H**.

L, F and **H** in vertical sequence make up a *mor* humus horizon. From it soluble mineral and organic substances excreted by soil fauna and released during micro-organism activity and cell lysis are washed down into the mineral soil, and from the **H** layer humus substances may infiltrate into the mineral soil. Within the mor horizon some herbaceous plants are able to obtain all their nutrient needs, and their roots contribute to its processes.

If the organic horizon remains wet so that soil fauna are absent and micro-organisms are short of oxygen, comminution stops and humification is inhibited. The residues accumulate to form peat, for which the letter **O** is used.

Mixed Mineral and Organic Horizons—A

The mixture of mineral and organic material produced by earthworm activity is known as *mull* humus. Humification is rapid and takes place within the soil rather than at its surface and the humus substances are in intimate association with clay minerals and calcium.

Moder humus is the form found in soil where, in the absence of earthworms, a simpler and less complete mixing process is involved. It is the result of the washing down of amorphous humus from an overlying mor horizon into the mineral soil, and of the excretion by arthropods of comminuted and partly humified material among the mineral particles, without the development of a close association of humus substances and clay.

Mull and moder are known as *humose* materials and the horizons they form are designated **Ah**. In soils which have been ploughed the humic horizon mixed by the plough is called **Ap**, and is derived from a mull humus, perhaps with the incorporation of non-humic soil from the horizon below but not altering the nature of the worm-produced association of mineral and organic material, or produced by ploughing-in a mor horizon to form a moder type of mixture, and in this case the incorporation into the mineral soil of the residues which would otherwise remain at the surface can greatly modify soil processes.

In **A** horizons minerals are altered under the influence of the products of organic processes. Roots of grass and herbaceous plants take nutrients from the **A** horizon close to their point of release from mineral and organic substances. Mobile substances are washed out of the **A** horizon into the soil below.

The definitions of mor, moder and mull used here conform to recent usage, but borrowings between languages in the course of evolution of terminology have resulted in shifts of meaning. The English term *raw humus* and the American word *duff* correspond to mor (Jacks, 1954). Moder is sometimes used in a sense equivalent to the **F** layer of mor, with no implication of eventual mixing by worms to form mull, and sometimes to mean a form of organic material which very vigorous arthropod and micro-organism activity is in the process of degrading but which is so abundant that the worms have not yet achieved incorporation into the mineral soil. Some deciduous woodland has a leaf mould which corresponds to this usage.

Eluvial Horizons—E

These are horizons of the mineral soil from which products of weathering have been lost by downward movement. The potentially mobile constituents are the iron and aluminium oxides released by alteration of primary minerals but not incorporated into secondary ones, or released by subsequent alteration of secondary minerals, and the secondary minerals themselves. Where the iron and aluminium are removed by the chelating activity of organic substances passing down the profile from the humic horizons the eluvial horizon is denoted by **Ea**. Where part of the soil colloid becomes dispersed and the finer clay minerals together with associated iron and aluminium oxides and humus substances move down the profile the eluvial horizon is denoted by **Eb**.

Because of the poverty of plant nutrients in eluvial horizons, particularly **Ea**, plant roots tend to be very sparse in this part of the profile. Those plants not rooting entirely in the horizons above send roots through the eluvial horizon to ramify in the horizons below.

In some systems of horizon nomenclature the eluvial horizons are grouped with the humic horizons, on the principle that both are characterized by loss of some of the products of weathering to the soil below. The organic, mineral and organic, and eluvial horizons are then called **0**, **A1** and **A2**, or **A0**, **A1** and **A2**, or **0**, **A** and **Ae**. The organic horizons may be subdivided into **01** and **02** equivalent to **F** and **H**.

Horizons of Accumulation—B

These are subsurface horizons where an absolute or relative accumulation of products of mineral alteration or of humus substances has developed as a result of alteration *in situ* or as a result of translocation from overlying horizons.

A "weathered **B**" horizon, **Bw**, sometimes called a *cambic* horizon, is formed by alteration of primary minerals and accumulation of secondary ones. A soil colloid composed of clay minerals and hydrous oxides is formed, and the soil structure develops, predominantly by the shrinking and swelling of the clay minerals, and obliterates the rock structure or stratigraphic arrangements of the parent material. The soluble products of alteration provide nutrients for plants, and roots are abundant, their exudates and micro-organisms contributing to soil processes and their residues adding organic matter which is humified within the soil. In some systems of nomenclature **(B)** is used in place of **Bw**.

Where clay mineral formation is slow in relation to the rate of alteration of primary minerals, or where kaolinitic minerals are formed rather than 2:1 minerals, a relative accumulation of iron and aluminium in the form of hydrous oxides develops. Such a horizon is called a "sesquioxidic B" horizon, **Bs**, the term *sesquioxide* referring to the hydrous oxides of both metals. Since the strong colour of the iron hydroxide dominates the soil colour this type of horizon is sometimes called a "colour B" horizon.

Illuvial **B** horizons are formed by an absolute accumulation of colloidal material, hydrous oxides or humus. A "textural B" horizon, **Bt**, occurs below an **Eb** horizon, beneath an **Ap** horizon where the **Eb** and the **A** horizons have been combined by plough mixing, and sometimes beneath an **Ah** horizon alone. The fine clay, hydrous oxides and humus substances removed from the horizons of eluviation are deposited by flocculation against the walls of pores and gradually incorporated into the soil structure as shrinking and swelling proceeds. The texture is modified by the increase of fine material, hence the name *textural* **B**. The **Bt** horizon may be superimposed on an earlier formed cambic **B** horizon, or it may form concurrently with the weathering processes. It is relatively rich in plant nutrients compared with the eluvial horizon above, and, holding more moisture, has more root development within it. Its higher humus content than that of the eluvial horizon is partly due to the greater amount of root material and partly to the humus component of the colloidal material which is derived from the humic horizon.

In soils where chelation processes have taken iron and aluminium from the **A** and **Ea** horizons, hydrous oxides and humus substances are deposited to form horizons of accumulation. **Bh** horizons are those where humus substances accumulate, **Bfe** horizons those where the hydrous oxide is predominantly that of iron, and **Bs** where both the sesquioxides are involved.

A **Bfe** horizon is usually in the form of a *thin iron pan*, formed where the mobility of iron is due to its reduction under conditions of waterlogging, rather than or as well as due to chelation, so that it becomes separated from aluminium. Because of the possibility of confusion between **Bs** horizons formed by two entirely different processes, that of relative accumulation during alteration *in situ* and that of absolute accumulation by translocation, particularly where the second is superimposed upon the first, **Bfe** tends to be used in the latter case even when considerable amounts of alumina are also present. An alternative is to use **Bw** to cover all horizons of accumulation produced by weathering *in situ* and keep **Bs** for the illuvial sesquioxide horizons.

Bhs and **Bhfe** indicate accumulations of the illuvial materials in association, where they have not separated to form distinct horizons.

C Horizon

The **C** horizon is the lowermost part of the profile, the rock undergoing alteration under the influence of moisture percolating from the soil above and the deeper roots of plants, which has not reached the stage at which rock structure is obliterated. Hydration and oxidation of iron is the main visible effect of alteration, and may pervade the fabric of the rock or appear only adjacent to cracks and structural planes, particularly those penetrated by roots. The rock is often softened by alteration of the more susceptible minerals and disrupted by the volume changes during alteration and by the shrinking and swelling of the secondary minerals formed.

The **C** horizon may have a sharp upper boundary or it may merge into the **A** or **B** horizon. A mixture of humic soil and rock debris is an **A/C** horizon, and a broad zone of merging from **B** to **C** in which the soil has characteristics of both is a **B/C** horizon. The **C** horizon merges downwards into unaltered rock.

Gleyed Horizons—G, g

Soils which periodically become waterlogged acquire a characteristic mottling produced by changes in the oxidation state and distribution of iron. The phenomenon is known as gleying, and a gleyed soil horizon is denoted by the letter **G** if gleying is its predominant characteristic, or by a subscript **g** to the main horizon letter if gleying is present but waterlogging not sufficiently continuous to inhibit the processes of formation of the horizon.

Waterlogging may be due to ground water affecting the lower part of the soil while the upper part remains aerated, or it may be confined to the upper part of the soil above a textural **B** horizon or a thin iron pan whose formation has impeded percolation. The pattern of mottling is produced by the fact that iron is soluble when it is in reduced condition. Reduction occurs when the oxygen in the soil water is used up by the micro-organisms faster than it is replaced by entry of air or infiltration of more oxygenated water or by diffusion from an air–water interface. The ferrous ions diffuse outwards from the points at which they are released into solution until they reach a zone where the oxygen has not yet been used up or until the soil drains and air again has access. There they are oxidized and the iron immobilized again as hydrous ferric oxides. The form of oxide which crystallizes under these conditions is thought to be lepidocrocite rather than goethite, and its colour is particularly bright. Patterns of gleying range from small flecks and patches relatively depleted of iron surrounded by thin zones of concentration in an otherwise uniform oxidized matrix, to wider areas of grey in a network of oxidized zones, based on the structural planes where percolation is more rapid and access of air more frequent, or to complete greyness. Overall iron content may stay the same, or iron may be lost altogether from the gleyed horizon if percolating water carries it away in the reduced condition. Movement of water through the soil superimposes a pattern of iron distribution on that produced by diffusion alone, and since in a groundwater gley water may carry iron up towards the oxidized zone, the concentration may be above the gleyed zone or may be related to topography. Iron pan can be produced by slow downwards percolation, the iron being oxidized at the lower limit of intense organic activity or at some change in texture which causes a change in rate of flow. The formation of such a pan can itself increase the degree of waterlogging to the point at which organic activity slows down and organic matter begins to accumulate.

The presence of roots and channels left by decayed roots often controls the distribution of iron oxides in soils subject to water-logging. There is a strong tendency for iron to accumulate in a zone around a root channel, and it can form a pan lining the channel to produce a pipe, or may fill it completely.

Subsidiary Horizons of Deposition

The increase with depth of the concentration of the soil solution can reach the point at which the less soluble substances leached from the **A** or **B** horizons are precipitated. Deposition may be within the **B** horizon, at its junction with the **C** horizon or within the **C** horizon, and its occurrence is signified by a subscript to the horizon in which it occurs or with which it is most closely associated.

The great abundance of calcium in the mineral and organic world, the ubiquity of CO_2 in the soil moisture and the low solubility of calcium carbonate result in frequent occurrence of horizons of accumulation of calcium carbonate. B_{Ca} and C_{Ca} horizons occur in soils on rocks containing calcium carbonate even under conditions of relatively high rainfall, and where the soil moisture commonly drains freely from the soil. In drier areas they occur on increasingly less calcareous rocks and at decreasing depth, and are very abundant where the rain usually only wets the profile without any loss to drainage.

Calcium sulphate, gypsum, is more soluble than the carbonate, and forms soil horizons, which can be designated by the subscript **cs**, in arid regions. Soils forming on parent materials with a high content of gypsum may have a gypsic horizon even in cool and humid regions.

Catena Relationships

A *toposequence* or *catena* (Milne, 1935) is a range of soils in a section of landscape covering hill top, slope and valley bottom in which profiles have developed differentially according to drainage conditions, percolation of soil solution downslope through the soil and the effects of soil movement. The rates of percolation into the soil, through its horizons of varying texture and structure and into the underlying rock relative to the length and intensity of rain showers control how much water runs off downhill at the surface, and how much runs obliquely through the soil in any horizon. Tree roots forming paths for percolation running laterally through the soil, flat rock fragments lying with their greater axes parallel to the slope, and the structural planes in the underlying rock may all contribute to lateral drainage. The effect of it is that at any point in a soil on a slope, soil solution can be received from all levels in the soils further up hill which are higher than that point. Thus, whereas in soils in top positions the **B** horizon receives moisture only from the **A** and the **C** from the **B**, on a slope the **A** horizon may be receiving moisture from the **A**, the **B**, or even, if the slope is very steep, the **C** horizons further up hill. The concentration of the soil solution increases downwards in the profile, so soils in lower slope positions may receive concentrated solutions even in their **A** horizons, and be not only totally unleached but sites of accumulation of soluble materials. In general, the depth in the profile to which leaching is effective decreases downslope. In certain circumstances, groundwater may come to the surface in lower slopes, creating "seepy footslopes", where soils affected by water-logging occur.

Classification of Soils

Soils are not very amenable to classification. Increase in understanding of soils has gone hand in hand with the development of systems of classification, since every attempt to arrange soils by visible and measurable characteristics in a way that is meaningful in terms of genetic relationships has revealed inadequacies in the understanding of genetic processes. The problem is increased by the common occurrence of relict and polygenetic soils whose formation took place under environmental conditions whose time span, nature, and variation are not fully known, and by the occurrence of partial profiles of eroded soils and multiple profiles of redeposited soils which are not always easy to identify as such. Nevertheless, most of the systems of classification in current use have genetic connotations, and in order to use soil surveys and maps which utilize these systems the archaeologist needs to have some understanding of them. Classification is inherent in mapping as soon as the scale becomes too small for single stage mapping units to be used. Whereas in large scale maps, discussion of classification is only necessary to set the soils of an area into the wider context, though even the choice of related colours for different units and the arrangement of the map legend in one order rather than another are classificatory activities, in small scale maps it is essential to the mapping process.

The emphasis on the various factors of soil genesis tends to differ with the nature of the country in which the soil scientist is working. In countries such as Russia, where soil studies were pioneered, crossed by several zones of climate and vegetation, uncomplicated by major topographic interruptions, soil types are clearly seen to be related to these zones, and systems of classification lay emphasis on these factors. In smaller countries, confined to one climatic zone but having great local variations in relief and geology, emphasis is placed on factors such as parent material and hydrological conditions. As soil science has become international in outlook, progressive synthesis and compromise between the different national systems has resulted in composite classification schemes, which change their emphasis at different levels of subdivision and which incorporate purely descriptive features.

In the attempts made in recent years to produce world wide classifications by rationalizing the various regional and national schemes much has been learnt about soils and about the difficulties inherent in trying to classify them. Things amenable to classification consist of discrete entities which can be described by a limited number of parameters whose range of variation and mode of combination allow division into significant classes. Soil is very complex and in most of its properties is a continuum. No universal system has yet been adopted, and those which have been put forward are still handicapped by the greater experience most soil scientists have of the soils of their own region and the fact that the more highly developed parts of the world, the temperate regions, have produced far more soil scientists than the tropical regions.

In trying to avoid the difficulties of classifying on a genetic basis before soil genesis is fully understood, several systems have been developed based on soil characteristics rather than on presumed processes, the ultimate development of this approach being the use of numerical taxonomy which is currently being studied by a number of workers (e.g. Rayner, 1966). Of the classification systems currently in use that of the United States Department of Agriculture (Soil Survey Staff, 1960 and Supplements, 1967 & 1968) is the most widely known, having been adopted in many countries which receive technical aid from the United States or send their soil scientists there for training. The system was published when it had reached its "7th Approxima-tion" and is familiarly known by that name. Since the primary use of soil classification is agricultural, and since the characteristics of surface soils are important for this purpose and provide the most accessible basis for survey, a high priority has been given to them as distinguishing criteria. The system uses a specially created language in which words are built up from syllables indicating specific charac-teristics. The syllables were chosen for their mnemonic qualities, but, as in learning any language, constant practice is needed before facility is acquired, and until it is the effect is largely gobbledegook. So far, in terms of international communication the result of adoption of the 7th Approximation by American soil scientists has been somewhat counter-productive, since people who could communicate perfectly well in one or other of their languages find that in the one field of their mutual interest the creation of a further language to be learnt from the beginning introduces a barrier. The 7th Approximation was the first of the non-genetic systems to be used, and so has been more readily adopted for use in regions where concepts of soil genesis were not far advanced. However, though intended for world wide use, it is often now being found to be inadequate in just those regions where the soils have been less fully studied and its classes less finely divided and closely defined. Presumably the 8th and subsequent approximations to the perfect soil classification system will be progressively better in these respects.

A committee working under the auspices of the Food and Agricul-ture Organization of the United Nations has produced a Soil Map of Europe (FAO, 1966) in which a correlation of the systems of nomen-clature and classification in use in European countries was made. The various European systems were already to a large extent interelated, many of the names of soil types having been borrowed, with or without translation, by one country from another in which they were first defined or more typically represented, but such borrowings have not always continued to be used in their original sense or within the same limits. Correlation was therefore a question of agreement on limits of soil types and on status of marginal and transitional types, and of standardizing the translation between the languages to be used, English, French and Spanish.

The Soil Map of Europe does not cover all of Eastern Europe but does include all the Mediterranean islands and the Asiatic part of Turkey. The map itself, in five sheets, is beautiful to look at but, at 1 : 2,500,000 is on too small a scale to be of practical use. However, the explanatory text is very useful to anyone wishing to become broadly familiar with the soils of the continent, and provides data on range of climate and types of vegetation associated with the soil types. The tri-lingual key is useful, and the text includes a list of bodies responsible for soil survey in each country, the types and scales of map they are producing and the stage reached by survey at the time of publication.

The Soil Survey of Great Britain (England and Wales) has been using a classification based upon a traditional hierarchical system in which soils are grouped according to parent material and arrangement of horizons. The basic units, the *soil series* or profile classes, are soils formed on the same type of parent material and having the same arrangement and degree of development of horizons. Similar series formed on different parent materials form *groups*, within which there may be sub-groups in which some characteristic of situation or parent material introduces marked variation without an overall change in profile type. Groups are arranged in *orders* which are differentiated on major differences in soil processes or in degree of maturity, so that soils in different orders differ markedly in the number and type of horizons making up the profile. The system has been continually undergoing modification in use, being adapted by different soil surveyors to the particular needs of the area in which they are working and taking into account the advance in understanding of soil processes. Since the nomenclature and the basis of classification are given in the memoirs or notes accompanying soil maps, it is easy to check that the terms and the limits and variations of classes used in a particular survey are fully understood. Avery (1972) introduced considerable modifications to the system, involving "tightening up definitions of existing concepts". The modified system is published in Avery (1973), and a monograph explaining it in more detail is expected. This will be the system used by the Soil Survey of England and Wales from now on. *Major Groups* and *Groups* are defined within the existing framework on grounds of nature and origin of materials, and presence or absence of the diagnostic horizons which distinguish the major modes of profile development. The *sub-groups* are given names compounded of words and qualifying prefixes drawn from terms referring to pedological concepts and familiar from previous usage, to make self-explanatory phrases. The use of soil series is to be retained for locally defined classes within the sub-groups.

FitzPatrick (1971) points out that almost all systems of classification applied to soils so far are hierarchical, while there is no basis for the assumption that soil profile types should fall into the relationships implied by a hierarchical ordering. He proposes a system based upon occurrence and position of horizons, which are designated by letter symbols with suffixes giving horizon depth. The horizons are listed in order of their occurrence in the profile, creating a formula which defines the profile type. Soils presenting the same formula, though horizon depth may vary, are grouped, and groups are assembled into sub-classes and classes according to whichever horizons or combination of horizons are particularly prominent or have significance in particular circumstances, allowing a flexibility of classification criteria and the creation of classes for different reasons as circumstances demand or as understanding increases, rather than requiring that every soil type fit into a previously defined class. Whether or not FitzPatrick's system comes to be widely adopted, his book provides a clear outline of soil forming processes and a very useful summary of horizon characteristics and relationships. He also gives a short summary of the history of soil classification, with an outline of earlier systems.

4
Disturbance and Displacement of Soil

Since soil is a relatively incoherent material it is much subject to disturbance and displacement. The agencies of disturbance are wind, water and animals, including man, and except in the case of wind, displacement downhill under gravity is the normal result of any disturbance. Some protection from disturbance is provided by the vegetation, and stability against displacement is provided by plant roots and by the structure of the soil itself.

Vegetation provides protection from direct attack by rain and wind. Rain is gathered by leaves and branches and reaches the ground by running down leaves and stems or by dripping in large drops from layer to layer of the canopy until it reaches the ground at a much lower velocity than that of the rain itself. A ground flora beneath trees or a grass cover in the open reduces the impact of rain on the soil to a minimum. The litter layer and humus horizons provide further protection. Although loose leaves of one season may blow about in quite gentle winds or be washed away by surface water, leaving some areas denuded and collecting in drifts in others, where they are present they intercept the direct impact of raindrops, and once they become part of a mat bound together by fungal mycelium they provide a coherent surface which protects the mineral soil from rain and wind. This is particularly important since the mor form of humus which provides such protection is best developed on types of soil in which structural stability is at a minimum.

Within the soil the mesh of roots of grasses and herbaceous plants binds the soil together, and the deep roots of trees can anchor the tree and the soil enclosed by its roots to the immobile rock beneath. Roots contribute greatly to the formation of soil structure, which itself provides stability against disturbance by the interlocking of peds and by holding the soil in aggregates which are more difficult to move than are individual soil particles.

The continuity of the vegetation and the humic horizons is very important in the maintenance of soil stability. Damage to forest by strong winds, which can uproot trees and tear bushy plants from the ground, by animals which break or fell trees or kill them by eating the bark, or by fire, opens up the cover and can disturb the soil or expose it to later disturbance. The hooves of animals break through the litter and humus layers. Grazing animals congregate in herds and tend to use particular paths and to gather at drinking places and in sheltered areas. In dry weather trampling breaks the soil peds and creates a loose soil which can be blown away or washed away in the next rain. In wet weather it puddles the soil, forming a layer in which physical disruption causes dispersion of peds which would otherwise be stable. Once the pore system has gone the soil has a low permeability and the puddled layer impedes infiltration and can cause removal of soil by surface run-off.

Other animals disturb the soil deliberately in their search for food or to make living places. Pigs rootle with their snouts and some birds break and scatter the humus layers looking for insects and worms. Burrowing animals undermine the surface so that the feet of larger animals may break through, and at the mouths of their burrows the continuity of the surface is broken, exposing the soil to further disturbance.

These factors affecting stability or disturbance of soil control the extent to which soil is displaced by creep, colluviation and erosion, and deflation. The first three are a sequence of increasing severity in which soil moves downhill under gravity. The fourth can happen at the same time as any of the others and on flat ground too.

Soil Creep

On sloping ground any movement of soil particles, brought about by changes in volume on wetting and drying, by growth or decay of roots, wind rock of vegetation, impact and percolation of water, or passage of animals through and over the soil, acquires a down-slope component. The nearer the surface the less is the restraint by the soil above and the less close is the packing of structural units. Whole peds and individual particles turn and jostle and gradually move downhill, those at the surface faster than those lower down and faster on a steep slope than on a gentle one. Every rain drop which falls on bare soil splashes up a little shower of particles, each of which falls down-slope of where it would if the ground were flat. Every tread of an animal's foot pushes part of the surface layer, whether it be soil or vegetation, a little way downhill.

Soil well stabilized by vegetation and by its own structure moves only very slowly even on slopes of moderate gradient, and the movement is predominantly in the humic horizons. Those parts of the slope which lose soil faster than it arrives from above suffer a thinning of the A horizon, and at the slope foot or in hollows there is an accumulation of humic soil forming a deep A horizon. At the other extreme, on very steep slopes no soil above the C horizon can remain since any rock whose coherence is reduced by weathering falls away under its own weight. Between these extremes the thickness of the soil horizons at any point on a slope depends on the rates of arrival and departure of soil material and on the rate at which weathering of the parent material and incorporation of organic matter contribute to the B horizon from below and the A horizon from above. As the A horizon loses material at its surface, it gains from the B horizon by incorporation of humus into the soil. The B horizon can be lost altogether under conditions of severe creep, not because it is moving more than the surface soil but because loss at the surface necessitates progressive incorporation of freshly added residues into more and more of the B horizon, turning it into A horizon. Eventually the humic soil lies immediately above coherent rock, since as soon as the weathering products become detached from the rock they become mixed with the organic material. In other circumstances, humic soil may move over the surface much faster than the rate of incorporation, in which case the A horizon may thin to the point of becoming discontinuous and the B horizon is exposed at the surface. If the net loss from the B horizon is greater than the rate at which the rock is weathering, the B horizon too breaks away and rock is left poking through. Fragments of rock become detached and are added to the creeping soil downslope, to remain at the surface or become buried as they move down, depending on the severity of the movement and the presence of worms.

The process of creep mixes soil from the different horizons, while in transit and in the final accumulation at the slope foot. The mixing may be fairly uniform, B horizon material being fully incorporated into the deep humic horizon, or stratification may occur. Rapid creep can produce stratification in the form of an inverted profile, in which a new A horizon has to form in the last deposited C horizon material. The effect of such mixing and inversion is to introduce fresh reserves of plant nutrients into the rapid weathering environment of the humic horizon, so that a soil which on flat land might be rather heavily leached or impoverished may have equivalents in sloping terrain of on the one hand thin and denuded aspect, and on the other rich and deep.

Colluviation and Erosion

The phenomenon of soil creep as considered so far takes place in dry or moist soils by movement of particles and peds individually. As the amount of water in a soil increases the rate of creep increases because of the greater weight of the soil mass and because pressures are transmitted through a continuous network of water filled pores to a very much greater degree than through a partially air filled system. In soils susceptible to dispersion an increase in water content may, by diluting the soil solution, reduce the strength of flocculation until peds become susceptible to break down on the slightest disturbance and the soil loses its structural stability. For one or other of these reasons, or a combination of them, a point can be reached at which the soil becomes semi-liquid and flows down the slope. The whole profile may move, stripping the soil down to coherent rock, or part of the profile may become unstable and carry the soil above with it, or the upper part alone may go. The flow may reach a stream bed and be carried away or it may come to rest at the slope foot and form a *colluvial* deposit. Colluvial soil which is carried into a gully and there transported in suspension may be deposited as the gully overflows or debouches onto a gently sloping terrace or foreslope. As the water spreads out is loses velocity and the soil is deposited as a sheet over the ground. This type of deposit is also included under the definition of a colluvial deposit.

Saturation of a soil or of part of its profile occurs when water arrives by infiltration from above or by lateral percolation faster than it can get away. The porosity and structure of succeeding horizons and of the underlying rock control how much can infiltrate and how much be held before a particular horizon at a particular gradient gives way. The cause may be torrential rain in the immediate locality or earlier in an area above, or prolonged rain over the whole hillside, or it may be an event such as overflow of a gully.

Colluviation is not the inevitable result of the arrival of too much water. Many soils are sufficiently stable to tolerate saturation on a slope, and surface flow of water may occur instead. Again, it can be the result of rain falling faster than it can infiltrate or it may be the result of seepage from soil or rock. Surface run-off carries particles of the surface soil and vegetation with it. The water may run into an existing stream or it may be gathered into a declivity in which it carries away more and more of the soil until it creates its own gully, and once a gully is formed, the break in continuity of the soil exposes it to lateral erosion. Gully erosion and hill wash as well as colluviation can occur at considerable distances from the point of origin of the surface run off and long after the rain has stopped.

Erosion by a stream or river is more localized than that by surface flow, being generally confined to deepening of the bed or a nibbling away at its banks, depending on the stage of development of the valley. Lateral erosion results in redistribution of relatively small areas of soil from the banks, where, since the edge represents a break in the continuity of the stabilizing surface layers and vegetation, the soil is particularly susceptible to attack. Water undermines the soil, which falls in chunks and is then dispersed and carried away. Rock from the C horizon forms the bed load and may be rolled and sorted and redeposited time and again as gravel banks, while the finer materials are carried further and may be deposited as floodloam, again to be re-eroded and redeposited until it is either left high and dry as a terrace or reaches an estuary and is deposited there.

Soil creep is a more or less continuous process, which if it continues at about the same level without interruption produces a landscape in which rates of movement and rates of soil formation achieve a balance at a level appropriate to each situation. Colluviation and erosion, on the other hand, are spasmodic, occasional or seasonal, and they can exhaust the supply of material they transport in the course of a few episodes. Deposits formed by these processes are stratified, and their characteristics vary with the power and duration of each particular event and with the amount of material still available After a mature soil has been stripped away, succeeding events can only draw upon the immature soil or weathered rock made available by further soil formation since the last event.

In colluviation the proportion of water to soil, the speed of flow and the distance travelled vary from episode to episode. The degree to which soil peds are dispersed or mechanically disintegrated depends on the amount of water and the turbulence of flow. The degree of sorting of whole peds and pebbles and of the constituents of disrupted peds depends on the speed and distance of flow. Thus the deposit formed by each episode has its own characteristics of proportion of soil to coarser material, proportion of whole peds to disaggregated soil, degree of sorting and degree of stratification of sorted materials. If the soil reaches a river, the amount of water becomes overwhelming and complete dispersion occurs, the soil materials are sorted and distributed in river deposits according to their size grade. Colloidal material remains in suspension until it reaches quiet water, such as the sheet of water left on a floodplain as the flood subsides, or the water of a lake, where it slowly settles out, or until it reaches an estuary where saline conditions cause flocculation and it settles more rapidly to contribute to the estuarine mud.

Deflation

The wind can pick up and carry particles whose size depends on the wind speed at the ground surface. Sand sized particles are carried clear of the ground only by very strong winds, usually by the greater speeds achieved in turbulence in an air stream whose overall speed is much lower. Much sand travels by *saltation*, in a series of leaps, being picked up by a swirl of wind and dropped again, to be picked up again later. Sand and small pebbles can roll along the ground, and soil peds of pebble size, being less dense than rock, frequently do so, particularly on a slope with gravity helping the wind. Silt sized particles can be maintained in suspension by a wind of moderate turbulence and are carried until wind speed drops or the turbulence is smoothed. Such particles frequently accumulate in hollows in the ground, such as ditches, particularly those with banks to windward, and particularly if there is vegetation in the ditch to prevent the material, once settled, being picked up again. Vegetation is effective in trapping deflated material on flat ground too. Soil picked up in open country is deposited at the edges of more vegetated zones. The formation of *loess* is discussed in Chapter 5.

Particles of clay grade can be carried indefinitely once they are picked up, since only slight turbulence, or updraughts, such as those induced by temperature differences at the ground surface even on still days, suffice to prevent them settling. This material is generally brought down by rain, after travelling distances dependent on constancy of wind direction and frequency of rain.

Dry organic material is very light, and this too contributes to windborne dust. Leaves and finer particles from woodland are distributed over much land to leeward.

There is always dust in the air, derived from soil surfaces, sediments, rock debris and vegetation, and many of the activities of animals and particularly of man increase the quantity. Not only does it contribute to soils already forming but it is also a major contributory factor in the formation of new soils on rock surfaces, in sterile sediments and in the abandoned buildings, ditches and other features of archaeological importance. The fine particle size of much windblown mineral material allows rapid alteration and release of nutrient substances, and the micro-organisms which will contribute to release of nutrients from organic residues are often carried in the wind on the residues or as spores or cysts and so are available to establish the population in the newly forming soil.

Volcanic Ash. An important wind blown material in a different category from those already discussed is volcanic ash. The glassy nature of most ash allows rapid alteration, and all but the most acid ashes contain a sufficient abundance of plant nutrients to be potentially very rich soil, and ash layers are very porous, allowing good drainage. However, a loose ash deposit is highly susceptible to movement by both water and wind, and even in environments where soil formation is very rapid a period of instability follows each ash fall. Partially weathered ash and its products of alteration are mixed and redistributed many times before a surface of sufficient coherence to maintain stability is formed, and the deep deposits of these very immature soil materials, are characteristic of volcanic regions. Stratigraphy produced by successive falls of ash and the buried surfaces of soils of greater stability are blurred by the continual movement, and may be finally obscured by the rapid formation of a deep soil in the highly permeable material once a surface is maintained for long enough (Cornwall, 1968; Limbrey, 1975).

Away from the main area of ash deposition small quantities of fine ash can contribute minerals rich in plant nutrients to soils over a wide area. Unless the ash contains some minerals resistant to weathering it can be difficult to establish the presence of these foreign materials in a soil, though its nutrient status may be higher than that which is normal in similar soils on the apparent parent material.

Effects of Soil Instability on Distribution of Plant Nutrients

Most of the processes which move soil affect the humic horizons, in which most plant nutrients tend to become concentrated by the cycle of growth and decay, to a greater extent than the rest of the profile. Wind is unselective in the areas from which it takes soil, except where other forms of erosion have created incoherent surfaces, but it concentrates its load in hollows of various sorts and where vegetation thickens or increases in height, particularly at its windward edge. Since thicker vegetation often implies higher fertility, deflation contributes nutrients where there is already relative abundance.

Creep affects the humic soil very much more than the subjacent horizons. Nutrients taken from the **B** horizon by plant roots and deposited on the surface in plant residues to be incorporated in the **A** horizon move off downhill. If this loss is greater than can be replaced by release of nutrients from fresh mineral material the vigour of the vegetation may be reduced, and the stabilizing effect of vegetation cover with a good system of plant roots and a strong soil structure become less effective in preventing further loss.

In receiving areas, accumulation of crept soil can be so great as to exceed that needed for full vigour of plant growth and only the upper part of the deep humic soil is then effectively used. The nutrients contained in the lower part are wasted, though the more mobile of them may be gradually exposed to leaching as humus substances are numeralized and may provide benefit to other soils in the drainage basin, or may be lost to the ocean.

More drastic erosion, involving more water, not only redistributes the soil, to the impoverishment of higher and more steeply sloping ground, but also takes away nutrients in solution. Again, they may be utilized further downhill in the terrestrial system, but much goes out to sea. In these circumstances, the soil colloid is susceptible to removal into the rivers, and not only the nutrients it carries but also the power of retaining nutrients is transferred from the source soil to any floodplain which may receive it or to an estuarine mud or sea bottom sediment.

Processes of soil movement therefore, as well as those of leaching, generally transfer a nutrient reserve from high and strongly dissected land to flat or hollow areas, impoverishing the soils of the steeper slopes and enriching those of the receiving areas, and exposing a proportion of the nutrients to total loss from the land surface of the time.

Soil movement blurs the boundaries between soils on different rocks. Wind redistributes mineral material irrespective of its origin. Soil creep and colluviation spread minerals from rocks cropping out upslope over those downslope, and blurr and shift downhill the boundaries between soils on contiguous rock types. Where a rock upslope provides an abundance of plant nutrients the effect is to spread a fertile soil over an infertile area. On the other hand a poor parent material or leached and impoverished soil upslope may contribute material which covers or dilutes the richer soils below.

Where soils are forming under a mild, humid climate on rocks which are rich in minerals which weather easily and contain an abundance of bases, removal of soil from areas exposed to deflation and from sloping ground may be easily compensated for by continuing soil formation. The supply of plant nutrients is maintained at a high enough level for the vegetation to persist without change in vigour or composition arising from nutrient shortage, and for regeneration of soil and vegetation on erosion scars to take place. Only on very steep slopes where trees may fall before they reach full size will the mosaic of disturbance and regeneration inherent in any forest with a normal age range and animal population not contain all stages of the complete cycle of forest growth, including the full climatic climax. Where any of the conditions necessary to maintain the supply of nutrients in the face of repeated or continual loss fails, the whole system may be kept short of completion of the full cycle, regeneration being slower, the risk of disturbance higher under less vigorous vegetation and structural stability of the soil being lower, so that in any area further disturbance is likely to happen before full regrowth has occurred. The more frequent or prolonged the periods during which the soil is exposed to disturbance the greater is the risk of nutrient loss, so those situations in which supply is limited to start with are liable to suffer losses in a vicious spiral leading to soil degradation.

In marginal situations where soils and vegetation are near to their limits of flexibility because of low availability of weatherable minerals, inadequate or excess moisture, low temperature or short growing season, or position on a slope, a change in climate or a single episode or season of extreme climate, the advent of a new disturbing factor, such as man, or a change in the habits of the people already there, can induce soil changes which are not noticeable in more stable situations. Marginal situations occur where climate limits the growth of important groups of plants, where the soil parent material has a low content of one or more nutrient elements, where climate limits the rate at which a nutrient can be made available and where high or low porosity of the parent material leads either to excessive leaching or low water retention or to a tendency to waterlogging. Where marginality of climate and that inherent in the parent material coincide sensitivity to change in climate or to extra disturbing factors is particularly high.

PART II

Soils of the Humid Temperate Zone

5

Background to the Development and History of the Soils

The Nature of the Parent Materials

Processes of the Periglacial Zone

The present periglacial zone is that of the tundra and taiga vegetation of arctic regions, where the development of soils and vegetation is limited by low temperature and short seasons of biological activity, and where soil formation and geomorphological processes are strongly influenced by the processes of freezing and thawing within the rock surfaces, sediments and soils.

Frost shattering at the surface of solid rock takes place where there is sufficient moisture for its expansion on freezing in cracks and crevices to develop high pressures and disrupt the rock. In the humid parts of the periglacial zone and during humid periods of glacial phases large quantities of rock debris are produced in the form of scree on and below steep rock faces, and on the shallower slopes the rock is broken and disrupted so that further infiltration of water is facilitated. These processes imply not just availability of moisture but temperature cycles passing through freezing point. Where cold is so intense that water rarely melts, frost shattering is reduced. Aspect may be important here, sunny faces experiencing daily cycles through freezing point when those in shade all day remain frozen, and such differences would be greater in low latitudes during glacial phases than in the arctic at present, where in summer the low sun makes a circuit of the horizon.

Glacial meltwaters and snow melt redistribute over the landscape the rock debris gathered up by glaciers and that produced by frost. The sediments formed and the weathered layers and primitive soils of the less eroded surfaces are also subjected to the effects of the freezing and thawing cycles. The summer thaw only reaches to a certain depth, below which is *permafrost*. The layer which thaws is known as the *active layer*. Because water cannot percolate through the permafrost the active layer in many situations becomes saturated with water, and on slopes greater than a few degrees only it may sludge downslope. The process is known as *solifluction*; the difference between solifluction and colluviation is really only a matter of definition, but if the word solifluction is restricted to the periglacial process, it is distinguished by its intensity and by its inevitability, colluviation generally occurring

Figure 9. Cryoturbation structures seen in section in chalky coombe deposits at Pitstone, Buckinghamshire.

in isolated episodes, due to climatic accidents or drastic interference with soil stability.

Solifluction deposits fill declivities and blanket lower slopes, and are subject to erosion by meltwaters and to further processes of freezing and thawing. Within the active layer as it freezes from the top downwards pressures are set up. Materials of different texture and different moisture contents freeze at different rates and expand different amounts on doing so. Movements occur in equalizing and relieving pressure and soft, incoherent materials are squeezed and injected into more solid layers. The resulting contortions, which can be seen in section in Fig. 9, take on characteristic flame and festoon shapes, with some size sorting within the mobile components. On level ground the regularly spaced points at which pressure is relieved result in polygonal patterns of such convolutions. On slopes the polygons elongate into stripes. Polygonal and striped patterns are also produced by size sorting within a soil or sediment which is not so water saturated as to form a continuous mass when it freezes. As freezing proceeds centres of crystallization grow into ice lenses within the soil, and when the ice melts, water percolates away and leaves a space into which fine particles can trickle. Over successive seasons large stones are heaved upwards as the finer particles trickle downwards. Combination of differential movement of size grades with lateral pressures set up by freezing among the partly sorted material result in development of cells in which the larger particles move upwards and outwards until they meet those of the next cell, forming polygons bounded by coarser material with finer material in the middle, which on sloping ground become stripes.

The formation of ice within a sediment of fairly uniform character produces a number of features of textural redistribution and structural arrangement. As ice lenses develop, expansion on freezing compresses surrounding materials into platy shapes of very compact consistency. Lateral contraction as water moves towards the centres of crystallization produces a network of vertical partings, down which water can run during the next thaw and which then themselves form centres of ice formation. A polygonal network of *ice wedges* develops within the lenticular layer. Stones become coated with ice and drop in the space left when it melts, leaving a space above them into which fine particles can trickle. Stones thus acquire silt cappings. Formation of needle-shaped ice crystals is thought to be responsible for the fine pores which are characteristic of the compact, platy material of the lenticular layer. Needle ice is also thought to contribute to the formation of stone polygons at the surface, growing crystals pushing stones in front of them.

While ice-wedges on a fairly small scale are a feature of the materials which show compaction and lenticular form, much larger ones, extending over long distances and reaching to several metres depth, are thought to be initiated by contraction of the mass of sediment as a whole as temperatures fall to very low values in the colder parts of the periglacial zone, rather than by the effects of movement of water movement during freezing. Once a crack has opened, water runs into it during the thaw and on freezing forces it wider open, distorting the surrounding materials. These wedges can grow upwards through accumulating sediments, or may be buried by them and fossilized.

When an ice wedge finally melts, it fills with material from its sides and from the surface or overburden, producing an "ice wedge cast".

The processes considered so far are largely dependent on the peculiar habit of water of expanding on freezing, and they are expressed to varying degrees depending on abundance of water and the intensity of the cycle of freezing and thawing. Deflation in the periglacial environment depends on dryness and results in the deposition of *loess*, material predominantly of silt grade which is picked up and carried by high and turbulent winds but not maintained in suspension when wind speed drops or turbulence is reduced. Frozen surfaces of glaciers and of glacial and periglacial deposits suffer ablation, ice evaporating without an intervening liquid phase, leaving loose particles on the surface which can be winnowed by the wind, the finer ones being carried away. Loess is mainly picked up in the dryer parts of the periglacial zone, from the surfaces of glaciers and moraines and from the sediments distributed by meltwaters but rapidly drying out and not maintaining a cover of tundra vegetation, and where snow cover is thin enough to be blown away or lost by ablation. It is deposited where the wind is calmed by features of the landscape or by vegetation or where rain or snow bring it down.

Soils which form on the materials of the periglacial zone are limited in their development. Few earthworms live under arctic conditions and there is no mull humus formation. The quantity of organic residues produced each year is low, though since rate of decomposition is also low, on stable surfaces fairly dark peaty humic horizons can form, but the amount of profile differentiation and the depth to which soil processes reach is low. Because of the presence of permafrost and the sudden abundance of water in summer, waterlogging of tundra soils is a common feature. For classification purposes, the tundra soils can be broadly divided into well drained and badly drained forms, together with the *raw* soils, that is, those in which horizon differentiation is not established, which are so common in a landscape of surface instability. In so far as horizon differentiation does occur, well drained soils on poorer parent materials may show incipient mobilization of iron and aluminium by chelation, and on base rich materials calcareous horizons are common where the active layer is not continuously waterlogged but free percolation from the soil is prevented. Study and classification of soils of the arctic regions is in its early stages, but under the pressure of strategic and economic demands it is developing fast, in Russia, Canada and Alaska, with the result that more variety of pedogenetic processes and profile development is being found than had been suspected.

Tundra soils support the grasses and herbaceous plants, ericaceous shrubs and other woody plants, including dwarf forms of birch and willow, which can withstand the long seasons of frost and complete their cycle of growth in the short summer. In the *taiga* zone the summer is long enough for coniferous trees to grow, and the active layer is deep enough for their roots to develop. The movements in the active layer cause trees to fall about in drunken attitudes. Where drainage is too poor for any soil aeration to be achieved, and in particular in low lying areas which receive water percolating laterally through the active layer, bogs develop in the tundra and taiga zones.

Where there is no permafrost, on the fringes of the arctic region or in dry regions where there is not enough moisture at depth in the soil for it to develop, soils still show features of periglacial activity. In the dryer parts, there may be enough moisture to produce heaving and formation of stone polygons, and the size sorting effects of ice lenses within the soil, and in the wetter parts solifluction occurs in the upper part of the soil as melting begins at the surface and the soil becomes saturated before thawing has reached the lower parts. Where scree or the active layer on a slope is not completely churned by cryoturbation and solifluction, stratified deposits may develop as the annual cycle allows fine material to be washed down during the thaw and the coarser rock fragments left behind to slide down over it during the frozen period.

The Pleistocene Periglacial Zone

In the periglacial zones of the Pleistocene ice sheets many of the processes now going on in arctic regions occurred. Their distribution has to be inferred from the pattern of relict evidence in the form of casts of ice wedges, convoluted layers and patterned ground, and loess deposits, and deduced from theoretical considerations as to the climatic manifestations associated with ice sheets at mid-latitudes. In general, more oceanic areas show greater evidence of processes associated with abundant water in the freezing and thawing cycles, and the more continental areas show greater evidence of wind activity, but local variations abound, and the differences in climate associated with different stages in growth, maximum extent and decline of an ice sheet, complicated by the phase differences between neighbouring or contiguous ice sheets resulting from their different sizes, have produced considerable complexities in the pattern of occurrence of periglacial features.

In applying knowledge of present day glacial and periglacial processes to the Pleistocene periglacial zones in order to understand the nature and distribution of the parent materials of the post glacial soils, important differences have to be borne in mind. The degree of weathering of mineral materials in the arctic is low. The soils formed there during the Tertiary epoch have long been removed by the successive advances of ice across them or by solifluction, and the rocks now being abraded by glaciers and shattered by frost to provide fresh surfaces and sediments have been cut back far beyond the depth of penetration of Tertiary weathering, and the restricted development of tundra soils during interglacial periods left insufficient weathering at depth to survive subsequent glacial or periglacial erosion. The periods of ice cover have been progressively shorter and less numerous as one proceeds further from the centres of the ice caps, and the rocks scraped by the ice retain more of the depth to which Tertiary weathering reached, and in interglacial periods soil formation contributed further alteration of rocks at depth, even though the ice subsequently removed the soil itself. The amount of material from the upper horizons of soils which is incorporated into glacial sediments is probably very low in relation to the amount of rock debris, but much of the rock debris was already somewhat weathered.

The present extent of permafrost is partly relict; effects of cryoturbation and solifluction now extend beyond the limits which they would reach ahead of an advancing ice sheet or achieve during glacial maxima. Also the temperature gradient away from the ice margin, and the length of the season of thawing would have been much greater at lower latitudes, so the effects of periglacial activity must have been confined to a narrower belt during glacial advance and still-stand than during retreat stages, and much narrower than the present tundra and taiga zones.

In the Pleistocene periglacial zones the processes of cryoturbation and solifluction acted upon the interglacial soils, and in successive phases of a glaciation upon the less mature soils of the interstadials. There may have been much disturbance of soils by erosion during the early parts of glacial phases, as the weather became more stormy and the vegetation cover unstable, so that the upper horizons of many soils may have been much reduced by the time the depth of frozen ground exceeded the depth of summer melting and the specifically periglacial processes set in. Sediments formed from the eroded soils would in many places have been buried by solifluction deposits. In the early stages, solifluction deposits would have contained much material from the upper parts of the soils, where these were still in place; gradually, as the soil thinned, the active layer would have worked down to the more coherent rock, disrupting it and contributing it to the moving material, so that the deposits became more stony upwards. The effects of cryoturbation acting on the fluviatile sediments and solifluction deposits would have been to squeeze masses of the incoherent material of higher moisture capacity predominating in the lower levels into the overlying more stony layers. On level ground the effects of cryoturbation included the upper parts of the soil profile. Where the profile was shallow enough for the active layer to reach the **C** horizon the churning process incorporated rock material, sweeping it upwards. Figure 10 shows such an effect. Where the depth of profile was greater than the summer melting the effect would have been slighter, since the development of differential pressures in the more uniform soil would have been less, and the mixing of **A** and **B** horizon would have modified the characteristics of the soil much less than incorporation of relatively unweathered material. However, areas of ground sufficiently level to have suffered no erosion and to keep the entire depth of interglacial soils through glacial phases are not extensive. Such relatively level areas as do exist in the temperate zone are mainly river floodplains and river or coastal terraces, and the soils on these were particularly likely to be eroded by glacial meltwaters or buried beneath river deposits or material brought down by solifluction from the valley sides.

An important difference between the arctic and the Pleistocene periglacial zones lies in the extent and amount of loess deposition. During glaciations an enormous amount of debris was gathered up and transported by the ice, deposited on land in moraines and redistributed by summer meltwater, and the area of loose materials exposed to deflation was further increased by the exposure of coastal and estuarine deposits as sea level fell. On the continental borders of the ice sheets, which have no equivalent in the arctic regions today, the climate was probably very dry, and during stable periods of glacial maxima the areas of dryness may have extended well into areas which at other times experienced moister conditions. The cold deserts exposed extensive areas of soils and deposits to deflation. Moraines and fluviatile deposits would have dried rapidly from the meltwaters, snow cover would have been thin and easily lost by ablation in the dry air, and growth of protective tundra vegetation restricted by lack of moisture as well as by intense cold. Thick blankets of loess were deposited in a belt lying to leeward of the dry areas, heaviest deposition occurring where the cold desert gave way to tundra, the vegetation reducing wind speed and perhaps snow fall helping to deposit the dust. Some weathering of loess probably took place concurrently with deposition in most places, the tundra grasses growing on it contributing

Figure 10. Part of a cryoturbation structure involving disrupted rock, seen in the side of an Iron Age ditch, at Embury Beacon, North Devon (excavation by J. Jeffries).

to soil formation. The very large surface area presented by the small particles exposed large quantities of mineral material to hydration, oxidation and solution, and the porous nature of a deposit formed of uniform sized particles initially containing very little clay allowed free access of air and moisture. The grey colour of rock debris gave way to the yellows of hydrous oxides of iron, and some downward movement and secondary crystallization of calcium carbonate occurred. Much of the loess was redistributed by meltwaters, and deposited in river valleys as loess loams, or mixed and interstratified with sands and gravels, and here too partial oxidation and the solution and recrystallization of calcium carbonate would have occurred (Lukashov *et al.*, 1970). Loess falling in areas where intense solifluction and cryoturbation took place at the same time or subsequently would have been mixed with the locally derived materials, modifying the mineralogy and texture of the deposits and soils but often not remaining visible as a distinct layer.

Initiation of Soil Formation

As the ice sheets shrank during decline of a glacial phase they left a landscape in which a great deal of already partly weathered materials were distributed: the loess sheets in which finely divided mineral material was already undergoing alteration and already contained small amounts of organic matter; the cryoturbated and soliflucted remains of former soils, already containing secondary minerals to provide a capacity for retention of water and nutrients as well as disrupted and partly altered rock and probably often a loessic component to provide fresh supplies of bases; weathered rock surfaces with some hydration and oxidation inherited from Tertiary and interglacial soil **C** horizons providing ready access for roots and for moisture to continue the soil forming process; boulder clay containing abundant fresh and partly weathered minerals in all size grades; sands and gravels composed of the harder and less weathered rock debris, but much of it derived from base rich and easily weathered rocks; and lacustrine deposits containing the finer residues. The soil parent materials of the post-glacial were therefore largely composed of materials which were either already partly weathered or finely enough divided for alteration to proceed rapidly as temperatures rose and the vegetation developed. Many of the materials contained debris from base rich and calcareous rocks. Where surviving interglacial soil horizons had been strongly leached, or where well sorted sands and gravels were poor in basic materials the incorporation by cryoturbation of base-rich materials from less leached horizons below or its addition from above in the form of loess, provided the bases which might otherwise have been in short supply. The glacial meltwaters running over and through moraines and the rainwater falling on them in the rapidly warming environment produced surface and ground-waters rich in bases. Fluviatile and lacustrine deposits retained the base rich moisture as the waters subsided, and secondary crystallization of calcium carbonate was probably widespread. In coarse grained materials which weathered slowly or were low in their own base reserves, and which initially lacked the clay component with a capacity to retain bases and to prevent rapid throughput of leaching water, the presence of free calcium carbonate would have provided a buffer against leaching until the secondary minerals began to accumulate, and provided an environment in which 2:1 minerals could form and remain stable, rather than the kaolinitic minerals characteristic of more leached environments.

Temperature probably rose rapidly as the ice sheets receded and, as the anticyclones associated with them became unstable the weather probably became wet and stormy, and these conditions may have extended into regions where continental stability and relative aridity held sway during glacial times and again when the pattern of inter-glacial atmospheric circulation became established. Soils would have been rather susceptible to erosion until the development of soil structure and a forest cover was achieved.

After the maximum extent of the Last Glaciation, recession was interrupted by periods of readvance of the ice margins, and periglacial processes buried or disturbed the soils beginning to develop on the more stable surfaces. The most marked climatic oscillation of late glacial times was the last, the Allerød oscillation during which birch forest began to cover large areas of the temperate zone, followed by reversion to tundra conditions as the ice again advanced. The Allerød soils, where they have been preserved by burial beneath subsequent periglacial deposits, show considerable accumulation of humus and formation of secondary minerals, with development of soil structure, giving a distinct A horizon. These soils were buried or were involved in cryoturbation and solifluction. In those places where soil cover survived relatively undisturbed, development could continue from a well established but immature soil when climatic amelioration was resumed. Elsewhere the Allerød soils contributed a small amount of humus and mineral soil materials to the mixed deposits on which profile development had to begin again.

Forest and its Soils

The Spread of Forest

As the forest vegetation spread from the refuges in which it survived glacial phases the tundra gave way first to open forest, or park tundra, with trees, initially juniper, birch, willow or pine, which act as pioneer species, being able to grow in shallow and immature soils in which the pool of available nutrients has not yet built up in the exchange complex, growing scattered among the grass and herbaceous vegetation or establishing small, open stands. Which pioneer species begins forest development depends on the distance and the rate of travel from refuges and on temperature and local factors such as soil moistness and groundwater level. As the forest cover thickened the light-demanding species began to disappear and the vigour of the grass and herbaceous vegetation was much reduced. The development of the soil was associated with changes in rooting depth, changes in the amounts of nutrients participating in the cycle of growth and decay, and changes in the environment at ground level. Because of the widespread distribution of superficial deposits it seems probable that even the most base-poor and porous sands and gravels had sufficient contribution from loess or from base-rich groundwaters to enable them to support a full deciduous forest cover wherever climatic factors did not limit the forest growth to conifers.

In the present temperate zone there are only a few mountain tops projecting above the theoretical climatic limit of tree growth and there a modified form of tundra persisted. Whether other high slopes and summits within the forest zone were kept bare of trees by exposure to high winds is not known. Since a developing forest provides its own protection as it grows it seems unlikely that exposure alone can prevent its growth, though the nipping off of buds and growing points by frost on windward sides and the top of trees might distort their shape and limit the ultimate height of the canopy. Within the mountain forests, refuges of light demanding herbaceous plants are provided by ledges on precipitous surfaces, where the weight of any tree which does grow, and the disrupting effect of its roots, brings down a rock fall, and by any erosion surfaces and screes which are too steep for the stability of soil and forest to build up.

Within the zone where climate permits the growth of deciduous trees, the establishment of mixed oak forest during the post-glacial period appears everywhere to have been achieved. The growth of tree root systems and the growth of successive generations of trees in the spaces between the older ones would have brought the whole soil within the influence of the roots and gradually smoothed out the local variations due to positions of major roots, dryness and lack of organic matter immediately beneath tree boles, and the effects upon the upper part of the soil of the shallow rooted ground flora between the trees. On parent materials of different texture the rate of development was probably very different, soils on hard, non-calcareous rocks with little cover of superficial deposits taking much longer to provide sufficient rooting depth for the deep rooted trees and to develop a **B** horizon than those on incoherent deposits and on soft or easily weathered rocks. Soils on valley bottoms would have been continually subject to the effects of river erosion and deposition and to flooding. Here the forest succession and soil development have to begin afresh on every newly deposited gravel bank in a river bend and on every abandoned meander and terrace. These areas provide refuges within the forest for the light demanding species and for those plants which flourish in conditions which are wet but not continuously deprived of fresh supplies of oxygen.

During the post-glacial period there have been fluctuations in climate, and throughout the period during which forests and soils became established man has been present and has been using fire. Evidence of change in climate, such as that derived by study of changes in behaviour of mountain glaciers, does exist, but interpretation of these changes in terms of temperature and amount and seasonal distribution of rainfall which might affect the vegetation of neighbouring regions relies on evidence which in many cases is likely to be confused with the effects of interference by man.

Sources of Evidence

Pollen analysis, the principal tool for study of past changes in vegetation, is limited in its application by the distribution of situations in which pollen grains are preserved. Information concerning the complete succession of vegetation during and since late glacial times is preserved in lake sediments and basin peats. In the formerly glaciated areas, and particularly in upland regions, overdeepening of valleys by glaciers and formation of hollows among moraine have provided numerous situations in which deposition of lake sediments, or growth of peat over lake sediments as the smaller basins filled, has occurred, and these deposits have been much studied. The pollen in the sediments and peats represents the vegetation of the region, and the local vegetation of the lake margin or the bog surface can be picked out in the analyses. Changes in the vegetation can sometimes be associated with changes in the amount of mineral material coming down the slopes of the basin sides or to changes in the rate of peat growth and its degree of oxidation, providing evidence of disturbance of soils and of changes in groundwater level. However, these changes cannot be tied directly to evidence either of change in climate or of the activities of man. Correlation between the pollen record and the archaeology of the area must be made by radiocarbon dating except in the rare instances of a direct relationship between the stratigraphy of the deposits and an artifact or an occupation site at the lake or bog margin.

In the lake sediments and basin peats there is no direct evidence of the way in which the changes in condition of soils are associated with changes in vegetation. Presence or increase of soil material show the physical results of increased erosion, but not the state of the soil being eroded. Direct correlation with the history of the soil itself is provided by pollen analysis of upland peats and of mineral soils and deep mor humus horizons, but here the evidence only begins to exist when soils have reached a sufficiently acid state, or waterlogging has become sufficient, for pollen to be preserved. The path of soil development to the forest soil profile and its decline from it to the degraded conditions under which pollen is preserved are not documented. Inferences can be drawn from relict horizons within the soils themselves, which can be observed but whose time of formation and period of persistence as part of the active profile cannot usually be determined. Further interpretation can be based upon the distribution of soils of different forms according to present climate and parent material, and the timing of the beginning of pollen preservation in a mineral soil or the beginning of peat growth, determined by correlation with the full sequence of vegetation worked out from the basin situations, in relation to soil parent materials and drainage conditions. However, these means, besides involving several interrelated factors whose relative importance is unknown, give no direct correlation either with climatic change or with human behaviour. Buried soils provide much more hope of direct observation, but profiles preserved intact are not common. In the earlier part of the period in question, when the beginnings of soil disturbance and degradation may have been taking place under the influence of hunting peoples, there are no earthworks whose construction preserves a soil surface at a time datable in terms of human culture, and the chance preservation of a soil by burial by natural processes usually involves erosion of the surface by the agency transporting the material which eventually buries it. In the periods for which we do have plenty of buried soils beneath mounds and ramparts the evidence they present is too strongly correlated with certain human groups and certain of their religious practices or strategic needs for it to provide the sort of overall picture of soil conditions through the full range of time and local conditions which we need. For instance, we do not know why the people who built burial mounds lived in some areas rather than in others or why the sites of the mounds were chosen. Any argument about their distribution or the agricultural needs or practices associated with them which are based upon the soils of the area as we see them now or the condition of the soil at the time of its burial introduces circularity into interpretation of the buried soil in terms of human activity.

The Natural Forest, its Clearings and its Margin

A forest undisturbed by man, in which fire started by lightning is a localized phenomenon and in which numbers of browsing animals are kept in check by predators and by the normal functioning of their social interactions, contains trees of all ages. Openings in the canopy caused by fire or storm damage or by death of a tree from age or disease are closed by the natural processes of regeneration at a rate which varies with climate and soil. The nearer the situation is to the climatic limits of forest growth or to limitations of moisture or nutrient supply, drainage or aeration, inherent in the soil or topographic situation, the slower will be the growth of young trees. The mosaic of full canopy and clearings in various stages of closure will contain higher proportions of the earlier stages as any of these limits are approached, and where limits due to two or more factors coincide their effects are reinforcing. A forest well within its climatic limits, on base-rich soil parent materials in gently rolling country has few, small and short lived openings, many of which can be closed simply by the neighbouring trees extending branches into the opening. Towards the climatic limits of tree growth, the better parent materials might still be able to maintain a more or less continuous forest where on the poorer ones, whether it be those which limit nutrient supply or remain wet and cold in spring at the cold limits or those which limit moisture reserve at the dry margins, clearings coalesce and forest gives way to scrub or grassland.

Beneath the closed canopy growth of grasses and herbaceous plants and low growing shrubs is reduced for lack of light. In deciduous forests there are herbaceous plants and shrubs which come into leaf early in the spring and produce most of their foliage before the leaves of the trees overhead have fully developed and cut off the sunlight. Many of them propagate vegetatively under these circumstances and do not flower. The early-growing plants are the only source of food for browsing animals in the closed forest, since the trees put their edible growth up in the canopy where there is plenty of light. Some of the shrubs and trailing plants of the understory have spiny leaves or thorns which protect them and anything growing beneath them from browsers, cutting down the food supply still further.

When a tree fall or other accident creates a clearing the brambles and thorny bushes put out vigorous new growth and begin to flower. Their seedlings and those of other bushy plants participate in colonization of the open space where grasses and herbaceous plants flourish and tree seedlings sprout up. Those bushes and tree saplings which grow among the thorny thickets around the edge of the clearing are protected from browsing animals and survive until the trees overtop and again shade out the shrubs. The clearing closes in from the edges as the thickets develop and the colonizing trees grow up under their protection. The presence of herbivorous animals is a major factor in determining the length of time it takes a clearing to close, and therefore the proportion of clearing to closed forest. Near to the limits of forest growth the proportion of clearing in various stages of regrowth becomes so high that adjacent clearings begin to coalesce. The forest margin takes the form of a transition from continuous forest with large, almost merging, clearings to open country with islands of trees surrounded by belts of bushy growth. As the margins are approached the contrast between the clearings and the forest in terms of soil regime becomes more marked. The soil beneath the low vegetation away from the shade or shelter of the trees experiences greater extremes of moisture and temperature, and snow lies more frequently and for longer in the clearings. The chances of tree seedlings surviving in the open areas decreases the larger these areas are and the closer they are to the climatic limit of tree growth.

The belt of vegetation around the edges of a clearing benefits both from the greater light of the open space and from the protection of the trees. In dry areas the atmosphere is more humid and the soil moister beneath the trees, and the roots of the vegetation at the edge can spread under the forest. In cold areas the plants here are protected from wind. Leaf fall from the forest trees contributes nutrients taken from the soil at depth to a vegetation using a predominantly shallow nutrient cycle and so compensates for losses due to leaching under high rainfall conditions. This belt of vegetation is therefore particularly vigorous and provides an abundance of food for grazing and browsing animals, more than the open space, and much more than beneath the closed canopy. The herbivores gather in clearings, grazing in the middle and browsing around the edges: if it weren't for the thorny bushes, closure of a clearing would be very much slower.

The animals trample the soil in clearings, wearing the ground flora away along their paths, breaking down soil structure and allowing it to be blown away, puddling it in wet weather and exposing it to surface erosion. Animals also transfer nutrients from the soil of the open spaces to that of the forest, eating in the open and dropping some of their dung in the forest where they retire for protection or shelter. A forest in full vigour is able to repair these localized areas of damage and nutrient loss, but as limitations due to climate or soil inadequacy are approached the losses become more serious, and the longer a clearing is open the more opportunity there is for such losses to occur.

The more clearings there are the greater is their total circumference and the belt of highly productive bushy vegetation. As clearings increase in number and extent the length of this belt increases, until the situation is reached at which adjacent clearings begin to coalesce, and then as this happens the total circumference decreases (Fig. 11). The animals help to maintain conditions in which their food supply is greatest by delaying the closure of clearings, until their numbers increase beyond the point where the clearings begin to merge. Animals which remain entirely within the forested zone live in small herds and their numbers generally remain low. The open country beyond the forest margin can support larger numbers of animals and they generally live in more numerous herds, perhaps for greater protection against predators from which it is more difficult to escape in the open. The animals of the open country converge on the forest margin during winter or the dry season, for shelter and food, and their browsing of the belts of bushy growth probably keeps the forest margin well on the forest side of the climatic limits of tree growth, with the result that growth there is more vigorous than it would be were climate alone the limiting factor.

Effects of Climate Change. Changes in climate change the positions of the climatic limits of tree growth, but since the forest margin is not at these positions anyway, because of the activities of the browsing animals and because the forest itself protects plants within it from the full extremes of aridity or cold, a direct response of the distribution of forest to a worsening of climate is not to be expected. The response of the forest to a change in climate is more likely to be by way of a change in soil characteristics than a direct effect of the climate itself on the vegetation. Within the body of the forest zone, within the extreme positions to which the margins may swing, there are marginal situations related to drainage conditions, slope, or extreme poverty of

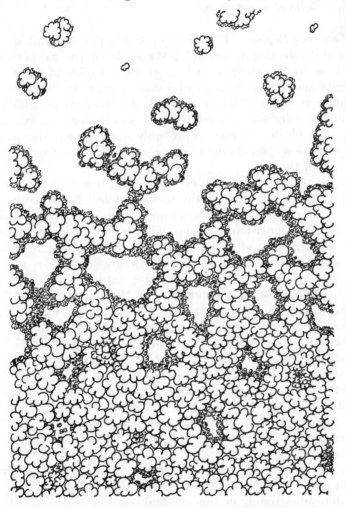

Figure 11. The transition from forest to open country, showing how the amount of bushy growth increases as the size and number of clearings increase, until clearings begin to merge, when it decreases again.

soils, and here too the changes in climate will be reflected, perhaps as much as at the forest margin on better soils. These parts of the forest margin where these other limiting factors already play their part in determining where the margin is, are particularly sensitive to climatic change. The behaviour of the animals too plays its part: a change in the length of the season of snow cover or a change in the seasonal distribution of rain may change the pressure of browsing on the marginal belts of vegetation.

Soil Leaching Regimes. The effect upon the development of the soil profile of the shifting mosaic of forest and clearing is that of temporary reversions to the shallow and short term cycles of moisture and nutrients and a somewhat greater risk of exposure to the extremes of climate, together with increased risk of soil disturbance and erosion. Within the fully forested zone such episodes are transient and the effect on profile development is probably slight, though the cumulative effect over a long period might be significant in controlling the eventual condition of the soil.

The overall trend in the forest soils of the temperate zone is one of leaching. Return of bases to the surface by leaf fall is inadequate to make up for downward movement and soils become increasingly acid, at least in the upper part of the profile. On base rich parent materials in the rainier parts of the zone, and on progressively less base rich ones towards the warmer and drier regions, leaching is not carried right through the profile. Fully base saturated horizons persist beneath neutral or acid conditions in the upper part of the soil. Calcareous horizons occur on parent materials rich in calcium and do so increasingly in warmer and drier areas. On base-poor parent materials the degree of leaching in the lower part of the profile depends strongly on texture. Fine textured soils have a greater capacity for retaining those bases which they do have, and also slow down the percolation of water through them, so base saturation may remain high in the lower horizons even under relatively high rainfall. Soils on coarse textured materials of low base content readily become leached throughout and there is little retention of bases in the lower horizons.

The effect on the leaching regime of removal of the protective canopy and deep roots of trees and their replacement with low vegetation, depends strongly on amount and seasonality of rainfall and the length of the growing season, as well as on the depth and texture of the soil. Where winters are cold and wet, leaching may occur to below the rooting zone while plants are not growing, and the rapid utilization of plant nutrients within the rooting zone during spring and summer is inadequate to compensate for this loss. Soils may therefore be more strongly leached in clearings than in woodland, and the effect is increased by the greater amounts of water reaching the soil, though the difference will be greater in coniferous than in deciduous forest because deciduous trees afford less protection to the soil during the winter. Where winters are mild and wet growth continues through most of the year and low vegetation may be able to make fuller use of bases released from residues in the upper part of the soil than the deeper roots of trees can do at depth, and leaching may be reduced in open areas. Even if the amount of water reaching the soil is greater, under higher temperature it is more rapidly used from the upper horizons and through-put may be lower. The risk of soil drying during the summer becomes an important factor in profile development. In areas where winters are cold and summers warm and dry, the degree of leaching is again reduced in clearings and soils there are dry for long periods during which they may remain moist under forest. Here too soil drying may be the dominant factor in profile development.

The transient occurrence of clearings, therefore, will delay the overall progress of leaching in the warmer and drier parts of the temperate zone and hasten it in the cooler and wetter parts. Where the transition between these regions occurs, soils of different texture respond differently, there being areas where the already intense leaching of coarse texture soils is increased in clearings while that of finer ones is reduced, and areas where summer drying affects the coarse textured soils but not the finer ones. Changes in climate would alter the relative distribution of areas where leaching increases and decreases in open areas and where summer drying of the profile becomes significant. To generalize broadly, the first effect would be important at the cold margins of the forest zone and on coarse textured parent materials within it, and the second would be important on the dry margins and on the finer textured soils within it.

Modification of Forest by Man

The balance established between the growth of trees, shrubs and ground flora and the animal population which forms the food chain based upon it, is likely to be upset by the presence of man. Without fire, early man would have had little effect except in so far as his use of tools and weapons and his ability to plan and co-operate in hunting gave him competitive advantage. His modification of the vegetation might be different in distribution or in detail from that of the other predators, but would not have been on a different scale. His use of timber for shelters, traps and barriers and smaller artifacts would be relatively slight, perhaps damaging the forest no more than do the larger quadrupeds, particularly elephants, or those which kill trees by eating their bark or fell them by gnawing. His paths and gathering places, his trampling at stream banks and crossings, would disturb the soil no more than do those of other animals.

Fire. Once he had fire, and it is probable that he had throughout the Pleistocene, man's effect upon the vegetation and soil was on an altogether different scale. A considerable amount of wood is needed just to keep a small domestic fire going, and one may postulate that fires would be kept going because of the inconvenience of relighting, and because of the value of fire in keeping away dangerous animals, in providing smoke against mosquitoes, and as a focus of group life. Once a fire gets out of hand, or if it is deliberately used to open up forest or drive game, its effect upon the vegetation can become a dominant factor in subsequent behaviour of the soil–plant system. Fires may often be started in forest by people using smoke to subdue bees and get at honey in holes in trees.

When vegetation burns, wood and foliage are converted to carbon, carbon dioxide, oxides of nitrogen, water, hydrocarbons and ash. The proportion of unburnt carbon depends on the degree of aeration in the fire. Most fires do not achieve complete combustion and charcoal remains among the ash. The carbon dioxide and most of the hydrocarbons go off into the atmosphere as smoke, and in the updraught caused by the heat of the fire much fine charcoal and ash are carried away too, to be distributed over a wide area. Charcoal is inert in the soil, but its absorbtive properties may participate in the exchange complex, and hydrocarbons adsorbed on it initially may contribute to organic processes. Ash contains the mineral components of plant material in the form of more or less soluble oxides. After a fire these mineral nutrients are washed into the soil, increasing their concentration in the soil solution and the exchange complex. The result of fire is therefore an immediate increase in the amount of nutrients available for growth of plants. However, with this increased availability is associated loss in smoke and in ash blown away or washed away by surface water before the new growth can protect the soil surface, and those nutrients not used quickly by the new growth are exposed to loss by leaching. The loss of the vegetation cover increases the risk of erosion not just of the ash but also of the soil itself. Humus may be dried and partly burnt by the fire, and the soil may be disturbed by the fall of burnt timber. The death of plants with deep roots exacerbates the risk of loss of nutrients by leaching. Though most forest and grassland fires pass too quickly for much heating of the soil to take place, some litter and superficial humic material may be destroyed, and micro-organisms may be killed even if the amount of burning is low. The burning of humus contributes to the immediate availability of nutrients and their ultimate loss, and the death of micro-organisms releases their materials into the available pool.

The most obvious immediate effect of damage to forest by fire is the lush new growth of herbaceous vegetation using to the full the abundance of nutrients in the full light of the open space. Herbivorous animals would gather and grow fat and the hunter would take advantage of this situation. Gathering activities benefit too, since the fruiting shrubs flower abundantly where there is free access of sunlight, and edible roots and rhizomes store more food when the plant gets plenty of light, as well as benefiting from the extra nutrients. Hunting and gathering peoples must always have found more food in clearings, and migratory or wandering groups probably deliberately returned to their earlier camping areas where their activities had extended a clearing and opened up the forest around it as they collected fuel and timber and where their refuse and the ash from their fires contributed to plant nutrition in their absence. We do not know when people began to open up the forest deliberately; they may have tended to live for preference in the forest margin areas, finding more vegetable food there as well as more game, and having access to plenty of timber for shelters and other artifacts and fuel for their fires. The effect of the presence of man in these areas would be to push back or widen the zone of the open mosaic of forest and clearing constituting the forest margin, bringing the zone of greatest extent of bushy growth into more climatically favourable areas and so increasing its productivity still further. The productivity of open country created in the former forest zone would also be greater than in the areas to which it is confined by the climatically controlled extent of forest.

These advantageous effects of man's activities would be immediate. The deleterious effects on the soil attendant upon repeated burning and the more frequent and longer lasting exposure of parts of the forest floor to greater leaching, in those areas where that is the effect of clearance, with permanent exposure in the marginal areas beyond the retreating forest, would not be apparent from season to season. Even if the overall productivity of the system were declining, the marginal belts and the recently burnt areas would still be relatively lush and productive, and appear so advantageous as to encourage their continued exploitation.

Hunting and gathering people, we may therefore assume, either deliberately or accidentally extended and kept open clearings and drew back the forest margins. In doing so they exposed the soil to the risk of nutrient loss by leaching and erosion and as a direct result of fire and soil disturbance, and caused an intensification of the effects upon the paths of soil development which in any particular area were attendant upon the reversion of moisture and nutrient cycles to the shallow mode characteristic of low vegetation.

It is a small step, though not necessarily an easy or an inevitable one, from opening clearings to encourage herbivores to come and keeping away other predators, to constructing barriers to protect and control the animals and avoid having to chase after them, carrying food to them when growth is inadequate or driving them from one clearing to another as herbage is used up. Some form of semi-domestication of red deer may have been practised in the temperate forests, equivalent to that of the reindeer which persists in the tundra today (Jarman, 1972), and there may be little difference in the effect upon the vegetation and soil between groups of hunters following and driving game and opening up the country to increase their numbers, and the herding within the forest and at its margins of domesticated sheep and cattle.

There is considerable incompatibility between the harvesting of the wild food grains and other edible herbage which would flourish in a forest clearing and the exploitation of the same clearing by way of increased meat supplies. For both purposes, and particularly for the successful pursuit of both at the same time, fencing becomes a necessity. The prickly bushes which contribute to the successful regeneration of forest can be most usefully deployed to delimit the area over which it must not encroach and to separate the activities of the incipient herder from those of the incipient cultivator.

Herding. Herding practices must vary with the need to reach grazing grounds throughout the year or to gather and store winter feed, and the need for shelter for the animals in bad weather. In the temperate zone there are areas where plant growth is almost continuous, and, provided the numbers of animals are not too high, shifting them about, up and down hillsides and to different parts of a river valley, is enough to keep them fed all the year. There are other parts where animals must be fed on stored food for many months and housed or gathered into sheltered folds. Between these extremes an enormous variety of patterns of behaviour have been established, based either upon taking the animal to the food or upon bringing the food to the animal. From the point of view of the soil the difference between these approaches is that if the animal goes where the food grows it drops most of its dung in the area from which it obtains its nutrition, whereas if the food is brought to the animal little if any of its dung is returned to the area from which the food came.

Animals concentrate nitrogenous and phosphatic materials and calcium relative to plants. The large quantities of plant food they need to obtain energy provides them with sufficient protein for their metabolism to involve a high rate of turnover of these materials as well as to use them for all structural purposes and even for wasteful excrescences such as antlers. Animals' bodies and excreta are therefore very rich in these nutrients. It is just these nutrients which are most likely to impose a limitation on plant growth. Calcium is abundant in many soils, but as soon as a soil becomes acid even those on parent materials containing plenty of calcium are likely to lose it by leaching. Phosphorous is often not abundant in parent materials, and may be present in minerals which alter only slowly. Nitrogen is particularly susceptible to loss from vegetation by burning, much of it going off with the volatile substances (Lloyd, 1971), and it is the nutrient most intensively concentrated by animals. It has to be replaced from the atmosphere and not from mineral reserves in the soil, and its replacement depends on a high rate of organic activity in the soil.

Wild animals transfer considerable quantities of nutrients from place to place. Migratory animals take them from their summer or wet season grazing ground, which they leave fat and in large numbers, and carry them to their winter or dry season refuge, where they become thin and where the old and the weak die. Dung is dropped and nutrients concentrated wherever animals go for shelter, and many such places, hollows and bushy places, are already receiving areas for plant materials and humic soil being blown about or moving downslope, so further concentration there may be unproductive. In rocky country animals go into cave mouths and rock shelters, where accumulation may be unproductive for lack of moisture or light.

People have the same effect as other animals, concentrating nitrogenous and phosphatic residues around their settlements and in their burial grounds. Their exploitation of animals extends and increases the unequal redistribution of nutrients, not only by the consumption of animal protein but by the discarding or utilization of bone, antler, horn, hoof, hide, hair and feather. Domestic animals kept near the settlement or brought in at night contribute further to the accumulation. Only extremely mobile people living in small transient camps return nutrients to the soil in a fairly uniform manner and in situations where they can be fully utilized by plant growth. Permanent settlements and the establishment of regular burial grounds produce concentrations which may not be available to plant growth for a very long time or which become available over a long period, during which the more mobile nutrients are susceptible to leaching. Loss by leaching is an immediate effect of collection of dung or rubbish in heaps and pits. The forms in which nitrogen is found in soils, those in which it is released rapidly early in the processes of decay of organic materials and those in which it is excreted by animals, are all soluble. The liquor draining away from dung heaps carries high quantities of nitrogen, and it often runs into the nearest water course or drains into the soil in swampy places where its concentration is too high and aeration too low for plant growth. Rubbish deposited in pits loses nutrients in percolating water which may enter the soil too deep in the profile for it to be used by the vegetation growing over or around the pit.

Some scatter of dung and rubbish around a settlement is beneficial to grazing land and crop growth while occupation continues, and people are likely to contribute their own excreta to productive land, at least during the daytime and in clement weather. But even where deliberate manuring of fields with the rubbish and dung of houses and byres is carried out, the labour of carting tends to restrict it to the nearer fields, which may receive a superabundance of manure and remain highly productive, and the outlying grazing grounds or source areas for gathered fodder receive little or nothing. However much manuring is practised, unless there is a source of plant and animal residues external to the area of food production and collection, no community can put back into the soil all that it takes from it. Export of organic material from an area, in the form of food or organic raw materials, or emigration of people, increases the loss of nutrients, with no return.

Since cattle were forest animals, like deer, before domestication and are browsers by preference, they can be kept in largely forested areas without the creation of grassy meadows. Open forest with small glades and much bushy growth among mature trees may have been deliberately maintained in some areas and at some stages during the development of animal husbandry. Cutting leafy branches to store as winter fodder would let in the light and allow vigorous growth of vegetation within reach of the animals browsing there in summer. This system is potentially very productive, since a combination of deep rooting and shallow rooting vegetation is maintained, and the trees, while thinned or lopped enough to let in light, provide some protection from wind and some shade in summer to prevent soil drying. It is equivalent to the maintenance of the bushy growth of the forest margin for long periods over the same tract of land, rather than allowing any particular spot to experience repeated oscillations between the extremes of open country and closed forest. It must require careful control of the numbers of domestic animals kept, and elimination of wild herbivores as well as predators. Continual exploitation of land in this way, since it makes maximum use of the potential productivity of the soil carries high risks of soil degradation, since much is being taken away and nothing put back; soils which can replace nutrients as fast as they are taken by drawing on the mineral reserve can maintain the rate of plant growth, but if the mineral reserve is low, or if the rate of mineral alteration is not high enough to supply nutrients as fast as they are taken, productivity must fall. Under this system, nutrients taken from depth by tree roots are returned to the soil surface in leaf fall and used by the low vegetation soon after release from the organic residues, reducing the risk of loss by leaching but also reducing the amounts percolating back down to the tree rooting zone. The nutrient reserve can therefore be exhausted throughout the depth of the profile if the low vegetation is browsed over long periods.

Herding practices which involve the creation of more open meadows, whether for grazing or for production of hay for winter feed, rely on periods of forest regeneration and the opening of new clearings to maintain the supply of food. Here, the upper part of the soil may be depleted of nutrients while a reserve builds up in the lower horizons by continuing mineral alteration and by retention in the exchange complex of those brought down by leaching. When forest grows again it brings about a return of part of this reserve to the upper part of the soil. However, repeated episodes of clearance and regeneration, with the loss of nutrients during the meadow phase, may entrain a lower level of productivity in a descending spiral. Whether this method of land utilization is more or less deleterious to the soil than that of permanent open forest depends on the rate at which the mineral reserve can release nutrients into the soil, and on the leaching regimes under open and forested conditions. In the long term the effects may not be very different, but one advantage of the meadow system is that meadows are more likely to receive some deliberate manuring than are the more diffuse grazing grounds and areas of fodder collection of open forest. It is difficult to scatter dung among trees and bushes, and since the decline in productivity is likely to be slower than in any one meadow phase of the cyclic system it is less likely to be thought necessary.

Cultivation. Cultivation requires a more deliberate and consistent policy of forest clearance than does herding. To facilitate sowing and reaping and the protection of seeds and the ripening crop from birds and animals, to prevent too much encroachment of seedlings and suckers from the forest into the cultivated area, and to ensure enough sunlight for ripening, regularly shaped fields of a manageable size must be created, rather than using scattered and irregular patches among the trees, and the work that goes into such clearance demands a number of years of cropping to justify it. The method of clearance depends on whether it is mature forest that is being cleared or the thicker and bushier growth of regenerating forest, and it also depends on the habits and needs of the community, the labour and type of tools available, the amount of timber needed, and the relationship between cultivation and herding. Large trees can be burned by piling brushwood around them, or cut down, chopped up and carted away, or they can be lopped, killed by ring barking, and the trunks left standing until they are needed for timber or fuel. The larger stumps can be left in the ground unless ploughing in long straight furrows is to be carried out, and this practice implies a source of tractive power which can be used to help in grubbing them out. Clearance of bushy, immature woodland is most easily achieved by cutting down and grubbing out the bushes and saplings and burning the smaller brushwood on site, since it is troublesome to transport, not thick enough or straight enough for construction of artifacts and much of it too twiggy for most domestic fuel needs. If cultivation and sowing follow quickly on clearance the ash provides nutrients for an abundant first crop, and the growing crop helps to protect the disturbed soil and the ash from being blown or washed away. Roots and stumps decaying in the soil provide a longer term reserve of nutrients, fed slowly into the soil and for a number of years compensating for the losses attendant on clearance and cropping.

In some tropical areas today, crops are grown by planting among the heaps of brushwood resulting from forest clearance. The planting method, that of dropping individual seeds into holes prodded with a stick, makes it possible to work through the brushwood layer. Broadcast sowing implies a prepared, that is, cultivated soil, so that seeds drop down into it. The tradition of sowing cereals rather than planting appears to be very strong in Europe and Western Asia, and it may be that where planting is used for cereals the tradition stems from the development of agriculture based on tuber crops. If the cleared soil is not cultivated and the brushwood is left lying on it, the soil is not exposed to erosion, and agriculture in this form is carried out on slopes whose steepness would disqualify them from use under the plough or the hoe. However, one of the purposes of cultivation is to control weeds, and it may be significant that it is the growth of weeds which causes abandonment of these fields, rather than falling yields resulting from nutrient loss alone.

Cropping involves removal of a large part of each year's growth, with almost complete removal of the more nutrient rich parts of the plants, the seeds. Reduction in the amount of organic matter deposited on the surface each year causes a reduction in the humus content of the soil and a decline in micro-organism activity. Since much nitrogenous material is taken away and since the nitrogen fixing organisms in the soil need an abundant supply of mineral nutrients and of organic material to supply their carbon requirements, nitrogen fixation is reduced just when it is most needed, and lack of nitrogen may be the first cause of falling yields in soils where the mineral nutrients are not yet in short supply. Manuring the fields affords some replacement, but even the most diligent collection and use of dung and domestic refuse leaves behind the high concentration of protein represented by the bodies of people and animals, so replacement can never be adequate over the whole of the land used for food production. Manuring involves the labour of carting; it tends therefore to be adequate or even excessive in the fields nearest to the settlement or the byres, and tails off away from them, particularly uphill. Domestic animals may be folded on the harvested fields and fed on carted fodder, involving more, but perhaps easier, carting, and lower losses of nutrients from the dung than is inherent in its collection and storage in heaps. However, animals trample and puddle the soil and damage the structure, which is already likely to be weakened by the loss of organic matter and of bases, and increase the probability of erosion.

Cultivated soils are always more likely to suffer erosion than those under grass or forest. They are exposed to wind and rain when these are at their most intense in winter, their structure is less stable, and they have no permanent root system binding them together. The process of cultivation involves soil movement on even the slightest slope, each disturbance of the soil by hoe or plough shifting it a little way downhill, unless deliberate attempts are made to move it uphill. The purpose of cultivation is to create a suitable environment for the germination of the seeds and for the growing seedling. A recently cleared soil which still has its natural structure should have a system of large and small pores holding a sufficient moisture reserve to prevent the seedling drying out, and sufficient drainage to allow plenty of oxygen to reach the roots. As the natural structure of the soil becomes less strong, it tends to settle down into closer packing, and cultivation is needed to open it up and create an artificial structure of plough clods. Cultivation early in the winter, though it exposes the disturbed soil to higher risks of erosion, is carried out to allow the frost to break down the plough clods into a finer structure. Cultivation is also practised to keep down weeds which compete with the crop for moisture and nutrients.

In comparison with the effects of more intensive agriculture, the amount of soil erosion resulting from shifting cultivation within a forested zone is perhaps not great. Small fields surrounded by forest are not badly exposed to wind, and any movement of soil by surface washing, by the increased rate of soil creep in the absence of binding roots and strong structure, and by the process of cultivation, would be halted at the edges of the field and soil would accumulate there. Redistribution of soil over short distances would have occurred, and as successive clearances used different areas the soil eroded from one patch would be used in the next. Wholesale stripping of soils followed, and still follows, the establishment of large fields with inadequate windbreaks, and the use of ploughs which cut deeply and have a blade or coulter which severs the remaining roots beneath the clods, breaking the coherence between the plough layer and the subjacent more stable soil. The hoe or the simpler type of plough scratches the soil, but does not turn it over, and between furrows the structural continuity is maintained, and sufficient roots remaining from the former crop pass down the soil profile to add some binding effect to that of the soil structure.

6

Soil Groups and Profile Types

Incipient and Immature Soils

The earliest stages of soil formation are somewhat different on solid rocks and incoherent materials. In the latter, plant roots can grow and nutrients are available for them to absorb as soon as dissolution of mineral materials begins, or, in many alluvial situations, as soon as nutrient-bearing waters subside to leave the soil moist. The limiting factor in soil development is likely to be shortage of nitrogen; some micro-organism activity is necessary to fix atmospheric nitrogen and make it available to plants. Colonization of sediments by bacteria starts with their introduction in dust or flowing water and can proceed as soon as moisture is available and there is a small amount of organic matter present. Incipient or "raw" soils on incoherent material in which alteration of minerals can begin throughout the rooting zone and organic matter begin to accumulate within the upper part, but in which as yet no horizon differentiation is apparent are known as *regosols*, or (Avery, 1973) as *raw sands*, *raw alluvial soils*, or *raw earths* depending on the nature and origin of the parent material. Where soil formation is beginning in a shallow accumulation of hard rock fragments, the soil is a *skeletal raw soil*.

Hard rock surfaces provide no root hold for plants and initially no moist and porous environment for soil fauna and micro-organisms. Soil formation begins with colonization by algae or by lichens. Lichens are symbiotic associations of algae, which photosynthesize to provide energy, and fungi, which exude substances which attack mineral surfaces and make available nutrient elements. The macro-nutrients are released from the rock by simple solution; the micro-nutrients are taken from the minerals by chelating substances in the exudate. Nitrogen is fixed by blue-green algae as components of lichen. The rough surface of the lichen-encrusted rock traps airborne dust, contributing to accumulation of weatherable minerals in finely divided

form and organic materials, which include the spores and eggs of the soil population. Contraction of the gummy exudates when the lichens dry plucks fragments of rock from the surface, and with the accumulation of these and the lichen residues a primitive soil develops which can hold enough moisture for mosses to grow. The cushions of moss contribute to creation of a moist environment in which the soil fauna can live and higher plants can take root. The solid rock begins to yield to hydration and oxidation as acid moisture percolates from the organic layer at the surface into cracks and along crystal boundaries, softening the rock and providing channels for penetration by plant roots. Cycles of freezing and thawing, and of heating and cooling, contribute to fragmentation in appropriate climatic situations.

Ranker

With the establishment of a distinct **A** horizon, either, in the case of the hard rock, predominantly organic, or, in the incoherent material,

Figure 12. A ranker profile.

a crude mineral and organic mixture, the soil is known as a *ranker* (Fig. 12). Rankers and ranker-like alluvial raw soils are the immature, **A/C**, profile type in freely draining situations on all parent materials except highly calcareous ones. On non-calcareous rocks, and on very hard calcareous ones, the **A** horizon develops initially by the activity of arthropods, which deposit their faecal pellets of finely comminuted plant residue among rock fragments initially present or loosened from the surface, or within the upper part of incoherent sediments. Earthworms cannot participate until there is enough calcium in the organic cycle to satisfy their requirements and until there is sufficient depth of soil to provide them with a moist refuge during dry periods.

Rendzina

On calcareous rocks the dissolution of calcium carbonate under the influence of rainwater and the acids of organic materials proceeds

6
Soil Groups and Profile Types

Incipient and Immature Soils

The earliest stages of soil formation are somewhat different on solid rocks and incoherent materials. In the latter, plant roots can grow and nutrients are available for them to absorb as soon as dissolution of mineral materials begins, or, in many alluvial situations, as soon as nutrient-bearing waters subside to leave the soil moist. The limiting factor in soil development is likely to be shortage of nitrogen; some micro-organism activity is necessary to fix atmospheric nitrogen and make it available to plants. Colonization of sediments by bacteria starts with their introduction in dust or flowing water and can proceed as soon as moisture is available and there is a small amount of organic matter present. Incipient or "raw" soils on incoherent material in which alteration of minerals can begin throughout the rooting zone and organic matter begin to accumulate within the upper part, but in which as yet no horizon differentiation is apparent are known as *regosols*, or (Avery, 1973) as *raw sands, raw alluvial soils*, or *raw earths* depending on the nature and origin of the parent material. Where soil formation is beginning in a shallow accumulation of hard rock fragments, the soil is a *skeletal raw soil*.

Hard rock surfaces provide no root hold for plants and initially no moist and porous environment for soil fauna and micro-organisms. Soil formation begins with colonization by algae or by lichens. Lichens are symbiotic associations of algae, which photosynthesize to provide energy, and fungi, which exude substances which attack mineral surfaces and make available nutrient elements. The macro-nutrients are released from the rock by simple solution; the micro-nutrients are taken from the minerals by chelating substances in the exudate. Nitrogen is fixed by blue-green algae as components of lichen. The rough surface of the lichen-encrusted rock traps airborne dust, contributing to accumulation of weatherable minerals in finely divided

form and organic materials, which include the spores and eggs of the soil population. Contraction of the gummy exudates when the lichens dry plucks fragments of rock from the surface, and with the accumulation of these and the lichen residues a primitive soil develops which can hold enough moisture for mosses to grow. The cushions of moss contribute to creation of a moist environment in which the soil fauna can live and higher plants can take root. The solid rock begins to yield to hydration and oxidation as acid moisture percolates from the organic layer at the surface into cracks and along crystal boundaries, softening the rock and providing channels for penetration by plant roots. Cycles of freezing and thawing, and of heating and cooling, contribute to fragmentation in appropriate climatic situations.

Ranker

With the establishment of a distinct **A** horizon, either, in the case of the hard rock, predominantly organic, or, in the incoherent material,

Figure 12. A ranker profile.

a crude mineral and organic mixture, the soil is known as a *ranker* (Fig. 12). Rankers and ranker-like alluvial raw soils are the immature, **A/C,** profile type in freely draining situations on all parent materials except highly calcareous ones. On non-calcareous rocks, and on very hard calcareous ones, the **A** horizon develops initially by the activity of arthropods, which deposit their faecal pellets of finely comminuted plant residue among rock fragments initially present or loosened from the surface, or within the upper part of incoherent sediments. Earthworms cannot participate until there is enough calcium in the organic cycle to satisfy their requirements and until there is sufficient depth of soil to provide them with a moist refuge during dry periods.

Rendzina

On calcareous rocks the dissolution of calcium carbonate under the influence of rainwater and the acids of organic materials proceeds

much more rapidly than does attack on other rock forming minerals. Solution widens cracks sufficiently for plant roots to penetrate and contribute further to development of channels reaching deeply into the C horizon. In the organic horizon calcium is available for earthworms from the beginning, and dark coloured stable associations of humus substances and minerals are formed; earthworms can retreat down cracks and root channels when necessary. A *rendzina* (Fig. 13) is a soil consisting of an almost black calcareous mull humus formed entirely of worm casts lying on relatively unaltered calcareous rock.

Progress from the ranker or rendzina stage depends on the establishment of a B horizon. On steep slopes, where erosion decapitates the

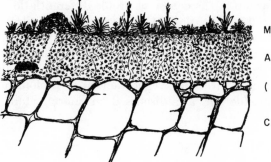

Mat of grass stems and stolons

A - mull humus

(A/C)

C

Figure 13. A rendzina profile.

profile from time to time, or where continuous and rapid soil creep occurs, the immature stage may be maintained indefinitely. Rendzinas can persist on any surface of pure limestone, such as chalk, provided there has been no contribution to the soil from superficial deposits, or after soils developed partly within such deposits have been eroded away. Much of the chalk contains so little mineral material other than calcite that there is nothing to form an appreciable B horizon. Solution of the chalk lowers the land surface without contributing more than a very small percentage of its volume to the soil, and development of a B/C horizon cannot proceed beyond oxidation of any iron present in the chalk at the surface of structural planes and in root channels. As the calcite goes, the mineral residue is immediately incorporated into the A horizon by worm activity. On the purer chalks it would take many thousands of years of soil formation without any loss from the surface by erosion to build up a depth of soil greater than that normally mixed by worms. Loss by erosion, even under stable

vegetation, can easily exceed the rate of growth of the **A** horizon from below.

From a ranker or a rendzina, various paths of profile development are possible, and the direction taken depends on a complex interaction between the characteristics of the parent material and environmental factors. The soils formed under the deciduous forests of the temperate zone are the Major Group of *Brown Soils*, of which the *brown forest soil* is the central character. The Brown Soils merge into the *Podzolic Soils* which are characteristic of the coniferous forests, and into the *Chernozems* of the steppes. Associated with all the soils are the immature soils and the gleyed soils. Gleyed forms are characteristic of low lying situations, or of clayey parent materials and impervious horizons which may develop in the course of profile differentiation in any of the main soil groups. The Major Groups of Brown Soils and Podzolic Soils will be described here, before the history of their development and the relationship between them, under the formation of the post-glacial forests, under changing climate, and in particular under human interference with the vegetation and exploitation of the soil for food production, are discussed in more detail in Chapter 7.

Brown Soils

The Major Group of Brown Soils as defined by Avery (1973) contains soils which have been called *brown earths* and *brown forest soils* under other classification systems. *Brown earth* has previously been used in the British system to cover the greater part of the Major Group, but is now introduced at the first level of subdivision, as a group. *Brown forest soil* is equivalent to brown earth at this level and corresponds to the usage of the F.A.O. Soil Map of Europe. It will be used here because of the way it draws attention to the forest history of these soils, and thus to their ancestral status with respect to soils in this and in other Major Groups, which is argued here.

Brown Forest Soil

Under a deciduous forest growing on a brown forest soil the vegetation is well nourished, and the leaves are rich in bases and poor in phenolic compounds. Rainwash from the leaves and litter and a high rate of microbial attack on plant residues return bases to the upper part of the soil and tend to keep the exchange complex saturated. Abundance of divalent cations prevents dispersion of the colloidal component of

much more rapidly than does attack on other rock forming minerals. Solution widens cracks sufficiently for plant roots to penetrate and contribute further to development of channels reaching deeply into the C horizon. In the organic horizon calcium is available for earthworms from the beginning, and dark coloured stable associations of humus substances and minerals are formed; earthworms can retreat down cracks and root channels when necessary. A *rendzina* (Fig. 13) is a soil consisting of an almost black calcareous mull humus formed entirely of worm casts lying on relatively unaltered calcareous rock.

Progress from the ranker or rendzina stage depends on the establishment of a B horizon. On steep slopes, where erosion decapitates the

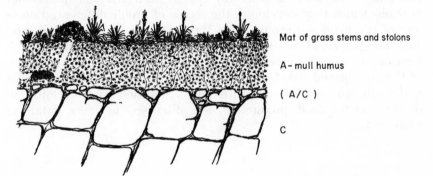

Mat of grass stems and stolons

A - mull humus

(A/C)

C

Figure 13. A rendzina profile.

profile from time to time, or where continuous and rapid soil creep occurs, the immature stage may be maintained indefinitely. Rendzinas can persist on any surface of pure limestone, such as chalk, provided there has been no contribution to the soil from superficial deposits, or after soils developed partly within such deposits have been eroded away. Much of the chalk contains so little mineral material other than calcite that there is nothing to form an appreciable B horizon. Solution of the chalk lowers the land surface without contributing more than a very small percentage of its volume to the soil, and development of a B/C horizon cannot proceed beyond oxidation of any iron present in the chalk at the surface of structural planes and in root channels. As the calcite goes, the mineral residue is immediately incorporated into the A horizon by worm activity. On the purer chalks it would take many thousands of years of soil formation without any loss from the surface by erosion to build up a depth of soil greater than that normally mixed by worms. Loss by erosion, even under stable

vegetation, can easily exceed the rate of growth of the **A** horizon from below.

From a ranker or a rendzina, various paths of profile development are possible, and the direction taken depends on a complex interaction between the characteristics of the parent material and environmental factors. The soils formed under the deciduous forests of the temperate zone are the Major Group of *Brown Soils*, of which the *brown forest soil* is the central character. The Brown Soils merge into the *Podzolic Soils* which are characteristic of the coniferous forests, and into the *Chernozems* of the steppes. Associated with all the soils are the immature soils and the gleyed soils. Gleyed forms are characteristic of low lying situations, or of clayey parent materials and impervious horizons which may develop in the course of profile differentiation in any of the main soil groups. The Major Groups of Brown Soils and Podzolic Soils will be described here, before the history of their development and the relationship between them, under the formation of the post-glacial forests, under changing climate, and in particular under human interference with the vegetation and exploitation of the soil for food production, are discussed in more detail in Chapter 7.

Brown Soils

The Major Group of Brown Soils as defined by Avery (1973) contains soils which have been called *brown earths* and *brown forest soils* under other classification systems. *Brown earth* has previously been used in the British system to cover the greater part of the Major Group, but is now introduced at the first level of subdivision, as a group. *Brown forest soil* is equivalent to brown earth at this level and corresponds to the usage of the F.A.O. Soil Map of Europe. It will be used here because of the way it draws attention to the forest history of these soils, and thus to their ancestral status with respect to soils in this and in other Major Groups, which is argued here.

Brown Forest Soil

Under a deciduous forest growing on a brown forest soil the vegetation is well nourished, and the leaves are rich in bases and poor in phenolic compounds. Rainwash from the leaves and litter and a high rate of microbial attack on plant residues return bases to the upper part of the soil and tend to keep the exchange complex saturated. Abundance of divalent cations prevents dispersion of the colloidal component of

the soil and the lack of polyphenols prevents chelation and the mobility of iron and aluminium by this means.

The high pH and low tannin content of the litter encourage worm activity and the litter is rapidly comminuted and mixed with the mineral soil. The mixing helps to overcome any tendency to downward mobility of bases or colloid. Worm holes and the strong crumb structure developed under the influence of worms encourage good drainage and aeration, so sites of anaerobic conditions are short lived and confined to the interiors of peds; continued worm mixing soon obliterates any effects.

Clay minerals formed under these conditions are the 2:1 minerals, illite, vermiculite, and interstratifications of vermiculite with chlorite and montmorillonite. Kaolinite may occur as the initial alteration product of felspars, and may be common in soils on igneous rocks rich in felspar. Montmorillonite will form in soils on basic igneous rocks, particularly basalt, when secondary mineral formation is going on inside a coherent rock fragment or in the C horizon. Once free in the soil it tends to be altered by substitutions in the lattice and the development of a magnesium hydroxide layer within the expanded water layer, giving interstratifications of chlorite or vermiculite. The 2:1 clay mineral suite provides a high cation exchange capacity, so that bases are not lost by leaching during prolonged rainfall and the nutrients are kept within reach of plant roots. The changes in volume of some of these clay minerals on wetting and drying help in the production of a good soil structure.

Iron is in the form of hydrous oxides in gel form or as tiny crystals in clusters associated with the surfaces of clay minerals and forming coatings to gains of primary minerals. Any iron reduced during temporary and localized oxygen depletion cannot move far before being re-oxidized. Humus substances are of the more condensed forms, in association with clay minerals and cations in the immobile colloid. The high level of organic activity and the high quantities of organic residues added to the soil by leaf fall in the deciduous forest ensure that much material is humified, and the stable forms in which the humus substances are held ensure that it provides a long term reserve of nutrients.

A diagrammatic profile of a brown forest soil is shown in Fig. 14. The A horizon represents the normal depth of worm mixing. It has a strong and well developed structure of porous crumbs which are largely old worm casts, usually a loam or a clay loam texture and a dark to very dark brown colour. The Bw horizon, a cambic horizon, consists of soil produced by alteration of parent material *in situ* beyond the point where the original structure or stratification of the rock or sediment on which the soil has formed has been obliterated and superseded by soil structure. The soil structure in the **Bw** horizon is formed under the influence of roots, the drying of the soil as roots extract water causing shrinkage of the colloid, which expands again after rain, and decaying roots and organisms of the rhizosphere providing organic substances which act as binding agents. The result

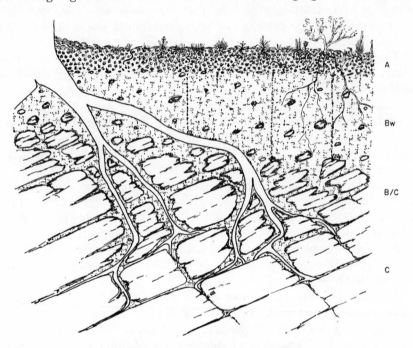

Figure 14. A brown forest soil profile.

is a granular structure of irregularly facetted peds, which may break to a second stage of structure of fine peds which, within the larger ped, expand on wetting until they press against each other, producing blocky shapes. The colour of the **Bw** horizon lacks much of the dark reddish browns of humus substances since its humus content is only that provided by decaying roots, and is dominated by the yellowish browns of hydrous iron oxides. The texture is very similar to that of the **A** horizon.

The parent material undergoes alteration in the environment of well-aerated, base-rich moisture percolating from the **Bw** horizon, and under the physical and chemical influence of growing and decaying tree roots. Softening and disruption by hydration and hydrolysis allow penetration of moisture and roots; the thickening of roots causes further disruption and the organic activity associated with them contributes organic materials in the form of exudates, micro-organisms and their products, and finally the decaying root itself. Clay minerals form *in situ* and their swelling causes more fragmentation. Iron is released and oxidized, and the hydrous oxides colour the whole of the softened rock in its upper part, and are present in root holes and on structural planes to great depth.

Calcareous Brown Forest Soil

On base-rich and calcareous rocks calcium saturation of the soil in part of the profile may be so high that calcium carbonate is precipitated. The secondary carbonate appears as a diffuse whitening of the soil in the **B** horizon or as concentrations in soil pores and in cracks and crevices in the upper part of the **C** horizon. It may be so concentrated as to form a discrete horizon within the lower part of the **B** horizon or at its base, or in the upper part of the **C** horizon. The **A** horizon and part of the **B** horizon may be decalcified, even acid in reaction, leaching having been sufficiently active in the upper part of the soil to remove all the free calcium carbonate there. The presence of a calcareous horizon and the depth at which it appears depend on the amount of calcium in the parent material, on the texture of the soil, and on climate. In areas where leaching is intense even limestone soils may be decalcified throughout the **B** horizon and be undergoing active decalcification in the **C** horizon. Where the soil is relatively impermeable, for instance in those formed on calcareous clays, the calcareous horizon develops even under high rainfall, and is often associated with gleying in a zone in which percolation is very much reduced. In increasingly dry climates it appears in soils on increasingly porous and less calcareous materials.

In Avery (1973) these soils are classified as *brown calcareous earths*.

Acid Brown Forest Soil

Acid brown forest soils occur on non-calcareous materials under high rainfall, where the rate of leaching is high in relation to the base content of the parent material or to the rate at which bases are released. Nutrient supply to plants is restricted, so that forest composition may be reduced in variety and its growth is less vigorous than on soils of higher base status, and the litter is poor in bases.

Alteration of minerals under the acid conditions is rapid, but the intensity of leaching inhibits formation of clay minerals, and especially of the 2:1 forms. The clay is kaolinitic, and its low cation exchange capacity contributes to the low nutrient status of the soil. Since no iron is incorporated into kaolinitic minerals, and since the amount of clay formed is lower than in less acid soils, the proportion of hydrous oxides of iron is high in relation to clay, and gives the **Bw** horizon, where it is not masked by humus colour, a stronger colour. Because of the low content and low capability for expansion of the clay, the **Bw** horizon may have a poorly developed structure.

The **A** horizon may be of mull form, but thinner and paler than in less acid soils. A frequent and distinctive characteristic of these soils, however, is extreme thinness of the **A** horizon, with little or no development of mull but without the accumulation of litter at the surface which would lead to the development of mor. It would appear that under certain conditions there is sufficient activity by earthworms and micro-organisms to achieve rapid degradation of residues, but insufficient calcium or 2:1 clay minerals to form stable complexes with humus substances, so mineralization is almost complete and there is little reserve of plant nutrients in organic form in the soil. The low organic content of the thin **A** horizon together with low clay content leads to poor structure, and high susceptibility to erosion when these soils are cultivated or otherwise disturbed.

Sol Lessivé

Dispersion of colloidal material in the upper part of the soil, its migration down the profile, and immobilization further down produce a horizon of absolute accumulation of clay, the illuvial clay horizon, or argillic horizon, characteristic of the *sol lessivé*. The French term has been used in English soil science because no English word or short phrase expresses the process involved so clearly. In Avery (1973) the term has been dropped, and the diagnostic horizon rather than the process is expressed in the term adopted, *argillic brown earth*, which has group status in the Major Group of Brown Soils.

Like the acid brown forest soil, the sol lessivé is the result of failure of one or more of the conditions required to maintain the high base status throughout the profile of the brown forest soil. Whether such failure, in either case, characterized the soil/plant system throughout the development of the soil under forest or whether it follows the development of a brown forest soil of high base status is not always easy to determine. However, in the case of the sol lessivé, if the soil parent material contained no clay initially one has to postulate a period of clay formation in the horizon from which it later began to be lost, implying a period of higher base status and probably the existence of a brown forest soil. Where the parent material was clayey to start with, or where alteration of very fine grained primary minerals of layer lattice type to form soil clays proceeds as soon as hydration begins, in parent materials such as slate or shale, a brown forest soil is not a necessary fore-runner of the sol lessivé, and it remains to be determined whether a tendency to dispersion of the colloid was a feature of the upper part of the profile throughout the history of the soil or whether it was the result of disturbance of the soil/plant system which had previously maintained a stable state.

Lack of bases, and particularly of divalent ones, in the upper part of the profile allows the colloid to begin to disperse. The low base concentration may pervade the whole soil fabric, or it may be confined to the larger pores: in fine textured soils mixing and diffusion between the soil solution in the larger and the finer pores may be so slow that percolation through the larger pores reduces base concentration in them to the point at which colloidal materials at their walls disperses, while the base status of the main body of the soil remains high. Where earthworms are abundant their mixing activity is sufficient to counteract the low degree of mobility of the colloid which probably occurs locally and transiently in many brown forest soils. Where this is not so the colloid moves down the profile. The finer the colloidal particles the more easily they are dispersed, so mobilization is selective, the fine clay and the amorphous components moving more readily. Immobilization occurs as the base concentration increases further down the profile, and the colloid flocculates against the walls of pores and on the surfaces of peds, producing skins and infillings.

The humic horizon of the soil is left somewhat greyish-brown in colour by the loss of some of its strongly coloured components, and there may be a light grey eluvial zone below the humic horizon. The argillic horizon is, unless it becomes gleyed, a strong brown colour and has a blocky structure. As the moisture content of the soil changes, the peds shrink and swell, and as the amount of material increases and pore space decreases, considerable pressure develops on swelling. The peds press against each other, producing simple facetted forms, and pressure may be relieved upwards by sliding, which produces shiny, fluted surfaces to the clay skins on ped faces: these are called *slickensides*.

Deposition of clay in the illuvial horizon tends to block up the pores and impede drainage. Oxygen depletion by micro-organisms in the A horizon and to a lesser extent in the illuvial horizon results in localized gleying. A soil in which this has occurred is known as a *pseudogley*, or, Avery (1973), a *stagnogley*. When a soil is forming in very clayey material pores can become blocked before much clay migration has taken place, and stagnogley conditions then dominate soil development rather than further clay migration.

Under very high rainfall, with an annual regime in which soil never dries and leaching predominates through most of the year the increase in base concentration with depth usually appears to be insufficient for an argillic horizon to form. As dispersion of the colloid

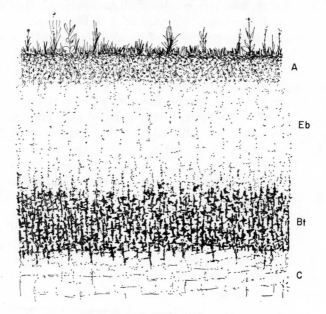

Figure 15. A sol lessivé profile.

occurs fine clay particles together with amorphous components break away from the ionic atmosphere of the flocculated material and in the base poor, acid environment of the descending sol the clay minerals are subject to surface attack. Since the rate of diminution of a particle increases as its size decreases, if they are not to disappear altogether they have to reach an environment in which attack stops and flocculation of the now even smaller particles can occur. If leaching is very strong, the increase of base concentration with depth is not rapid enough to prevent destruction of the mobilized clay. As the fine clay is lost, cation exchange capacity decreases and porosity increases, so leaching becomes even more intense and progressively coarser clay particles begin to degrade and follow the same path to destruction. For a sufficient increase in base concentration to occur to flocculate the colloid before clay is destroyed it appears to be necessary for downward movement to cease as the soil dries seasonally. As water is withdrawn from the pores the colloid flocculates against their walls.

A sol lessivé therefore develops where soils can dry to some extent for part of the year but are subject to strong leaching for another part. It is formed on fine grained parent materials, but not so markedly on those which are very fine and impermeable, and in soils whose previous history has produced a moderate to high clay content. A sequence of changes in leaching regime might result in a brown forest soil first developing acid characteristics and then becoming a sol lessivé.

Podzols and Podzolic Soils

The Major Group of Podzolic Soils includes the fully developed *podzol* and a number of profile types in which the expression of features due to the process specific to podzolization, that of mobilization of iron and aluminium in association with organic compounds as chelates, is limited by some other process or characteristic of the parent material or situation.

Podzol

A podzol has an almost black organic horizon, mor humus, overlying a horizon from which iron, aluminium and most of the weatherable minerals have gone, leaving a loose, bleached mineral soil. Below this horizons of deposition of iron and aluminium oxides, and in some cases humus, develop. Figure 16 shows a podzol profile.

Poverty of plant nutrients leads to a restricted flora, and those plants which can tolerate low nutrient levels produce large quantities of tannins in their leaves, and condensed tannins in particular. The litter is very poor in bases and is protected from micro-organism attack by the tanning of proteins in the cell walls (Handley, 1954). There being no earthworms in the acid and polyphenol-rich soil, organic residues accumulate on the surface; comminution by mites and collembola goes hand in hand with slow attack by fungi, whose mycelium binds the residues into a moist, compact mass. Below a loose litter layer the residues become increasingly reduced, and as their structure is obliterated they become darker in colour, until at the surface of the mineral soil a black amorphous material forms a distinct layer. The mor humus is thus composed of three distinct horizons: **L**, the loose litter; **F**, the mass of comminuted and fungus-bound material in which plant residues are still recognizable; and **H**, the amorphous material. There is often a zone at the top of the

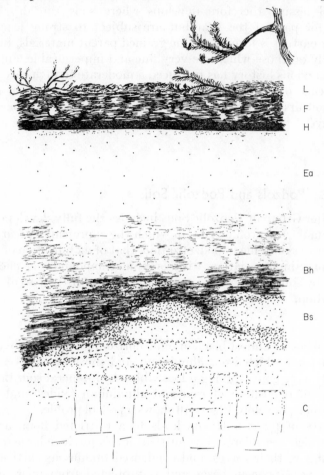

Figure 16. A ferri-humic podzol profile.

mineral soil which is coloured by the amorphous humus as it is washed downwards, producing a form of moder horizon.

The polyphenols produced by the poorly nourished plants are not all involved in tanning proteins; only the larger molecules with numerous hydroxyl groups suitably disposed act in this way. Some polyphenols are leached from damaged living leaves, and when the leaves die and lysis occurs more are released than are used in tanning, and they too are washed into the soil from the litter. During passage through the organic layer and the upper part of the mineral soil they enter into chelating reactions with iron, aluminium and other metals, and the soluble complexes formed are washed down the profile.

The translocated materials may remain in close proximity on deposition, humus and iron oxides together giving a dark reddish-brown colour, or they may form separate horizons of deposition, a black illuvial humus horizon lying above the horizon of hydrous oxides coloured by the ochreous iron colours, or be missing altogether. It is easy to forget about the alumina content of the illuvial horizon of a podzol, since it hardly modifies the colour, but in many podzols the relative accumulation of aluminium is greater than that of iron. The horizon of hydrous oxides is also the site of accumulation of other di- and poly-valent metals which participate in the chelation process, and of phosphate, which under the acid conditions in podzols can only move when the metals or humus with which it is associated become mobile.

Whereas the processes by which humus and the metals become mobilized in a podzol are fairly well understood, the causes of immobilization of the descending sol to form the illuvial horizons are still somewhat obscure. There are probably a number of different processes which come into play under different conditions, and the process which acted to produce a given horizon in a given profile may not be the one which appears to be appropriate at the time when the profile is being studied.

The location of the illuvial horizons in many podzols is clearly determined by a change to a finer texture of the soil or by the presence of the water table, either of which would slow down or halt the downward movement. In other situations the deposition is spread over the range of depth to which the boundary of saturation falls between showers of rain, again indicating a slowing up of movement. Once deposition has begun, porosity is reduced and the rate of flow through the horizons further reduced. As leaching becomes less severe with depth the pH of the soil remains higher and this alone may be enough to flocculate the sol. In this case the humus and the metals would remain closely associated and the mixed form of illuvial horizon develop. It seems probable, however, that some more active process which involves breaking the chelate is usually involved, and this would allow a physical separation of the humus and the metals to occur once they were released from chemical bondage. A catalytic action by iron oxides already present, once the depth is reached at which active chelating power has been used up, has been postulated.

The activity of micro-organisms is probably the most favoured hypothesis. The hydrous oxides can sometimes be seen, under the microscope, to have the form of an encrustation on the surfaces of colonies of cells, and it is suggested (Aristovskaya and Zavarin, 1971) that the colonies grow in spring at the level where the zone of contact of air and water in the capillary fringe provides a suitable environment for certain bacteria, which utilize the organic component of the chelate and reject the metals. The metals are oxidized immediately on release from the chelate and form a sort of slag at the cell surfaces.

Whether or not humus deposition occurs depends on the balance over the year between mineralization and accumulation of the products of metabolism of the micro-organisms in the illuvial zone. The humus molecules are synthesized within the illuvial horizon, using as structural units the phenolic and other compounds produced in microbial cells. Whether any of the humus substances which leave the organic horizons are deposited in the illuvial horizon without undergoing degradation and resynthesis is open to doubt. They may be broken up in the thin moder layer into which they are washed from the base of the mor humus, only their component parts travelling down through the eluvial horizon, picking up metals as they go.

The common situation of a humus horizon lying above the horizon of metal oxides and itself containing very little of the metals is somewhat puzzling. There is a tendency for this type of profile to occur in the wetter and colder parts of the podzol zone, and it may be that under wet, but still freely draining, soil conditions formation of the oxides is not so rapid and the metals have time to move a little way down the profile as ions in solution after release from the chelate. However, the complexity of the illuvial horizons of many podzols and the very irregular nature of their correlation with climatic factors suggests that a sequence of soil development is involved.

Multiple sequences of illuvial horizons suggest that changes in conditions at the soil surface may change the balance between mobilization and immobilization, and between mineralization and accumulation, so that horizons already formed may be destroyed and reformed lower down, or may move downwards and new horizons form above them.

Podzols are the characteristic soils of the coniferous forests at high latitudes and high altitudes. Rainfall varies greatly, from about 400 mm to over 2,000 mm per year, but since chemical and biological processes are slow under low temperatures, and may be virtually at a standstill for a large part of the year in the colder parts of the zone, the rate of leaching can be high relative to rate of release of soluble substances even where rainfall is relatively low. In the northern coniferous forests podzols form on coarse grained sediments, most of which are glacial and fluvioglacial deposits, and on the rocks of the palaeozoic shields, many of which are coarse grained igneous rocks. Where the rocks or sediments are very fine grained and where topographic arrangements or the presence of permafrost lead to accumulation of water, peats are associated with the podzols. The conifer zones on mountains are found at increasing height at lower latitudes, and as the duration and intensity of summer warmth increases so the rainfall necessary to produce the strong leaching which is the pre-condition of podzolization increases.

Podzols reach perhaps their most extreme expression in the lowland heaths which occur on base-poor and porous parent materials in a belt which stretches from southern and eastern England eastwards across northern Europe. The reason for their development under a climate very different from that of their zonal distribution is discussed in the next chapter.

In the mountain regions which fringe the western seaboards of northern Europe, the coincidence of high land and prevailing wind brings very high precipitation, increasing with altitude and not strongly seasonal in distribution. The mountains also provide steep slopes and soil parent materials dominated by fine grained metamorphic rocks. Under this conjunction of factors profile development is now very often limited either by the rejuvenating effects of continuous and episodic erosion or by waterlogging. Podzolization is widespread but rarely leads to the development of deep, freely draining profiles. Associations of *brown podzolic soils* on sloping ground and on the flatter ground gleyed profiles in which podzolization is expressed to varying degrees are very common.

Brown Podzolic Soils

Brown podzolic soils have characteristics which are transitional between those of acid brown forest soils and podzols, and the inadequacy of most schemes of classification to deal with transitional situations has been brought out with particular force in this case. These soils may be regarded as very weakly developed podzols, in which iron and aluminium are beginning to be mobilized in organic complexes but have not moved to a great enough extent, and nor has the level of organic activity fallen sufficiently, for the horizon sequence of the podzol to have become apparent. The profile has an **A** horizon of moder type, in which grains of resistant minerals, such as quartz, may be clean, the iron oxide which would have coated them in a brown forest soil having been stripped away. The humic horizon thus includes and masks the zone of eluviation of metals which in a podzol forms a distinct horizon. The **B** horizon has a highly characteristic colour and structure: the concentration of hydrous oxides of iron, high in a strongly acid soil in which relative accumulation has occurred by the destruction of clay minerals or by their failure to form, is increased as absolute accumulation by translocation begins. The **B** horizon therefore has an extremely strong ochreous colour. Its structure and consistence are influenced by the mode of aggregation of the hydrous oxides, which instead of being immobilized in stable form in close association with the surfaces of primary and secondary minerals remain to a greater extent in more easily mobilized forms, and are less closely associated with the other components of the soil. The structure is of a loose, extremely fine crumb or granular character and the soil has a consistence which Ball (1966) vividly describes as "fluffy".

The difficulty in classifying brown podzolic soils is associated with differences of opinion as to their genetic status. Their distinctive and consistent profile, characteristics and distribution suggest equilibrium in soil processes rather than active transition between soil types. Apart from those cases in which such transition is related to a recent change in vegetation or land use, the most satisfactory explanation is that the apparent stability is provided by a state of dynamic equilibrium, in which slope movements keep the podzolic profile immature. As rapidly as differentiation processes occur, loss of material in the surface horizon and mixing within the profile prevent their effects becoming established.

Gleyed Podzolic Soils

Where brown podzolic soils occur on slopes, flatter land both upslope and downslope often has soils which show the effects of waterlogging. In the lower situation the gleying is due to groundwater, and may affect the lower part of the profile or the whole of it. In the upper situations impedance of percolation is due to conditions within the soil profiles themselves. In the former case, the soils, which Avery (1973) assembles into the Major Group of Groundwater Gley Soils fall into groups according to their nature before gleying superseded other soil forming processes. In the latter, the soils are divided between the Major Groups of Surface-Water Gley Soils and Podzolic Soils.

In the Surface-water Gley Soils, a Major Group which has members in almost all climatic zones, gleying occurs in the **B** horizon or in both **A** and **B** horizons, as a result of impedance of percolation through a clayey **B** horizon, which may be a cambic horizon or an argillic horizon. Some of these soils have formerly been called *pseudogleys*.

The gleyed profiles in the Major Group of Podzolic Soils include some, a group of *gley-podzols*, in which gleying occurs below the podzolic **B** horizon, above which the eluvial and humic horizons remain well aerated, but most of these soils are strongly gleyed above an impervious horizon which may be a normal **Bs** horizon of a podzol but which is often a *thin iron pan*. These soils, which have also been called gleyed podzols, peaty gleyed podzols, and thin iron pan soils,

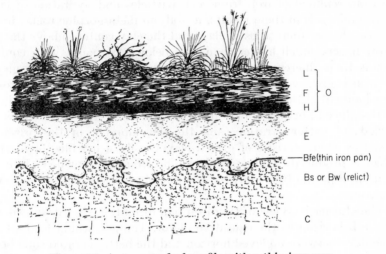

Figure 17. A stagnopodzol profile with a thin iron pan.

are now called *stagnopodzols*. Waterlogging in the upper part of the profile is usually associated with the accumulation of organic matter to form a very thick, peaty mor layer. Figure 17 shows the profile of a typical member of this group, in which an iron pan is well developed.

There is some disagreement about the mechanisms by which thin iron pans form, and this may reflect a variability in the circumstances in which they occur, in soils on different parent materials and under different distories of land use. Broadly it is thought that prolonged periods of saturation of soil arise, under high rainfall, either from the great capacity for water of mor humus, or from impedance of percolation through the lower part of the profile. In the first case, waterlogging depends on development of a thick mor layer, which absorbs so much water that seepage from it keeps the upper part of the profile wet between periods of rain. Iron is reduced as a result of organisms in the mor layer using up all the oxygen, and moves downwards in the seeping moisture until it reaches more oxidizing conditions in the **B** horizon and the oxide is deposited to form the pan. This mechanism would account for the persistence beneath the pan of the brightly coloured, freely draining **B** horizon of a brown podzolic soil, which is often observed even on very fine grained parent materials such as slate or shale. The pan is continuous, and even passes right through stones, and once it has become continuous it develops no further, since all downward movement ceases in the soil above it. Hence the other characteristic of such pans, thinness. In the gleyed horizon, removal of iron from rock particles and hydration of layer lattice minerals in them results, in soils on metamorphic rocks, in the release of large quantities of clay, and the clay content above the pan is often very much higher than that below. As soon as clay content begins to be increased in this way percolation through the gleyed horizon is greatly reduced, and this may be a contributory factor in development of thin iron pans. Once the pan has formed, residual iron in the gleyed horizon moves about by diffusion, and with occasional periods of better aeration, and the development of the prismatic structure characteristic of gleyed clayey soils, a marbled or mottled effect is produced. A humus sol seeping from the mor layer into the structural pores often contributes a dark brown veining to the gleyed horizon.

The alternative view of thin iron pan formation regards impedance in the **B** horizon as the causative factor. The pan would form at the boundary between a gleyed horizon and the better aerated soil above, the iron diffusing upwards in stagnant soil moisture. Once formed, the pan would permit no more moisture to penetrate into the soil below, which would drain and revert to a more fully oxidized condition, while that above in its turn became waterlogged and gleyed. The waterlogging would then lead to accumulation of the peaty mor layer, rather than being the result of it. In soils whose **B** horizon is so porous and fully oxidized as a brown podzolic soil, and in which after formation of the pan these features are predominant, it seems unlikely that this mechanism could have been responsible, but many soils show evidence of the development of an argillic horizon superimposed on the brown podzolic **B** horizon, in such a way that the illuviated clay tends to block up major pores, and induce gleying, leaving islands of more porous and fully oxidized soil isolated in a clayey network (Limbrey, in press). Once this has happened, impedance by the argillic horizon may lead to formation of an iron pan at its upper boundary. In soils with initially higher clay content in the **B** horizon than the brown podzolic soils, such as brown forest soils on clayey parent materials whose cambic **B** horizon develops a very high clay content, or in sols lessivés developed from a brown forest soil rather than from a brown podzolic soil, the stages in formation of a thin iron pan are less complex, and all that is required is a change in external conditions leading to increased water content in the soil, or a progressive increase in clay content in the horizon causing impedance. In some soils on parent materials which have compact layers of periglacial origin, particularly those of loamy texture, the compacted material is cemented by deposition within it of small quantities of a material which is probably alumina brought down in the incipient stages of podzolization, before iron had begun to move. This too can so block up drainage that further development of the podzol is taken over by iron pan formation and gleying.

Many soils have multiple and rather diffuse iron pans, suggesting that rather than forming rapidly and directly, as would those resulting from the thickening of a mor layer until it held enough moisture to induce pan formation, they formed by a series of changes in balance between oxidizing and reducing conditions. In such soils, the pan does not develop such continuity, and may wander up and down through a strongly gleyed horizon, which merges downwards into less gleyed soil at some depth. In this case, the iron pan does not form so marked a barrier to plant roots as it does where it is continuous, and some retrieval of nutrients from below the pan is possible. Where the pan is continuous, a mat of roots often develops just above it, but none pass through, and all nutrient reserve in the **B** horizon below is forever cut off from the vegetation. Vegetation is limited to shallow rooting plants of low nutrient demand which can tolerate shortage of oxygen in their rooting zone.

7

The History of Podsolized Soils in Britain

In the zone in which forest development was limited to conifers, as the forests became established in the post-glacial period, tundra soils and incipient soils on fresh rock and glacial deposits developed into podzols, even on materials which were initially base-rich or even calcareous. In the cold conditions of the conifer zone there could be only slow build up of secondary minerals to provide a means of retention of bases, and only slow alteration of minerals to replace bases leached away during snow melt and summer rain. The low level of base demand of conifers limits the amount returned to the surface from depth, and together with the high polyphenol content and resistant cuticles of the leaves limits the activity of soil animals and micro-organisms. Podzolization becomes inevitable, but the rate at which it occurred, and its timing in relation to the establishment of coniferous forest and to any climatic fluctuations which have occurred subsequently, are unknown.

Whether podzolization normally played any part in the history of development of soils as the deciduous forests became established is uncertain. Pine trees participated in the forest succession throughout the temperate zone, but at a time when soils would still have been relatively immature and unleached. Pine was associated with or followed birch woodland, and birch is a tree under which podzolization is not initiated and can be halted and reversed (Dimbleby, 1952). Locally, rankers or alluvial raw soils may have developed into podzols under pine, but nowhere that climate allowed it does further forest development to the full expression of the mixed oak forest appear to have been limited by soil factors.

A quite different question is that of whether progressive leaching of soils under undisturbed mixed oak forest, once developed, led to podzolization, and if so, whether the postulated wetter climate of the

Atlantic period was a contributory factor. Evidence from buried soils of brown forest soil profiles which must have persisted through the Atlantic in areas which later succumbed to severe podzolization as forest was cleared, or subsequently, would suggest considerable stability against podzolization even in extremely rainy areas, or on extremely base-poor and porous materials, so long as forest remained undisturbed. However, work by Valentine (1973) throws considerable doubt on the idea of such stability. Valentine studied a soil catena in a landscape buried by encroachment of fen peat as the sea transgressed over the low lying land of south Lincolnshire during the Atlantic. Pollen analysis shows that before the spread of fen reached that particular area mixed oak forest was growing there, and it continued to grow on unsubmerged land nearby. The soil is a catena ranging from freely draining iron-humus podzols on the higher land to gleyed podzols in lower situations, and had become well established during the Atlantic period on a parent material, gravels of glacial outwash origin, which was apparently originally calcareous. There is no evidence of the presence of man in the forests, and no evidence that podzolization was associated with changes in forest composition.

Evidence of retrogressive succession of vegetation without, so far as can be told, the participation of man, is presented by Iversen (1964, 1969) for the Draved forest in Denmark. Soils on coarse, base-poor materials supported mixed oak forest. Mor humus had begun to accumulate and the soils had podzolized. Episodes of tree fall could be picked out by the scatter of upthrown sand or by a layer of bark within the mor layer, and the pollen record shows that brief florescence of light-demanding herbage in the openings was followed by complete closure of the canopy again. It was not until the advent of man in this forest that openings became more permanent and heath vegetation was established, but before this had happened the succession of vegetation shows the elimination of the more base-demanding species as the soil became leached.

We do not know what the ultimate paths of soil development during an interglacial would be. The present interglacial has not lasted long enough and man has interfered almost everywhere with the natural course of development of forest and soil. In earlier interglacials man was present in the forests of the temperate zone, but we do not know the distribution and intensity of his interference with the vegetation. He had fire and must therefore have used considerable amounts of wood to keep it going and must have caused forest fires and soil disturbance. The pollen record for the Hoxnian Interglacial shows a phase of vegetational changes associated with occurrence of charcoal and of increased quantities of mineral material in the lake deposits at Hoxne and Marks Tey (West, 1965/6; Turner, C., 1970). Whether or not the fire and soil disturbance should be attributed to man's activities is still a matter for discussion. A single, widespread episode of forest fire is postulated and the changes in forest composition and the increase in light-demanding species and herbaceous vegetation which seems to have persisted for several hundreds of years can be attributed to the persistence of the effects of erosion and the destruction of the large forest trees. It seems equally probable, however, that this was a period during which hunting people were constantly or repeatedly present and taking advantage of increased game stocks and greater abundance of vegetable foods in the clearings, their fires contributing to the persistence of open conditions whether or not they were the initial cause.

At a later stage during the interglacial, modifications in the forest composition are attributed to the increasingly leached state of the soil. No change in climate is postulated at this stage and there is no direct evidence of the condition of the soil. Since instability of climate associated with the build up of ice sheets as glaciation developed is likely to have resulted in increased rainfall and increased leaching of soils, it is difficult to see how a retrogressive succession from this cause could be distinguished from one due to soil degradation as a result of time alone. More detailed knowledge of the flora and of the ecological requirements of individual species is necessary in the absence of direct information about the soil itself and about the fluctuations of climate during the interglacial and leading up to the subsequent glaciation.

Podzolization is widespread under deciduous forest today, and forest will survive, and even regenerate, on well-developed podzols provided burning is prevented and browsing animals kept away (Dimbleby, 1953; Dimbleby and Gill, 1955). The condition of the soil is sometimes closely related to the particular tree whose canopy is overhead. Oak and beech produce condensed polyphenols under conditions of low nutrient supply (Handley, 1954), and beneath them podzols are likely to develop or to be maintained. In the acid litter from these trees mosses flourish and contribute their own polyphenol production to podzolization, and much of the sparse ground flora roots entirely within the mor layer, using such nutrients as become available as soon as they are released from residues, so reducing availability to more deeply rooting plants. Birch trees too can spread their roots through the mor humus. They are extremely efficient in utilizing bases from extremely low concentration in podzolizing soil, and in doing so produce a mild litter free from polyphenols. Once birch litter has begun to accumulate, earthworms can colonize and mull humus begins to be formed. Thus, birch's function as a colonizing species in closing clearings is associated with a function in ameliorating soil conditions, providing a more fertile medium for seedlings of the forest trees than is provided by their own litter. Among other trees which can contribute to forest regeneration in this way are rowan and holly, and in the latter case the protection it provides to the saplings is an additional advantage when browsing animals are about.

In some oak and beech woods, the tendency to podzolization is expressed in the form of miniature podzols, in which the whole sequence of horizons is developed in a few centimetres. Such miniature podzols are probably entirely obliterated when openings in the canopy occur and such trees as birch or holly spring up in them. Since profiles of depths intermediate between these and the typical deep podzol of freely draining type do not seem to occur, it is unlikely that profile development in a podzolizing soil proceeds by downward growth; it is more probable that horizon differentiation intensifies throughout its depth, within a framework provided by the moisture regime of the profile, with only relatively minor adjustments of the level at which critical environments are established leading to a complexity of development of illuvial horizons which is characteristic of these profiles.

Soils of the Lowland Heaths

The present distribution of deep, fully developed podzols outside the zone of coniferous forests is apparently anomalous in respect of the climatic conditions one might suppose to be conducive to their formation. They occur not in the cool, wet uplands, but in the lowland heaths. These lowland areas are marked by low total annual rainfall and relatively warm, dry summers, with a strong risk of drought. The deep podzols occur in freely draining situations where soil parent materials are very poor in bases and porous. In England such materials are some of the sandstones and the Tertiary and Pleistocene gravels and sands.

The heaths on Tertiary sands and gravels and on the Greensand in southern and eastern England are within the area in which loess is detectable in soils on other parent materials (Hodgson et al., 1967; Catt et al., 1971; Perrin et al., 1974). Loess must therefore have contributed to these soils too, and at an early stage in their development would have lain as a distinct layer at the surface or been mixed with the underlying materials by cryoturbation and solifluction to a greatly variable extent.

The soils of the heaths now are deep iron-humus podzols and their vegetation is dominated by heather and bracken. They are largely common land today, and have long been used for pasture by the communities living around their edges on soils of higher cropping potential. Burning has been widely practised to produce a flush of growth in early spring from the *Calluna*, so providing a short period of abundant grazing.

In England, Neolithic monuments on the heaths are very few, and none have been excavated in recent years, but there have been several excavations of Bronze Age barrows, whose buried soils show podzolization developed to a slighter degree than subsequently, if at all (Dimbleby, 1962). In Holland, buried soils beneath Neolithic barrows have shown podzolization (Waterbolk, 1957).

Dimbleby's pollen analyses of the Bronze Age buried soils have shown that they formerly carried mixed oak forest. During the early history of these forests, the soils were not acid enough for pollen to be preserved, so the pollen record begins after soil degradation and retrogressive succession of vegetation had begun. Examination of the soil profile has sometimes shown evidence of former worm sorting, in the form of a stoneless upper few centimetres and stone lines, and in some cases artifacts, at a depth below the surface consistent with their having been lowered through the soil by worm action. The pollen profile then begins with a confused zone, such as would be produced had worm activity continued for a time after the soil became acid enough for pollen to be preserved. It is therefore reasonable to suppose that brown forest soils with mull humus existed there.

The pollen record at some sites shows that interference with the forest began, with an increase in pollen production by light-demanding shrubs, particularly hazel, and by grasses and herbaceous plants, before the first evidence of cereal production. The occurrence of mesolithic artifacts at these sites strongly suggests that the opening up of the forest was caused by the activity of mesolithic peoples (Dimbleby, 1965). What better food for such people than hazel nuts? Would it not have been obvious to them that by opening the canopy the hazel could be encouraged to abundant production?

Full scale clearance of forest had occurred by the time the Bronze Age barrows whose soils have been studied were built, and the usually undisturbed nature of the pollen profiles shows that cultivation was not widespread in the preceding periods. Sporadic occurrence of cereal pollen suggests that it was derived from some way away, and the paucity of Neolithic monuments in these areas suggests that the land was for some reason unattractive to people at that time. If podzolization had already been initiated by neolithic times, yields under cultivation would have been low; pasture may have been a better proposition, and later the heaths seem to have attracted Bronze Age people. However, under pasture on these soils degradation continued; to what extent the continued development of podzols and the extreme degree to which they developed is due to the attempts to maintain some usefulness as pasture, in particular by burning, is not clear: it seems probable that burning has been an important factor in the history of these lands, and in particular in the establishment and maintenance of the *Calluna* vegetation, which ensures by production of condensed polyphenols that podzolization is inevitable.

In these deep and well developed podzols erosion can become very serious. Initiated in paths or gullies which break the continuity of the mor horizon, erosion proceeds rapidly in the loose, incoherent eluvial horizon, which washes or blows away very easily. Burning greatly increases the risk of erosion, destroying a high proportion of the protective vegetation, drying, and in extreme cases burning, the humus and destroying its coherence. Once erosion begins, slight hollows act as gullies for surface water and as funnels for wind, and blow-outs quickly develop: the heath becomes the blasted heath.

If we now consider the processes by which the heathland soils reached so sorry a state, we have first to ask "what happened to the loess?". Here it lay upon and was mixed and convoluted with coarse siliceous materials which provided neither a chemical buffer against leaching, as did the chalk, nor a physical one, as did the clay-with-flints, whose soils are discussed in Chapter 8. It would therefore have suffered extremely rapid alteration, contributing its base supplies to the establishment of the forest and the brown forest soils, but in doing so being to a large extent destroyed, leaving in the soils only the most resistant minerals. Loess is particularly noticeable in chalk soils, since soil formation removes chalk and concentrates the loess, whereas the reverse is so in a siliceous context, and its residues are difficult to detect. Where sand predominates in wind blown material the term *cover sand* is used; the distribution of cover sand and loess has been found to be to a large extent complementary in Belgium (Paepe and Van Hoorne, 1967; Paepe, 1971). In England, cover sand occurs in East Anglia, and, material of sand grade not travelling so far as the finer silt of loess, much is derived from the local sandy deposits of the area in which it is found. The presence of a wind blown component is clearly particularly difficult to detect in these circumstances.

As the brown forest soils developed, clay minerals would have been formed as a result of alteration of the less resistant minerals of the loess, but with continual leaching the clay would have begun to degrade. Whereas the chalky loess soils became sols lessivés, the lack of a base-rich horizon at depth, and the extreme porosity of these soils, would have prevented the development of an illuvial clay horizon and degradation of the clay minerals would have been total.

The different time scale of development of degraded soils on the chalky and the siliceous materials may be significant. If the brown forest soils on chalk remained stable until a period of use by Neolithic cultivators, the possibly drier summers of the sub-Boreal period may have contributed to the formation of the illuvial horizons of the sols lessivés. The evidence that on the heathlands the soils were already podzolized by that time implies that their degradation began during the Atlantic period, when climate is thought to have been wetter, and in particular the summers less warm and dry; the degree of summer drying may have been inadequate to flocculate the colloid in the lower part of the profile. The earlier onset of degradation would then be all that had to be attributed to poverty and porosity of the autochthonous material. As the soil development under forest proceeded, the trees rooting into chalk maintained the base status of the soil, whereas those rooting into sand or gravel could not. If this be so, need we invoke the intervention of man? According to Iversen (1964), these siliceous soils will podzolize under closed forest without any interference by man, and Valentine's work (Valentine, 1973) supports this, and it may only be the extreme degree of development of the heath vegetation which is attributable to interference. However, where pollen evidence does show an opening up of the forest at a much earlier stage than in Iversen's example, the opening occurs before the onset of retrogressive succession, and though the soils must have been acid enough for pollen preservation to begin, there is no evidence to suggest that they were then actually podzolized. The effect of man may therefore have been to begin a process which would have happened anyway later on. The fact that he did so during a period in which possibly cool, rainy summers might have increased the degree to which leaching is enhanced in any clearing might have exaggerated his impact. Indeed, the situation of the heathlands and the porosity of their soils may bring them into a situation in which slight fluctuations in climate carry across them the boundary between areas in which soils become more leached in clearings and areas in which they become less so.

Moorland Soils

Figure 18 summarizes the present distribution of soils in much of

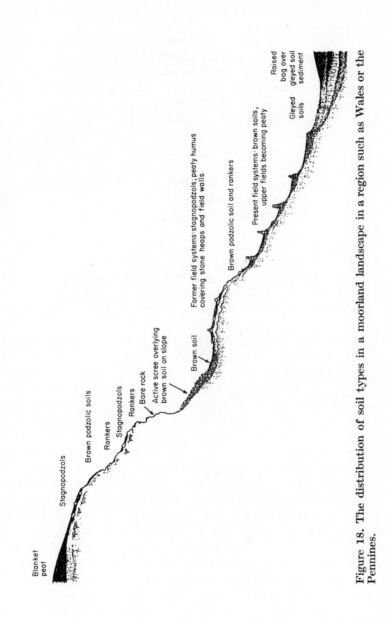

Figure 18. The distribution of soil types in a moorland landscape in a region such as Wales or the Pennines.

northern and western Britain. The stagnopodzols occupy large areas of gently undulating uplands, with blanket bog covering the higher rounded mountain tops and elevated plateaux. On slopes soil creep and erosion expose areas of rock and maintain brown podzolic soils, lateral drainage preventing gleying within their profiles. Where the percolating water emerges or rises into soil profiles in lower slope and bottom positions, groundwater gley soils occur, and in valley bottoms peat has accumulated on top of fully gleyed soils. Lower slopes and river terraces with colluvial soil accumulations may have fairly well drained soil profiles, or the whole accumulation may be gleyed and have a peaty capping. Some of the gentle slopes and undulations of the lowlands, such as hummocky glacial debris and river terraces which have not been overwhelmed by colluvial soils, and do not receive seepage from the uplands within their profiles, have well drained brown soils of neutral to acid character. Throughout the uplands there are areas of basic rocks which, provided they are not affected by gleying because of situation or high clay content, maintain soils of high base status and mull humus, and others which can do so if slope movements contribute to maintenance of base supply. As one moves northwards, the whole pattern tends to shift downhill.

The pattern of vegetation follows the distribution of soil types fairly closely, and in many details is strongly related to present or recent land use and management. Lowland fields in current use are given dressings of lime and dung and carry good pasture. Enclosures on the lower slopes often show strong contrasts between those under good management, limed, dunged and grazed at limited intensity, and those less well managed or abandoned, with gaps in their walls and left to uncontrolled grazing by sheep. In the former, good pasture is maintained, while in the latter coarse grasses and bracken or heather predominate. Whether grazing is by sheep or cows strongly affects the vegetation: cows are rather unselective feeders, munching steadily on and consuming all the grasses and herbs, apart from thistles, butter-cups, and the young fronds of bracken. They also have broad feet and great weight, and their trampling of emergent bracken fronds helps to keep it down. Sheep are pernickety feeders. They can survive on poorer pasture than cows, and do so by nibbling around selectively, taking the more tender and nutritious grasses and herbs and leaving the coarse, base-poor ones, and avoiding bracken. They have small feet and step between the tussocks in which the coarser grasses grow, leaving the growing bracken fronds undamaged and creating paths among the tussocks which provide runnels for surface water and so increase erosion. Bracken needs a moderate depth of soil for its rhizomes to develop, and does not tolerate waterlogging, but wherever soils are deep enough and moderately well drained it tends to spread into pastures grazed by sheep, and can only be eliminated by being cut each year before the rhizomes have stored enough food for the growth of the next year's fronds. Where bracken grows it shades out grass and smothers it with the dead fronds in the autumn, so that new growth is inhibited; it produced condensed polyphenols and so induces or enhances podzolization. Accumulation of mor humus beneath bracken may be one of the stages by which previously adequately draining soils reach a condition in which thin iron pans form, and once the wetness of the soil reaches a point which bracken can no longer tolerate, it is superseded by a grass or heather vegetation characteristic of iron panned soils.

Where soil is too shallow or too wet for bracken to grow, as sheep take the more nutritious grasses the base-poor ones are increasingly successful. These too are polyphenol producers and lead to podzolization. On already strongly podzolized soils, grazing by sheep helps to prevent the spread of heather into grassland, because they prefer its young shoots to the poor grass. Where heather is already established, it is often regularly burnt, and since heather is more resistant to burning than the grasses, the combination of burning and grazing probably helps to maintain heather moor. The distribution of heather moor and grass moor is thus strongly related to the intensity of sheep grazing and to the history of burning.

The pattern of vegetation is fully and clearly presented in Pearsall's book, "Mountains and Moorlands" (Pearsall, 1971). The types of vegetation he describes can be related to soil types as follows. The *Festuca-Agrostis* grassland, dominated by the nutritious grasses, sheeps fescue, and one or more species of bent grass, provide the mainstay of sheep farming, on brown soils and brown podzolic soils on the better parent materials and where slope movements are adequate to maintain the supply of bases against leaching and grazing losses. Where cows graze the *Festuca-Agrostis* grassland, bracken is kept at bay; where sheep graze it tends to spread wherever soil depth is adequate, and leads to degradation of pasture.

Where leaching is too severe to allow these more base-demanding grasses to persist, generally in a zone extending from the *Festuca-Agrostis* grasslands up into higher rainfall at greater altitude, more strongly podzolized soils carry mat grass, *Nardus stricta*, with some association of wavy hair grass, *Deschampsia flexuosa* on drier soils, and with a strong tendency to invasion by moor rush, *Juncus squarrosa*, in wet situations. The residues of mat grass and moor rush accumulate to form a peaty mor humus, but some contact with mineral soil apparently has to be maintained, and these plants do not grow on deep peat. Mat grass is very poor in bases, and it is this grass which succeeds where sheep take the more nutritious ones. It is thought that *Nardus* grassland has been produced by sheep grazing, in that part of the former extent of the *Festuca-Agrostis* grassland where the combined effect of leaching and grazing has reduced base content of the soil below that at which the more nutritious grasses would maintain their existence in the face of selection by sheep. The *Nardus* grassland is a·major component of the zone of stagnopodzols, and it is tempting to correlate the intensification of podzol horizon differentiation, and the build up of peaty mor leading to thin iron pan formation with the change in vegetation consequent upon intensive grazing of the uplands by sheep. That is, in many areas, these soils reached their present form late in the history of land use, after the development of the system of sheep farming dependent on free ranging in the uplands, that is, after wool production became the predominant use of the uplands under the influence of the Cistercian monasteries.

If many of the features of present soil distribution are of recent origin, what might have been their earlier history? There is plenty of evidence of former extent of forest in areas now quite devoid of large trees. Pollen analysis of peat and of lacustrine sediments carries the vegetational history of these areas back to Late Glacial times. It is clear from the pollen evidence and from the evidence of buried soils and relict horizons in present soils that brown forest soils must have developed throughout most of the upland regions. In Britain, only a few mountain tops in Scotland project above the theoretical climatic limit of tree growth; the more precipitous mountain slopes clearly could not support large trees, but bushes and small trees will grow on ledges wherever there is enough moisture and room for roots to develop in cracks in the rock or in a surface accumulation of fragments. Whether northern Scotland and the islands beyond its coast, and the most exposed mountains were in fact fully forested is open to question; where the shape of a mountain is such that there need be no breaks in the continuity of plant cover, the trees would have provided protection for each other as forest developed. However, it is under such conditions, where climate and soil factors approach values which impose limitation on tree growth that the effect of herbivorous animals has to be considered, and this entails consideration of the behaviour of any hunters who might have been imposing deliberate or accidental controls on numbers and distribution of animals. It may be that the forest was never able to close where soils had the lowest base reserves and climate was most severe. Does persistence of birch forest in the far north and in high situations during the Atlantic period, while mixed oak forest developed elsewhere, imply that the forest trees could not grow there, or is it a question of continual recolonization by birch in situations where further development is limited in part by browsing by deer? There is a certain amount of evidence to suggest that some of the areas which did not achieve complete cover by mixed oak forest succumbed soonest to soil degradation, whether or not they were the worst situated climatically, and it is in just these areas that abundant evidence for the presence of mesolithic people has been found. The areas most studied in this respect are the Pennine moorlands (Conway, 1947, 1954; Hicks, 1971, 1972) and Dartmoor (Simmons, 1964, 1969).

It is extremely difficult to establish a clear chain of cause and effect in the sequence of vegetation and soil changes. It is perhaps no coincidence that, apart from the highest mountains and the far north, the areas where forest cover does not seem to have been fully developed are those areas where the rock is not only coarse grained and base-poor, but also has not had all traces of former soils removed by glacial erosion, even though these are not the regions of shortest growing season and highest rainfall. The soils of Dartmoor and the Cornish granite moors, in particular, are formed to some extent on rock whose history of weathering reaches back to the intense leaching under tropical conditions in Tertiary times, and the southern part of the Pennines and much of the North York Moors, not glaciated in the Last Glaciation, would have retained some depth of interglacial weathering, presenting to post-glacial soil formation a parent material whose base reserves were already somewhat depleted. Perhaps there is a parallel here to the soils of the lowland heaths, podzolization having occurred much more severely on Tertiary and Lower Pleistocene land surfaces than on Upper Pleistocene river terraces.

These soils would have been rather acid in character from the beginning, and on such coarse grained rocks highly susceptible to leaching. Soil acidity may therefore have been a contributory factor in delaying forest development, and the inadequacy of tree cover in protecting the soil from the full force and volume of rain, together with the sparsity of deep rooting trees to retrieve bases from depth, would do little to prevent acidity increasing. By contrast, the soils on finer and richer parent materials would not have been acid to start with, indeed, they may well in many places have been calcareous where there was a contribution to their parent materials from base-rich rocks, and the forest would have been able to develop before leaching became too extreme, and in doing so slowed down the process of acidification. Texture, base reserve, rainfall and temperature must combine in such a way that a critical point was reached during post-glacial forest development at which forest cover would continue to thicken or would fail to do so. The presence of herbivorous animals and their predators, including man, has to be included in the factors affecting the situation, and its importance would increase as other factors approached values which imposed limitations on forest growth. Though in winter, cold and snow would have driven them down into the thick forests, in spring and summer a forest margin situation would have attracted large numbers of animals, and man might have driven them into open areas, and might have employed fire. Even without deliberate burning to control the abundance and distribution of browsage, it is unlikely that accidental burning would never have occurred, and much wood would have been consumed by domestic fires. In these regions of high rainfall, low weathering rate and rapid percolation, the loss of nutrients from the ash might very rapidly reduce the reserve in the soil to the point where the vegetation was limited to species which, while tolerating low nutrient levels, induce podzolization. The situation might have developed in which congregation of herds, by their own preference or encouraged by hunters, limited further an already marginal vegetation, contributing to its long-term impoverishment even though each spring the vigour of herbaceous and bushy growth, perhaps increased by the short-term availability of nutrients after fire, would continue to attract the herds. The animals themselves would remove nutrients from the summer feeding grounds to their winter refuges in the forests, and people feeding on them would do the same, increasing the nutrient loss already exacerbated by exposure to leaching.

The establishment of time scales for the degradation of the soil-plant system in any area depends on availability of suitable situations for pollen analysis and samples for radiocarbon dating. The simplest point to establish is the onset of formation of blanket bog at any site, which can often be placed both in the broad context of pollen zonation and in the absolute time scale by the dating of tree stumps buried by the encroachment of the peat. The earlier history of the vegetation can sometimes be followed in the downwash pollen profile in the soil buried beneath the peat, but the definition of pollen zones is less clear in this type of analysis, and there is no means of providing a timescale. The downwash profile generally does not extend back to Late Glacial times, though Godwin (1958) discusses an example where it apparently does, suggesting that usually there was a period when soils were not yet sufficiently acid for pollen to be preserved, and preventing study of the earlier stages in the response of vegetation to soil acidification. The evidence so far available suggests that the Pennine and Dartmoor peats began to accumulate in the highest situations during the sixth millenium B.C., that is, early in the Atlantic period. Mesolithic flints are frequently found at the base of the peat and in its lower few centimetres, that is, in and below the mor humus of the podzolic soil which preceded peat growth. In some areas, the lower part of the peat contains remains of trees. These are the birch trees, or, sometimes in Scotland, pines, which were the last which were able to grow in the impoverished soil before peat growth overwhelmed it. In other areas, absence of tree remains in the peat suggests that woodland had ceased to be there before accumulation began. On Dartmoor, pollen analysis in these high situations shows bracken spores in the mor humus below the peat (Simmons, 1969), which suggests that soils remained fairly well aerated, implying that woodland ceased to be able to regenerate because of a factor other than the waterlogging which later led to peat growth. Bearing in mind the efficiency with which birch trees can extract nutrients from very poor soils, it is tempting to blame their failure to grow on grazing pressure or fire. There is much evidence in these areas for the presence of mesolithic people.

It is very widely accepted that the onset of climatic conditions of the Atlantic period, involving higher rainfall throughout the year, or in summer, or greater cloudiness, was responsible for initiation of the spread of blanket bog. The evidence for wetter soil conditions lies mainly in the increase of alder in the mixed oak forest, and in the beginning of peat growth itself. In some parts of the uplands the soil was already strongly enough leached before the beginning of the Atlantic period, in the sixth millenium B.C., for pollen to have begun to be preserved in the mineral soil before peat growth began. If, where forest development has already been upset by progressive degradation of the soil or by mesolithic people, inception of peat growth is taken as indicating the beginning of the Atlantic period, without an independent means of dating, the argument is circular. There is no need to postulate increase in climatic wetness to account for signs of increased wetness in the upland soils. Opening of the forest canopy allows more water to reach the soil. Loss of tree roots, assisting drainage and taking water from depth to return it to the atmosphere via the leaves, increases soil wetness. Build up of a water-retentive mor layer and associated inception of iron pan formation keeps the upper horizons wet. All these can as easily result from progressive acidification and the effects of man's activities as from increased raininess or cool, cloudy summers. This is not to say, necessarily, that the climate did not change, but if inception of peat growth is regarded as simply a climatic effect, differences in timing remain unexplained, and if it is taken as a time-marker, they do undetected.

The increase of alder is also not simply explicable in climatic terms, and is not necessarily even an indicator of soil wetness. Pollen diagrams show it arriving and achieving high values everywhere very quickly, which would suggest a rapid spread into suitable territory rather than the more sporadic beginning which might be expected if climatic change had initiated soil changes which would reach the condition suitable for alder at different rates in different places, and in areas where soils remained well drained for reasons of situation whatever the climate, not at all. The present distribution of alder may reflect some limitation and restriction to wet habitats by other soil conditions rather than by wetness alone, and alder woods can be found on well drained slopes today.

From the time that the earliest peat growth began, areas at lower altitude and on finer-grained and more base-rich rocks have suffered progressive or intermittent soil degradation of a similar nature and variable severity. In considering their history deliberate clearance of forest and use of the land for pasture and for crop growth have to be taken into account. Pollen evidence in almost every study which covers the relevant period shows episodes of interference with the forest increasing in length, frequency and extent from mesolithic times onwards. The earliest episodes which can be correlated directly with activity of neolithic people or which coincide in time with their known presence in the area, are of very similar type to those which have been suggested as the effect of mesolithic activity, showing increase in pollen production by light-demanding shrubs and herbaceous plants but with no certain indication of the presence of open pasture within the forested areas. It is suggested that the forest was being used as a source of fodder, and the common occurrence of a drop in elm pollen frequency during these episodes suggests that it was elm foliage that was being gathered. If people were living in small groups, and their cattle were tethered or closely tended within the forest, as would be necessary while wild cattle were about, they need not have had any greater impact upon the vegetation than did their predecessors and contemporaries who had no domestic animals, apart from the selective use of certain foliage, and the effect upon the soil would have been comparable. Where the forest was far from the sensitive, marginal situation, every clearing or opening in the canopy, whether by deliberate felling for timber and to make room for habitation, clearance or lopping to improve browsage and to collect winter fodder, collection of fuel for fires or accidental fire damage, perhaps in the collecting of honey, would have achieved full regeneration when the people moved on or turned to a different area for their needs. Nearer to the marginal situations, in a zone extending downwards from that already affected by the activities of mesolithic peoples on the more sensitive parent materials and now perhaps beginning to include the higher and more northerly regions on better soils, as nutrient loss increased, each regeneration phase would have taken longer and included less of the full complement of species, until podzolization of the soil became widespread under a mosaic of poorly nourished trees and low vegetation containing heather, bracken and the poorer grasses. As areas of forest formerly used for browsing or fodder collection became less productive, they would join the already degraded areas as hunting grounds, either for the neolithic people themselves or for remaining groups of mesolithic hunters, and as neolithic activity within decreasing areas of productive forest increased, the wild herbivores would have been driven out into these degraded lands, increasing grazing pressure there.

There is little or no evidence of crop growing in upland areas during the period of brief and temporary openings in the forest indicated in the pollen analyses. If cereals were grown, with the number of pollen diagrams available for upland areas, at least one or two samples would have yielded a sufficient quantity of cereal pollen for it to be said with confidence that the arable fields were nearby, for however short a period, rather than the occasional appearance of cereal pollen which is interpreted as indicating derivation from some distance away. It is also uncertain when the open areas at high altitude began to be used for grazing by sheep. The effect upon the vegetation and soil of controlled exploitation of wild animals, perhaps with regular and deliberate burning, and the herding of domestic ones would be similar in type, and though it might differ in degree, the intensity and efficiency of the two systems would probably vary sufficiently to create considerable areas of overlap.

Disturbance and erosion of soils was sufficiently frequent at the stage of small and short-lived openings in the forest for lake sediments to show increased inwash of soil material. Once the increase had occurred, the higher levels of inwash were maintained thereafter (Mackereth, 1965–6). Any human activity is bound to cause soil disturbance, and it would be greatly increased once soils had begun to podzolize and any mineral soil exposed by a break in the humic horizon had little structural stability.

The transient clearings which mark the early phase of Neolithic activity gave way to longer episodes of clearance. An area of primary forest or secondary woodland would have been cleared and the land used until its productivity declined, when it was left to revert to woodland while other areas were used, either as a long term rotation within a constant area or in the course of a gradual expansion. This system of land use seems to have persisted in the highland zone from the Neolithic through the Bronze Age and well into the Iron Age. Turner, J. (1964) suggests 50 years as the period of each cycle of clearance and regrowth. What the mode of use of the clearings was remains in doubt. Bradley (1972) discusses the problems associated with the assumption that pastoralism was predominant throughout this long period, but the fact remains that it is a period for which pollen diagrams from upland and lowland sites are abundant, and in the great majority of them cereals and the weeds of arable land are not strongly represented. Such evidence as there is for cereal production is localized. To some extent this is a reflection of the types of site and deposit in which pollen is preserved, but if cereal production was widespread but of short duration at any particular place, for instance, as a catch crop on newly cleared soils before nutrient reserves had fallen too far (Fleming, 1971), surely some buried soil or peat deposit somewhere should have brought to light at least one thin layer or horizon in which cereal or weed pollen was sufficiently abundant for it to be said that the crop was grown in the immediate vicinity. Pollen of the weeds characteristic of arable land rather than of pasture appears more frequently than that of cereals most of which are self pollinated; most of these weeds do not necessarily imply crop growth, however, since the process of clearance would produce a transient stage of soil conditions similar to those of land ploughed for arable use.

Some cereals certainly were grown, and the localities might have been quite restricted. Those stretches of coastal lowland in which raised beaches and dunes provide calcareous soil parent materials or where such fertilizers as seaweed, fish residues and shell sand are available, together with a favourable climate under maritime influence, are clearly suitable. The evidence from the Cumberland coastal plain shows that once clearance had occurred there in the Neolithic period the land was kept open for much longer than in other lowland areas, and cereals were grown (Walker, D., 1966). Other suitable coastal areas have not yielded pollen diagrams; the most suitable soils for prolonged cereal growing are the least likely to preserve pollen, and it is only in the lucky circumstance of adjacent lacustrine deposits that the evidence is preserved. Most of the lowland sites which have been analysed are the sediments and raised bogs of estuarine areas, such as those around the head of Morecambe Bay, and such coastal geomorphology does not provide base-rich but well-drained situations nearby. Terraces further up river valleys might provide cereal growing sites; floodplains are nutrient rich but though well drained in summer do not drain early enough in spring for successful cropping. However, many of the rivers draining the uplands are steadily cutting down in

Figure 19. Plough marks in a podzol profile in Orkney. Most of the plough soil has been trowelled away so that dark grey soil in the bottoms of the furrows shows up in the white eluvial horizon. The plough soil was buried by Bronze Age buildings and occupation material. (Excavation at Skaill, Deerness, by P. S. Gelling.)

their middle courses by means of incised meanders. As each meander migrates downstream it leaves fragments of the floodplain abandoned as terraces a little way above the newly established flood plain, and these recently abandoned sections with their capping of unleached floodloam and good drainage, usually clear of the winter floods, provide ideal situations for cultivation. As each suffers nutrient loss under leaching and cropping, more become available. Areas providing such situations have not provided any sites for pollen analysis.

Indirect evidence for cereal growing, in the form of plough marks indicating cultivation, is more hopeful. Figure 19 shows plough marks in a podzol profile in Orkney. The land surface underlying Late Bronze Age, and possibly earlier, occupation had been extensively ploughed in spite of the extremely poor state of the soil. Where soil accumulations could be studied, it was apparent from the evidence described on page 331 that manuring had been practised. Seaweed, used for reclamation of moorland in the area in recent times, would have been available.

If cereals were grown only in such restricted areas, elsewhere the clearings on the floodplains and valley sides would have been used for pasture, winter folding and hay production. Floodplains, which if undisturbed would develop a woodland vegetation of trees such as alder and willows, which tolerate limited oxygen availability around their roots for part of the year, would be too wet for grazing in spring, but the grass would grow rapidly to produce abundant hay crops, the nutrient supply being replenished with each load of floodloam, and the aftermath could be graze in late summer and autumn. Clearings on the valley sides would be used as winter folds, and in spring and summer might either be grazed or used for hay production. It cannot so far be determined from either the archaeological evidence or the environmental evidence whether the lowland clearings were grazed in summer or whether the whole community with its cattle and sheep went up to the higher pastures. The mode of use of the upland clearings, and of the open ground, which was increasing in extent as the soils became poorer and forest regeneration failed, is not clear. Cattle may have been as important as sheep, or more so, and the sheep as well as the cattle were probably used for milk and meat, wool and skins sufficient for a community's needs being by-products. A system of herding based on free ranging over the upland pastures is unlikely. The danger from wolves would have been great, and numbers of animals beyond the milking need would be unnecessary until wool production for export becomes important since a milking herd will supply meat from the old and from the excess male young, whose meat and hide do not increase in value once they are full grown. While animals are in milk they have to be gathered fairly closely together within easy reach of habitation. The habitation might be a mountain bothy in summer and a lowland farm in winter, or a year-round settlement in an intermediate position, depending on the scale of the landscape and the relative extent of lowland and upland areas, and perhaps depending also on the relative importance of cattle and sheep. Sheep are hardier than cattle, can obtain nutriment from poorer pasture and can survive more severe conditions in winter.

The abundance of round barrows in upland areas is in notable contrast to the scarcity of Neolithic monuments; the pollen analytical evidence shows no consistent change which could be attributed to anything but more intensive occupation under the same system of land use in either the lowland or the upland areas from Neolithic to Bronze Age. The barrows are presumably incidental to the economic use of land, but it is possible that they were sited on the soils already most degraded or, bearing in mind the possible requirement of visibility, in the most open areas, and since the direct evidence about soil conditions comes almost entirely from soils buried beneath the barrows some caution is needed in generalizing too widely from it.

With this proviso, the buried soils do show that by the time the barrows were built, in these areas at least, soil degradation in the uplands had reached an advanced stage. The podzolic soils have an eroded aspect, suggesting either abandonment and development of heathland after a period of severe disturbance, and the possibility of cultivation does remain, or intense grazing, perhaps with regular burning, which prevented accumulation of a stable mor humus layer under continuous vegetation. That humic horizons did suffer erosion at this stage is shown by the arrival of organic matter, and pollen grains of cricaceous vegetation in lake sediments (Mackereth, 1965–6). If continuity of vegetation and humic horizon is not maintained, disturbance mixes organic material with the eluvial horizon and both organic and mineral soil are subject to erosion and deflation, so the depth of development of the podzol profile is restricted. Where parent material is stony, as the glacial residues covering many areas of upland are, the finer material is washed or blown away and stones stick up through the surface of the mineral soil or are left lying on it, the humus collecting in pockets among them. Here may lie the origin of the Bronze Age cairn fields, some of which are probably heaps of stones cleared from fields. It is not necessary to postulate cultivation to explain the necessity for such stone clearance. The area of pasture and the protective continuity of plant and humus cover are greatly increased by removing the projecting stones. This can be seen around present farms in the Highland Zone, where the stones from the fields have been used to build the walls around them, and the depth and continuity of soil within the fields is in part the result of such clearance and is in strong contrast to that of the stony moorland beyond the enclosed areas. In much of the Highland Zone there is little indication in the buried soils of this period that iron pan and gleying had developed to any extent. The profiles are thin or weakly developed podzols in which the thin eluvial horizon is often masked by the colour of a moder humus, and the **Bfe** horizon is so diffuse as to be scarcely detectable against the background of a strongly coloured **B** horizon of the brown podzolic soil which existed before the stronger differentiation began to develop.

In several profiles on sandy glacial deposits in Yorkshire, the earlier history of the profile can be partly elucidated (Limbrey, in press). In buried soils beneath round barrows at Irton Moor illuvial clay horizons are present, and show signs of having developed within a brown podzolic profile. The illuviated clay is present as fillings and linings to fine pores, and spreads from them into the matrix to engulf the brightly coloured, more open textured soil of the brown podzolic **B** horizon, which remains in the interiors of the peds. Above the illuvial horizon, its counterpart is seen in the sandier texture and paler colour, particularly on ped faces, of the upper part of the former **B** horizon. Gleying has followed development of the illuvial horizon and its effects spread from the major structural planes, formed as the prismatic structure developed, in the gleyed soil. The horizon thus presents a tricolour appearance, with a two stage marbling, of reddish-yellow sandy soil within a pinker and more clayey network associated with the fine pores, and both enclosed in a coarser network of grey associated with major structural planes. It is tempting to associate development of sols lessivés in these soils, as in those of the chalklands, and as Romans *et al.* (1973) do in the case of a soil in lowland Scotland, with opening up of forest cover. The acid brown forest soils or brown podzolic soils on these coarse grained parent materials would have been maintained in a stable form under closed forest, but where some clearance had already led to loss of nutrients, the combination of high leaching in winter and partial drying in summer in open areas led to clay migration and flocculation at some depth. Subsequent podzolization might have followed abandonment of a cultivated or heavily grazed soil which had suffered considerable erosion, leaving a stony surface upon which establishment of an impoverished pasture or heathland led to accumulation of mor humus in pockets among the stones. The podzol may thus have begun to develop after the erosive phase or contemporaneously with deflation and surface washing under sparse or disturbed heath vegetation. Buried soils beneath round barrows in other parts of the North York Moors studied by Dimbleby (1962) indicate that the barrows were probably built in clearings within a patchily wooded area, with hazel growing around the edges, and that any cultivation for cereal growing was some distance away.

The predominantly pastoral use of the upland soils during the Bronze Age and the early part of the Iron Age does not appear to have been the immediate cause of the spread of iron pan formation but rather to have pre-conditioned the soils for it. During this long period the bare bones of the uplands were gradually exposed, the forest and its soils wearing thin in patches and the frayed edges peeling back from the hill tops. The pattern of soil distribution which had developed from the generally acid brown forest soils would have been a complex of brown podzolic soils, sols lessivés, and profiles in which the characteristics of the latter were superimposed upon those of the former, distributed in the landscape according to variations in parent material, slope and climate, and probably strongly influenced by the frequency and persistence of phases of forest disturbance before and during the neolithic period. As grazing by the cattle and sheep of the Bronze Age people increased in intensity and time, the nutrient status of the soils would have declined, the clearings would have been slower to close, and the vegetation in the clearings and that which sought to close them would have been affected in vigour and composition so as to induce or increase the tendency to podzolization: heather and bracken would have been present in the undergrowth of scrubby woodland and would have encroached upon the grassland in abandoned clearings. Where sheep rather than cows grazed, their preference for the tender and more nutritious grasses and herbs and their avoidance of the coarser ones and of bracken would have allowed these polyphenol-rich elements of the vegetation to increase. Thus in both woodland and pasture shallow podzols would have developed in the upper part of sol lessivé profiles and the horizon differentiation in the brown podzolic soils would have been intensified. The mor humus would have been easily disturbed by the feet of the grazing animals and the heavily grazed vegetation provided poor protection from the rain and wind. These are the thin, stony podzols found under the upland round barrows.

The lowlands remained wooded through the greater part of the period during which the uplands were losing their trees; in these less sensitive situations forest regeneration could continue so long as only small areas of lowland pasture and arable land were needed at any time, and pigs probably rooted in the woods, contributing to disturbance of their soils, limiting the growth of herbaceous plants beneath the trees and removing nutrients, but also perhaps helping to ensure continuing tree growth by implanting acorns with their feet, so protecting them from birds and rodents.

The evidence from pollen analysis suggests that clearance of the forests of the lowland areas took place on a much wider scale than previously round about 400 B.C., and that the cleared areas remained open (Turner, J., 1964). It is supposed that this activity was associated with denser settlement, but still a predominantly pastoral way of life. The need for more lowland pasture and more hay could arise from a greater concentration on cattle, or from a difference in stock management practices, involving less dependence on the upland pastures, perhaps without the habit of transhumance, whether or not there was an increase in population.

Bog stratigraphy provides evidence of increased wetness, which is taken as marking the beginning of the sub-Atlantic period. In the lowland bogs a high degree of humification, indicating intermittent partial drying and aeration of the bog surface, allowing some activity of micro-ogranisms, gives way to a low degree of humification and rapid accumulation of peat, indicating saturation throughout the year. The upland blanket peat shows similar changes and also increased its rate of spread. Rather than a simple change, in many places conditions must have fluctuated, so that there are a number of horizons at which humification changes, and correlation is not always clear between different sites, and the change in environmental conditions appears to cover a wide range of dates within the first millennium B.C. If a simple climatic trend is involved, there may have been some flexibility of response of the peat bogs, so that increased wetness became critical at widely different times. Flexibility would be introduced by the variety of soil conditions in the drainage area feeding the basin on whose sediments a bog is growing, and in the soils of the area over which blanket bog was encroaching.

The acceleration of the spread of blanket bog over the upland podzols may be an indication of changes in moisture conditions which also lead to increased formation of iron pans in those soils not engulfed. The thin, eroded podzols of the upland pastures could have become wetter either because of an increase in climatic wetness or because of a relaxation of grazing pressure which allowed the mor humus to thicken up and hold more water. Thickening of the organic horizon as grazing was reduced, perhaps also as the frequency of burning fell, could be correlated with the increase in use of lowland pasture, late in the period over which climatic change is postulated. Such change in land use might itself be a result of climatic change, as the uplands became more unpleasant and as increased leaching and shortened growing season reduced the productivity of the pasture. Gleying in the **B** horizon could be the result of climatic wetness alone, or it could be the result of final collapse of drainage channels provided by roots when soils had reached so impoverished a state that no more trees could grow, or when grazing became so continuous, perhaps with regular burning, that no regeneration was permitted.

We do not know where to draw the boundary between the areas which became open pasture and moorland during the Bronze Age and those which remained generally forested until major clearances in the Iron Age. The boundaries would have been extremely irregular, taking account of differences in soil parent material, slope, drainage, aspect and the density and continuity of settlement throughout the earlier periods. In many of the more stable, forested areas, the soils would have been strongly leached by the time they were finally cleared, and would have degraded rapidly once forest regeneration was prevented and grazing became continuous; in other soils stability would have been upheld by their high base reserves or by their situation in receiving areas for base rich moisture or for soil material on a creeping slope, and once clearance had taken place these soils would have differentiated, some still able to maintain their nutrient reserve and structural stability, others succumbing to the increased leaching and grazing losses and the disturbance of clearance and trampling by animals, and others benefiting from the moisture or soil material derived from these.

Those soils which survived the forest clearance and grazing with their structure and nutrient status in good order, because of favourable position or parent material, were available for cultivation as the field systems were established which are now still visible as "celtic fields". We do not know whether some of the soils which did not survive so well were also cultivated. In other words, does the limit of the celtic fields, which extends beyond that of subsequent enclosures, approximate to the limit of soils with sufficient stability not to have podzolized severely after a period of intensive use as pasture, or to that of the soils which, though stable enough to have remained largely forested during the Bronze Age, began to degrade when permanently cleared and heavily grazed, or does it extend even beyond this into the open pastures and moorland which had existed since the Bronze Age? It is unlikely that the ploughmen would have tackled soils with iron pans. If the pan were below plough depth the soil would remain too wet for crop growth, and if it were shallower the pan would have to be broken, which is difficult even with modern machinery. Podzolized soils with no iron pans, under grass or heath vegetation, might have been enclosed and ploughed, provided manure was available.

Farming in a permanent nucleus of enclosed fields is only made possible by the application of manure. The advent of true ploughs, with their deeper bite, undercutting share and turning action of the mould board, makes possible cultivation not only of heavy soils but also of those which need the incorporation of the **B** horizon into an eluvial horizon to produce a good plough soil. Together, ploughing and manuring may be responsible for the spread of arable farming up the valley sides. The amount of manure needed where leaching is so high is very great, even on the most stable soils and, except where coastal resources were accessible, the stock management practices of the mixed farms must have been largely controlled by the need for dung. Using the arable fields as winter folds helps to keep up the nutrient and organic matter content of the soils, but the risk of puddling and damage to structure, causing fall in crop yield in spite of adequate manuring, and exposing the soil to more severe erosion than is anyway inherent in cultivation, is high where soils are very wet in winter. The stubble would have been grazed in late summer and autumn, but the animals would probably have to be taken off it as the water content of the soil rose early in the winter. The enclosed fields may have been used in rotation, with periods as pasture and as hay meadows. Either use, provided grazing is carefully limited, will allow a build up of ogranic matter and improvement of soil structure, but will contribute to loss of nutrients unless these fields too are manured. To make up for leaching losses and for that part of the rubbish and excreta of the settlement which does not get back onto productive land, much manure must originate from outside the enclosed area, brought in by animals which graze in summer on pastures beyond the enclosures and come in at night for protection and for milking, or brought in as winter fodder, from riverine hay meadows or from woodland or moorland. Maintenance of soil quality in the enclosed fields hastens degradation elsewhere.

During the periods of intensive forest clearance and of arable farming, movement of soil downhill increased greatly. Once fields are enclosed the lower boundary stops the moving soil, but the field walls may not have been built until sufficient plough erosion had taken place to expose large amounts of boulders and rock outcrops, necessitating field clearance as well as providing the materials with which the walls could be made, and on the lower slopes there may not have been so much enclosure as in the higher settlements, whose traces have survived subsequent land use. The soil which moved downhill, from the disturbed and degraded soils of the pastures and in great quantity from the plough, formed deep colluvial sediments on the lower slopes and in the valley bottoms.

More soil was carried into streams, and with the increased run off from the denuded slopes and the greatly increased load, sedimentation of the floodplains built up, merging and interfingering with the colluvium of the slopes to give the wide, flat bottoms characteristic of many valleys in upland areas. As the rivers silted up the bottom lands became permanently wet and peat growth began. At the same time, blanket bog was spreading, and so the upland pastures, the riverine hay meadows and the pasture or cultivable land on lower terraces were all reduced in area. This process may have played some part in stimulating the use of the hillslopes for cultivation, and must have influenced stock rearing practices considerably. Its continuance, particularly the loss of hay from continually enriched floodplains, which had provided an external source of fodder which then yielded manure, may have been important in the subsequent decline of the way of life represented by the celtic fields.

The celtic fields were eventually abandoned, and arable land never again extended so far up the valley sides. Pennington (1969) suggests that the growing of cereals in the Lake District at higher altitudes than before or since indicates a period of warmer, drier climate, that is, a climatic oscillation within the generally cool, wet sub-Atlantic, within the first few centuries A.D. The abandonment of the settlements is often attributed to worsening climatic conditions, with growth of peat over the enclosed fields as a direct cause of retreat, supported by the sheer nastiness of winter weather. Worsening climate would increase the leaching of the soils, necessitating more cartage of manure while upland pastures and sources of winter fodder declined in productivity and so reduced the availability of manure. This would hasten the approach of the point at which the farms could not be worked with the labour they supported. However, the decline in productivity is inevitable under constant climate, and the farms would probably have had to be abandoned sooner or later in any case. Once the land ceases to be cultivated, all the advantages of manuring and ploughing cease and the soil rapidly succumbs to the same processes of degradation as the soils outside the enclosures, and if the area is one in which iron pan formation and waterlogging are rife, encroachment of peat over the enclosures is likely to occur. Peat encroachment is the consequence of abandonment, not its cause.

Subsequent history of land use is dominated by the expansion of sheep farming initiated by the Cistercian monasteries. The monks cleared much of the remaining woodland and cultivated the lowlands extensively around their monasteries and granges, and they exploited the uplands for wool production. This is a major change in land use, with much greater concentration on sheep than ever before and the introduction of the practice of free ranging. With the hardy sheep on the mountain all winter, and no need to gather them daily for milking, importation of manure to the lowland fields is reduced, and with the

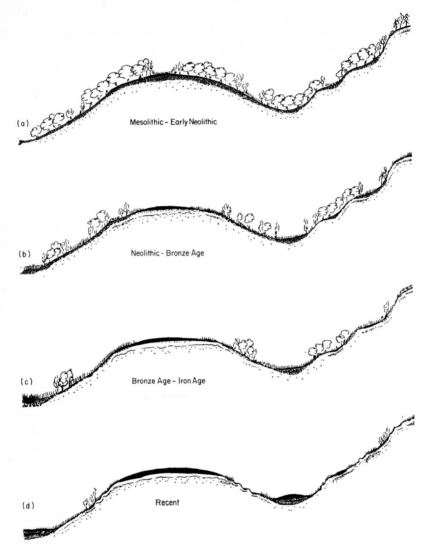

(a)   Mesolithic – Early Neolithic

(b)   Neolithic – Bronze Age

(c)   Bronze Age – Iron Age

(d)   Recent

Figure 20. A possible sequence of development of landscape and soils in the highland zone.

export of wool, nutrients, and particularly nitrogen, are continually exported, not just from the uplands but from the region as a whole.

The characteristics and distribution of soils have reached their present form largely under the influence of this mode of land use, which has continued into recent times, with fluctuations in intensity of grazing and in the area of land on the lower slopes which is now enclosed. The enclosed land is not used mainly for cultivation, as was probably the case with the celtic fields, but entirely for pasture, year round for cows and in winter for the less hardy but higher yielding breeds of sheep, and for lambing.

Clearly the history of the upland soils is complex, and the relationship between the different types of vegetation, the soils they occupy and the animals they feed, needs much further investigation from the historical point of view. More factors than have been dealt with here have to be taken into account. The behaviour of deer, their change of habit from that of forest creatures to that of moorland ones, and, much later, the effect of control of moorland vegetation and grazing so as to preserve them, and grouse, for the modern version of the hunting way of life, must be considered. The introduction of ponies, possibly the use of the uplands for pony production, which would be an exporting economy based on the uplands and beginning at a much earlier stage than sheep farming, and the continuing effect of grazing by ponies in some areas, are probably important.

Figure 20 shows a possible sequence of soil distribution in the highland zone. It is obviously very much over simplified, but perhaps provides a hypothesis which can be tested as more information becomes available from excavation and soil studies.

8

The History of Chalk Soils in England

Characteristics of Chalk

Chalk is predominantly calcium carbonate: in part the tests and shells of foraminifera and marine animals and in part precipitated in very fine crystal size. The fossils are better crystallized than the precipitated matrix and more resistant to disruption and dissolution; they are often found loose in chalk soils and the foraminifera are sometimes found concentrated as calcite sand where locally derived wind or water transported chalk residues occur. Some fossils in the chalk are made of flint: sponges and echinoids are particularly frequent in some localities and when found on archaeological sites are sometimes thought to have been deliberately collected.

Most of the flint which occurs in chalk is in irregular nodules concentrated in certain layers, or in tabular beds. Flint is silica in the form of very fine crystals, cryptocrystalline silica, and amorphous, chalcedonic silica. The outer part of the nodules has lost its more soluble, amorphous component, and the cryptocrystalline mesh encloses air spaces, giving the opaque white appearance of the flint cortex. Patination of flint artifacts is the same phenomenon, but in the context of soils and river deposits it is very variable in occurrence and in rate of development. The rate of dissolution of silica is much lower in the acid environment of a humic soil than in chalk: it is often observed that artifacts lying on a buried surface and covered by an earthwork of chalky composition, or at the base of a worm-sorted layer, have a different patination on their upper and lower surfaces. The surface in contact with chalk has a white patina while that in contact with soil appears quite fresh. In a soil or a river deposit iron or manganese oxides may infiltrate the patina and give a brown or black staining.

Other large inclusions in the chalk which appear in soils and may turn up on archaeological sites are pyrites nodules. They are usually approximately spherical and can be up to fist size. When fresh, the facets of the pyrites crystals can be seen at the surface, but in the soil the surfaces become oxidized and present a rusty, knobbly appearance.

The fine mineral component of chalk is clay, silt and fine sand blown or drifted into the Cretaceous seas from surrounding land surfaces. The clay is largely montmorillonite and is accompanied by iron oxides, and the silt and sand are predominantly silica. In certain areas, the silt and sand, probably derived from desert loess, comprise a considerable percentage of the chalk, but in much of the chalk the non-calcite constituents comprise no more than about 2%. Some chalk has a certain amount of amorphous silica, which makes it harder than usual, and this form, known as chalk rock, has been much used for building. It is also a good raw material for cement making, having the silica component already present, and is much quarried for this purpose.

Chalk is porous in itself and also has joints and bedding planes which facilitate percolation. The fine calcite is readily dissolved in slightly acid rain water and in soil acids: the whole mass becomes wetted as water percolates along all the boundaries of the tiny crystals. At the contact of soil and chalk, and in the joints, down which water flows freely, dissolution occurs. Roots penetrate cracks and joints, and their acid exudates, and those of the rhizosphere organisms, contribute to dissolution. The oxidizing environment of a freely draining chalk soil and the rhizosphere results in immobilization as oxides of iron released by alteration of impurities in the chalk: iron staining can be seen on the faces of joints to considerable depth.

The Parent Materials of Chalkland Soils

The amount of mineral material released by dissolution of chalk is usually very low, and the depth of soil which can form simply from chalk residues is very small. In many situations loss from the surface by hill wash, creep or deflation is likely to be at least as rapid as gain in depth by soil formation, and in the complementary situations of accumulation the depth is achieved by derivation from a wide area. However, the chalk of the temperate regions occurs in areas of deposition of till, outwash materials or loess during glacial periods, and few chalk soils can have been derived initially from chalk alone.

Clay-with-Flints

Before the beginning of Pleistocene glaciation much of the chalk must have had a cover of Tertiary sediments. During the Eocene period the uplifted chalk was eroded and complex marine, estuarine and fluviatile sediments, the Reading Beds, were laid down. The succession begins in many places with a basal gravel, largely of flint derived from the chalk but having pebbles of other rocks of the region as well, and continues with sandy or clayey sediments. Erosion later in Tertiary times and during the Pleistocene has removed these deposits altogether in some places and cut well down into the chalk, but in other places the basal surface is still represented in the chalk landscape and a layer of the deposits is associated with it (Jukes-Browne, 1906; Hodgson et al., 1967). Earlier in Pleistocene times more of this surface must have been present, but repeated cycles of weathering and erosion have brought the land surface closer to the chalk on hill tops and plateaux, initiated and extended the system of dry valleys and seasonal or perennial streams, and cut more deeply into the chalk in the main river valleys which cross it.

Periods of disturbance of the superficial materials by periglacial activity and additions to them of wind blown sand and loess, alternating with periods of soil formation under interglacial conditions, have produced a complicated assemblage of deposits in which Tertiary and Pleistocene sediments and their products of weathering are to varying degrees mixed and convoluted with each other and with the chalk residues. The variety of these materials and their disposition in the landscape have led to prolonged discussion about their nature and origin. *Clay-with-flints sensu stricto*, a firm, reddish-brown or yellowish-red, somewhat mottled clay with flint nodules, many of them broken but unworn, lies adjacent to the chalk, having a sharp and very irregular contact with it. *Plateau drift* often overlies clay-with-flints in the higher positions in the landscape, and contains pebbles of flint and other rocks, and has variable amounts of sand and silt as well as clay. Its colour is more varied and the mottling stronger than in clay-with-flints. The two layers are so mixed and convoluted together that it is often difficult to distinguish a boundary between them, and terminological differentiation is correspondingly confused. In places where the plateau drift is deep, its presence is clear, but the degree to which the Reading Beds of whose materials it is composed have been redistributed by fluviatile, glacial or even the marine activity of a postulated early Pleistocene high sea level, as well as contorted by cryoturbation, is still a matter for argument. In lower lying situations where a thinner remnant of Reading Beds is almost entirely mixed with clay-with-flints it has not always been identified as such, and it is often included in a broader usage of that term. Confusion is made worse by incorporation of loess by further cryoturbation, and the term *brickearth*, discussed below, enters the field.

Loveday (1958, 1962) summarizes the arguments and defines clay-with-flints *sensu stricto*. Avery *et al.* (1959) and Hodgson *et al.* (1967) investigate deposits in detail to provide an understanding of the parent materials of soils in areas covered by Soil Survey publications Avery (1964) and Hodgson (1967). As a result of these studies it becomes clear that the possibility of defining the deposits in terms of initially distinct layers is complicated not only by the effects of periglacial activity but also by the effects of clay migration, so that clay-with-flints has a fine clay component derived from the Reading Beds and from loess, and plateau drift is rendered loamy in its upper part by the loss of downward moving clay as well as by the addition of loess.

Formation of clay-with-flints is thus the result of the following processes. Dissolution of chalk at the lower boundary of the Tertiary materials has been going on ever since they were laid down. These siliceous materials are base-poor, and under tropical and interglacial soil forming conditions would have become leached and acid. The clays hold moisture, allowing continual slow seepage from the zone of contact into the more freely draining chalk, carrying away the calcium released and allowing dissolution to proceed. The contact of clay and chalk is very irregular, there being deep pockets and pipes of clay going down into the chalk. Where the overburden is relatively shallow some of the pipes may represent the positions of roots of trees of the interglacial forests, but much of the irregularity is probably the result of the solution process itself. Slight irregularities in the original surface would have developed into hollows and pockets as the deeper soil remained wet longer and the pockets held a concentration of the acid soil moisture.

Infilling of the volume left by dissolution of the chalk has apparently taken place partly by infiltration of fine clay, mobilized under conditions of degradation and dispersion of clay in the **B** horizons of the tropical and interglacial soils, represented by the overburden. The zone closest to the contact with chalk therefore has a mixture of chalk residues and illuviated clay. Where the contact has remained below the depth of periglacial disturbance, this zone can be seen as a dark clay horizon, which has been called a *Beta horizon* (Bartelli and Odell, 1960), and is described, in deposits in Hertfordshire, by Catt *et al.* (1970). Infilling also takes place by slumping and oozing of the clay, and flow structures can be seen in thin sections of the clay. However, in many places, the effects of cryoturbation have contributed to the mixing process and obliterated the evidence for the more gradual and gentle processes. Where cryoturbation reached into the chalk, the boundary is strongly convoluted and masses of clay occur within chalk, and of chalk within clay. The cryoturbated chalk is disaggregated, and may have become recemented by calcium carbonate as the solution processes at the zone of contact proceeded. Flint nodules released from the chalk during solution are fragmented by cryoturbation, and are often seen to be aligned vertically in the involutions.

Sarsens

In the context of materials of Tertiary origin it is perhaps appropriate to mention *sarsens*, the sandstone blocks which litter the chalk landscape in parts of Wiltshire and which provided the builders of Stonehenge and Avebury with large stones. These too are residues of the sediments laid down over the chalk. Under a tropical climate in later Tertiary times soil formation must have resulted in mobilization of silica, and a *silcrete* horizon developed. The more heavily silica-cemented patches resisted subsequent erosion and weathering and now lie scattered on the chalk surface, shifted by solifluction, so that they can be seen as if streaming downhill in dry valleys of the Marlborough Downs. The matrix of softer materials which must have participated in the solifluction has since been lost by erosion. The hypothesis of Kellaway (1971) that the sarsens are the remnants of ice-rafted sandstone blocks and that the Stonehenge bluestones were also brought to the area by glaciers extending much further across southern England than is usually supposed is hardly tenable in the face of work by Green (1973), who shows that gravels of river terraces, which would have contained pebbles of rocks foreign to the area had erosion removed most of the glacial residues, do not do so. The only rocks from beyond the watersheds of the areas drained by the rivers that are represented in the gravels of their terraces are those which were already present as pebbles in the Tertiary sediments.

Coombe Deposits; Head

Where Tertiary materials and accumulated chalk residues had been eroded away before the Last Glaciation, periglacial processes during that glaciation affected only chalk and loess, and in such areas solifluction deposits consisting of mixtures of these materials fill the bottoms of dry valleys and cover many slopes. The term *coombe rock* should be reserved for predominantly chalky deposits in which chalk lumps are cemented together by secondary deposition of calcium carbonate. The mobilization and recrystallization of calcium carbonate may have occurred beneath a soil, the coombe rock then being the C_{ca} horizon, or it may have happened soon after the deposit was formed, under the influence of calcareous ground or surface waters. Coombe rock can occur on hill tops, where cryoturbation involved little lateral movement, as well as in the form of solifluction deposits, and it can be difficult to tell from solid chalk. It can be even more difficult to tell from chalk redeposited by human agency, such as the packing of a grave cut into either solid chalk or coombe rock itself, or the primary fill of a ditch cut into these materials, and causes much puzzlement in excavation.

Where larger quantities of loess were present, solifluction and cryoturbation produced festoons consisting of layers and lenses of mixtures of chalk and loess in greatly varying proportions, from chalk lumps set in a sparse matrix of loess, ground chalk and recrystallized calcium carbonate to calcareous loess with only a scatter of chalk particles. Flints can occur throughout, or may be concentrated in pockets or strings in the festoons, and are often sharp, broken fragments. A deposit consisting almost entirely of flint gravel in a chalky matrix fills the bottom of some dry valleys and shows cryoturbation structures. Where there is clay-with-flints on higher ground, solifluction deposits contain material derived from it, and where the clay-with-flints lies in declivities on slopes it may be covered by chalky, loessic solifluction deposits from the slopes above.

A number of terms are in current use for these very variable and very widespread deposits. *Head* is a general term covering them all, and is preferable to *coombe deposits* since the latter would appear to wish to exclude deposits lying on open slopes. The word *drift* persists from the diluvial period of geological thought as a useful term for superficial deposits of almost any origin, and is used in the descriptive terms employed by the Soil Survey: silty drift, flinty silty drift, and so on. Where it is clear what the materials are and how they achieved their present disposition, it is preferable to be specific, and use phrases such as "chalky, loessic solifluction deposit".

Brickearth

The term *brickearth* has been applied to a great variety of materials suitable for making bricks by virtue of having a silty or fine sandy texture with enough clay to provide coherence on firing. Many such materials are fluviatile deposits, particularly the loams which cap many Pleistocene river terraces and probably contain large amounts of redistributed loess. Others are the loessic materials which overlie and are involved in periglacial mixings with chalk, coombe rock and clay-with-flints, and the term brickearth is used for those which are calcareous because of secondary recrystallization of calcium carbonate, those which became mixed with ground chalk during cryoturbation, and the products of decalcification and soil formation on either. Calcareous brickearths are usually yellowish-brown, decalcified ones brown or strong brown to reddish-brown. It is perhaps wise to avoid the term brickearth for these materials of such different history and characteristics, except where a deposit is locally known as such and the place name is specified. Otherwise, as the distribution of loess becomes better known it will be possible to speak of *loess loam* for the fluviatile materials, as is done in continental Europe, and use descriptive phrases for the other deposits and soils.

Development of Soils

While some of the loess which fell on the chalk was removed from slopes by solifluction during at least one period of periglacial conditions after the last episode of loess deposition, its contribution to the soils which developed together with the post-glacial forests is undoubted. Soil mineralogy studies such as those of Hodgson *et al.* (1967) attest its presence in soils on clay-with-flints (*sensu lato*), where less cultivation has been practised than on the more base-rich parent materials and soils have been less eroded, and elsewhere buried soils provide evidence of the former existence of deeper soils with more of the loessic profile intact. The present thin rendzinas and stony plough soils are the result of man's use of the land, and as in the case of the podzolized soils of upland regions and the lowland heaths, sufficient interference with the soil–plant system to cause irreversible changes in the soil profiles may well have begun with disturbance of the forest by mesolithic people. Full development of forest on the chalk, and its clearance are shown by the only depositional pollen sequences for locations close to the chalk, those of Godwin (1962).

Development of the forest soils involved decalcification of the chalky, loessic materials and would have proceeded faster in the less chalky parts, so that the boundary of decalcification followed the involutions of cryoturbation structures in an intricate manner. Clay mineral formation would have been rapid in the decalcified loess, leading to development of a **B** horizon of loam or clay loam texture and establishment of a brown forest soil of high nutrient status. However, at some stage in development of the profiles, base status fell and sols lessivés developed. In the soils in which loess was involved with a deep clay-with-flints layer or with plateau drift, base status would not have been continually maintained under forest by trees rooting into the chalk, but elsewhere it may be necessary to invoke clearance of forest before clay migration would begin. Loss of bases through cropping and grazing would have contributed to impoverishment. Exposure of the soils to rain would have been increased, especially in winter when arable soils have little plant cover. Loessic soils drain freely, and as a sol lessivé develops they do so excessively in the eluvial horizon, increasing leaching and enhancing dispersion and degradation of clay minerals. The eluvial horizon becomes loose and poorly structured. If the soil is cultivated and the organic content of the A horizon falls, the plough soil also loses its structure, and the plough soil and eluvial horizon become highly susceptible to hill wash and deflation. However, once these horizons have been lost and the plough begins to bite into the illuvial horizon, the soil can become more stable and has a greater base content and greater ability to retain bases and water, but the clayey soil is difficult to cultivate.

If cultivation is persisted with or resumed at a later period, and cropping reduces the base content of the plough soil, dispersion of the clay renders it sticky and liable to surface washing in wet conditions, and extremely hard when dry. The process of clay eluviation may continue, clay leaving the plough soil and settling in the undisturbed part of the illuvial horizon, so the plough soil continues to lose structure and suffer erosion while the illuvial horizon becomes more clayey and more difficult to work as the plough bites down into it. This type of soil might be unworkable with a simple ard, but with a more advanced plough the greater cutting power would make cultivation possible and the turning action would help to offset leaching and eluviation, so that the soil could continue in cultivation. However, whereas the easy cultivation of the clay depleted soil with an ard would render it liable to erosion by wind and water, turning the clayey soil with a plough would increase mass movement as well.

As cultivation continued the plough would eventually, on all but completely flat land, reach into the still calcareous C horizon, and the soil would again become base rich and well structured. However, the thin soil and the porosity of the chalk provide little capacity for retention of moisture, and dryness is likely to be the limiting factor in use of the land for crop growth, and would also restrict its productivity under pasture.

History of Soils and Land Use

How does this sequence of stages in profile development and degradation fit the evidence of their history that we have from buried soils? Variability in distribution and depth of loess deposits, in the amount of their mixing with chalk, and in the extent of erosion of the loessic materials before soil formation in the Post Glacial period got under way makes it difficult to establish an overall picture, but the general framework may be adequate as a basis for detailed local studies.

So far no soils or deposits in which archaeological evidence provides a means of dating stages of development in terms of human activities have been investigated by the mineralogical techniques which allowed Weir et al. (1971) to suggest a very complex history for a soil on the deep loess deposit at Pegwell Bay, Kent. They identify an early stage of some clay eluviation following decalcification and followed by erosion early in post glacial times. Development of a sol lessivé then proceeded in the truncated profile, but gave way to a stage in which drainage was inhibited, eluviation ceased, and clay minerals were degraded in situ instead of moving down the profile. The soil was buried by a colluvial deposit which probably came downslope as a result of cultivation, and a radiocarbon date determined on organic matter in the buried soil suggests, bearing in mind the unknown but possibly large mean residence time of the humus, that the soil was buried in neolithic times (cf. Kerney et al., 1964).

The Pegwell Bay loess is a deep deposit, and though some of it had been lost by erosion before post-glacial soil formation began, the soil parent material was calcareous loess, not the more chalky cryoturbation mixtures which provided the parent materials of the archaeological buried soils which have been investigated. It provides an example of one end of a range of loessic profile types susceptible to development of a sol lessivé, and it serves as a reminder that the soils studied in the field and not subjected to this type of study of the processes of mineral alteration and translocation have probably yielded too simple a story.

A soil which was available for study at Saffron Walden, Essex, while an area was being investigated archaeologically in 1973, illustrates a larger part of the range of profile development than is usually visible in a small area, but again its stages of development are undated, and it was only studied in the field. Figure 21 is a diagramatic representation of a cross section of the area. On a gentle slope chalky, loessic cryoturbation structures had elongated into stripes, and running downslope was a loess-filled hollow. Ploughing had cut across the

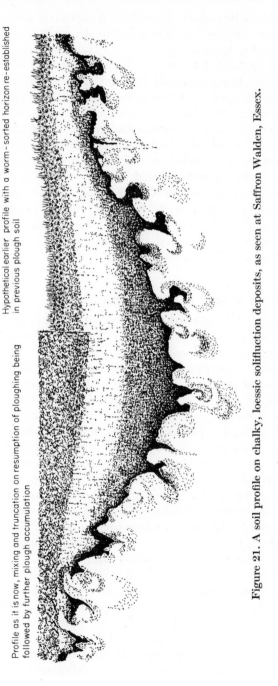

Profile as it is now, mixing and truncation on resumption of ploughing being followed by further plough accumulation

Hypothetical earlier profile with a worm-sorted horizon re-established in previous plough soil

Figure 21. A soil profile on chalky, loessic solifluction deposits, as seen at Saffron Walden, Essex.

convoluted lower boundary of the illuvial horizon of a sol lessivé where the profile was shallow, but where it was deeper it had truncated the profile at a higher level and part of the eluvial horizon was preserved. The illuvial horizon had followed decalcification down the more loessic parts of the periglacial festoons, and there were pipes and pockets filled or lined with the reddish-brown clay enriched soil, as can be seen in Fig. 29, which shows part of the area, at the edge of the loess filled hollow, after trowelling had removed the soil profile down to the point at which the higher parts of the C horizon had been exposed, and the cryoturbation structures and the pattern of pipes and pockets of clayey soil revealed in plan were being investigated in section. The character of the eluvial horizon here demonstrated the reason why it is only preserved in the chance circumstance of a protected situation and early burial. It was an almost structureless, incoherent, very pale coloured silt, which would not have survived attack by wind and rain for long once the humic horizon was disrupted.

The Condition of Soils in the Neolithic

If the line of plough truncation is drawn at different levels in Fig. 21, the various parts of the profile serve to represent the situation in many profiles buried beneath Neolithic and Bronze Age monuments, with in many cases superposition of a worm sorted horizon with re-establishment of soil formation under stable conditions on the truncated profile or on the plough soil. Of the soils beneath Neolithic monuments discussed by Evans (1972a), those beneath a henge bank at Marden, Wiltshire (Wainwright, 1971) and beneath a long barrow at Kilham, Yorkshire (Manby, 1971) show development of sols lessivés, though with eluvial horizons of less extreme form than that at Saffron Walden. These soils had remained undisturbed for long enough before burial for worm sorted horizons to have been well established, indicated at Marden by a line of stones, with neolithic pot-sherds, lowered to the base of worm sorting, and at Kilham by a thin iron pan which developed at the base of the humic horizon as the organic matter in it was mineralized after burial. At Kilham, pollen analysis indicated two periods of cultivation after forest clearance, but at neither site were what may have been considerable amounts of soil disturbance during clearance and agricultural activity detectable by examination of the profile. If a sol lessivé begins to form or continues to develop after disturbance has ceased, the movement of clay and the re-organization of structure may disguise the signs of disturbance in a soil which has few stones, and establishment of a worm sorted horizon will obliterate such traces, perhaps throughout the depth of ploughing.

It is thus established that in two instances sols lessivés had developed by neolithic times, and that at least one of the soils concerned had by then had a considerable history of interference by man. Other neolithic buried soils discussed by Evans have shallower profiles, and if they were sols lessivés of partly loessic origin they had been severely truncated and profile development re-established on a plough soil which involved the illuvial horizon or mixed it with the C horizon. Evans regards the soils beneath the Beckhampton Road and Ascott-under-Wychwood long barrows, the latter on limestone rather than chalk but having a complex of buried soils which are strikingly similar to those of the former, as not being significantly truncated. They do, however, invite a different interpretation. It would account for the shallowness of those areas of soil whose calcareous rendzina profiles are better explained as immature soils than as the result of several thousand years of soil formation under forest, particularly in the case of Ascott-under-Wychwood, where the solid limestone and clayey layers of the Inferior Oolite series, fragmented and mixed together by cryoturbation, which provide the parent material, have a much higher non-calcite content than does chalk. It would also resolve the problem of the "subsoil hollow" at Ascott-under-Wychwood, which contained snail faunas of forested conditions in soil materials which are only explicable in terms of a more complex history of profile development and disturbance than the more simple profiles over and around it would seem to demonstrate. Discussion during the excavation at Ascott-under-Wychwood and since has not reached any conclusions about these problems, but increased understanding of the features of prehistoric soil profiles has introduced factors which were not appreciated when these profiles were available for study.

The following history for these neolithic soils may be postulated. Interference with forest in mesolithic times initiated dispersion and eluviation of clay in brown forest soil profiles which had a high loess content in their parent materials. Neolithic people cleared wider areas, and cultivated the soils, increasing the degree of development of sols lessivés by increasing exposure to rain in winter and to drying by sun and wind in summer, by preventing retrieval of bases from depth by tree roots, and by removing bases in their crops. Cultivation led to widespread hill wash and deflation, so that the eluvial horizons of the sols lessivés were lost. Rotation of land use would have interrupted the process, so that it proceeded in a series of declines and recoveries of soil quality as periods of cultivation alternated with pasture or regrowth of woodland. Manuring was probably practised, as is strongly suggested by the finding of large quantities of bracken spores in soils whose snail faunas are incompatible with a bracken cover (Dimbleby and Evans, 1974), and the gathering of bracken for litter in the byres as well as the feeding of cattle on gathered foliage and herbage would have slowed down recovery of base status of soils not under cultivation while attempting to maintain that of the areas being manured with the muckings out of the byres.

Cultivation ceased, in some cases long before the monuments which buried the soils were built, and in some areas woodland had encroached over pasture (Dimbleby and Evans, 1974), though it cannot be said whether this demonstrates a general change in land use or simply that the existence of such areas in the pattern of rotating land use provided suitable sites for monuments. However, a general change in land use does seem to have come about by the time Bronze Age barrows were built, and the fact that by the time it did the sols lessivés had lost their eluvial horizons and cultivators were having to deal with the clayey illuvial horizons may have some bearing on the matter.

To account for Evans's molluscan profiles one needs to postulate that the shells of the forest populations found in the lower parts of his profiles had become incorporated deep into the soil before base status in its upper part had fallen sufficiently for them to be destroyed during transport down through it. They therefore represent forested conditions before leaching under undisturbed conditions, or accelerated by forest disturbance, had progressed far. The "sub-soil hollows" may be the result of uprooting or fall of trees, soil materials representing various horizons in the soil profile as it existed at the time collapsing into the hole and carrying with it snail shells, and sometimes mesolithic artifacts. There must be a hiatus in the snail profiles, corresponding to the period during which base depleted soil profiles existed, and shell preservation was only resumed when truncation had reached an advanced stage and the base rich illuvial horizon was reached. The snail populations indicative of a clearance phase would then be those of a stage of re-clearance in the cyclic pattern of land use.

It is appreciated that the positive evidence in favour of the hypothesis presented here is less conspicuous than the negative evidence which it seeks to explain, but this problem is inherent in a situation in which loss of soil materials and destruction at different stages of both pollen and snails are part of the process one is trying to elucidate. There is a great need for thin section and mineralogical studies to be applied to the buried soils beneath neolithic monuments, since these can provide evidence in the soil materials which remain of processes which can only have come about as a result of conditions in soil horizons which have been lost.

The Condition of Soils in the Bronze Age

Soils buried by Bronze Age monuments have snail assemblages indicative of the dry, unshaded conditions of short glassland (Evans, 1972a), and worm sorting appears to have been long established in soils which resumed development of the eroded remnants of sols lessivés. I. W. Cornwall has made a number of studies of Bronze Age soils and his observations of the soil characteristics seen in thin section are consistent in showing the orientation of colloidal material and the lining of pores with it, which show that the soils were sols lessivés. Cornwall's interpretation of these features was made in terms of the only soil studies then available which were based on thin section work, those of Kubiena (1953). Kubiena described the microscopic characteristics of soils of the mediterranean zone, in which illuvial clay horizons are very well developed, and often relict. He did not recognize the sol lessivé as a soil of temperate regions and suggests that horizons having the characteristics of an illuvial horizon are indicative of profiles relict from Tertiary times. The identification of sols lessivés was a curiously retarded pedological phenomenon, particularly in Britain. Cornwall was ahead of British soil scientists in using thin sections, but therefore did not have access to a sufficient body of studies appropriate to this region to enable him to interpret them. Thus, the interpretation of soils buried beneath round barrows as indicating a period of warm climate and dry summers in the Bronze Age is not substantiated. Cornwall's thin section studies are nevertheless valuable in preserving evidence of the soil type of the profiles he studied, and without them we would not know that they were, in fact, truncated sols lessivés. Development of an illuvial horizon, rather than further degradation of clay minerals does seem to require some degree of soil drying, but the existence of sols lessivés in the Neolithic shows that a dry period assigned to the Bronze Age will not do, and in any case the freely draining character of a loessic profile and the exposure of the soil to drying in summer once forest was cleared are probably all that need be invoked.

The clay enriched horizons which remained after erosion of the sols lessivés would have presented a problem to the cultivator, and this may have been part of the reason why they remained under pasture for so long. Their use as pasture must have been a very stable condition, since if the quality of the pasture had fallen either they would have been abandoned and regrowth of woodland would have occurred, or they would have been losing bases, further clay migration would have occurred, with the growth of a potentially unstable eluvial horizon and with the danger of overgrazing as pasture became poor and consequent disruption of the surface, leading to erosion. There is no evidence that either of these things happened. The soils were thin enough for even pasture vegetation to have access to nutrients in the C horizon, and the high clay content would have helped to prevent leaching and maintain base status, even in the face of losses of nutrients as meat and other products were taken.

Celtic Fields and Later Land Use

These are the soils which were eventually ploughed for establishment of the celtic fields over large areas of the chalk country, and the ploughing up of the well developed turf and the clayey soil may have become a feasible proposition only with the replacement of the ard by the plough. As ploughing proceeded soil movement downhill built up accumulations against the lower boundaries of fields on even the shallowest slopes, creating differences in level between fields which we now see as lynchets, and filled the bottoms of many dry valleys with deep soil deposits. These valley bottom accumulations, and those which formed the tertiary fill of many earlier ditches, often show a number of stages, which can be correlated with periods of expansion and contraction of arable farming. These changes in land use are probably attributable more to economic factors than directly related to condition of the soil, though the fact that plough land did not until recent times extend again over the full extent of the celtic fields may be due to the fact that in those on the steeper slopes and convexities the soil had become too thin to hold enough moisture for full growth and ripening of a crop to be reliable. Where soil accumulated rather than being stripped by the plough, cultivation could continue or be resumed without meeting this limitation.

Continued use of the celtic fields was made possible in part by the action of the plough in moving soil, since once it began to attack the C horizon chalk was brought into the plough soil and both stripped areas and areas of accumulation were ensured of a supply of calcium, providing an essential nutrient and helping to maintain soil structure so that other nutrients were not lost and so that soil was not stripped altogether by wind and rain. However, if manuring had not been practised, as organic content of the plough soil fell nitrogen deficiency would have become serious, and deficiencies of trace elements, some of which are in short supply in chalk, would also have become apparent, once the contribution of loessic material in the earlier soils had been lost. Rotation in use of the fields, between pasture and arable, would have allowed an increase in organic matter and nitrogen availability, and the use of the fields for growing beans rather than cereals would also have helped. It is not implausible that the value of green manuring was known. The vigour of the next crop would surely be observed, if ever a leguminous crop which failed for lack of water or because of attack by pests had to be ploughed in.

The expansion of sheep farming in mediaeval times initiated another long period when the chalk lands were predominantly under pasture, and the formation of a worm sorted horizon to a depth which could easily exceed that of the thin, stony plough soils obliterated their former character and established the profile of the typical chalk rendzina. Pasture could again continue as a stable system, without further degradation of the soils, since soil creep, though hastened under the feet of sheep, particularly when grazing pressure is high, and forming the "terracettes" characteristic of many slopes, is not too serious a problem where structure of the calcareous soils is stable and where thinning of the humic horizon does not deprive plants of access to further supplies of nutrients. Drought and deficiency of nitrogen and some trace elements, and interference by the abundant calcium with uptake of others, limit productivity to below that of lowland meadows, or that of modern systems of improved pasture, but a sufficient wool crop was maintained for sheep farming to be a profitable activity into recent times. Competition for food by rabbits, since warrening ceased and they became widespread, introduced a new limitation on productivity of sheep, and caused significant changes in the flora of the pasture because of their closer cropping of the turf. Their burrows caused erosion and reduction in pasture area by the spread of soil at their mouths. The reduced burrowing habit of the rabbit population becoming established since the myxomatosis epizootic has limited this aspect of their damage to pasture and soils in recent times.

Other Lowland Soils in England

The implications of the history of chalk soils for those on other parent materials have yet to be pursued in any detail. The soft Jurassic limestones provide the closest parallel to the chalk soils, as is illustrated by the similarities between the Ascott-under-Wychwood and Beckhampton Road profiles. Soils on gravels formed from these limestones provide support for the idea of the widespread development of sols lessivés on soft calcareous materials. It is difficult to detect a loessic component in these soils against the background of the large residue of decalcification of the limestone, and it is possible that soils are formed in a partly loess-derived sandy loam layer capping the gravel terraces. The gravels are usually strongly cryoturbated, and, as in the case of the chalky materials, the boundary of decalcification follows the forms of festoons. The illuvial horizon is often a strong reddish-brown colour, produced by concentration of the iron oxides already present in the limestones, and it is a distinctive feature of sections exposed in gravel pits. It provides difficulties for archaeologists excavating ahead of gravel winning, since large areas are stripped by machinery down to this horizon and it then dries very hard, and the convoluted nature of the boundary with the C horizon introduces the same problems of identification of man-made features as are encountered in the cryoturbated chalky materials.

An indication of development of a sol lessivé and its truncation in neolithic times was provided by the soil buried beneath the henge bank at the Devil's Quoits on the Upper Thames gravels near Stanton Harcourt, Oxfordshire (excavation by Margaret Grey, report in preparation). Only small patches of the buried soil, in the deeper cryoturbation hollows, survived levelling of the bank, but what remained was the illuvial horizon of a sol lessivé, and where traces of bank material remained in place it could be seen that the upper horizons of the soil had been lost before the bank was built. On the rather flat surface of a river terrace, the most important agent of erosion, once the surface was disturbed, must have been wind, and the wide expanse of the flight of terraces of that area provides little obstruction to wind. Gravel archaeologists are only too familiar with the replication of conditions which the Neolithic ploughman would have experienced, when they are working in winter on stripped areas, sand storms adding to the discomfort of the wind. The upper part of the sandy loam soil, perhaps itself initially the result of wind winnowing the unvegetated gravel under periglacial conditions, would have reverted to a loose and unstable character when it became the eluvial horizon.

More complete sol lessivé profiles, retaining traces of eluvial horizon, on limestone gravels around Bredon Hill, an outlier beyond the main Cotswold escarpment, in south Worcestershire, are buried by plough accumulations, in an area of Iron Age and later occupation at present being investigated. Stages of soil profile development prior to the Iron Age are not dated, but Neolithic and Bronze Age artifacts indicate the presence of earlier people in the area (Britnell, personal communication).

The sol lessivé is perhaps the commonest profile type in lowland Britain. In the archaeological examples discussed so far, subsequent events have truncated or buried early-formed sol lessivé profiles. It

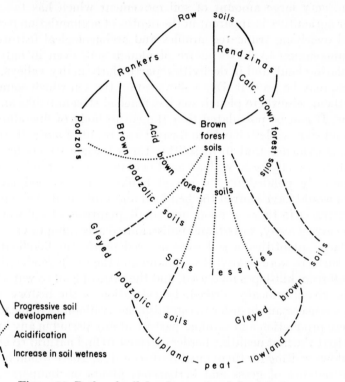

Figure 22. Paths of soil development and degradation.

remains to be determined whether the sols lessivés which give present day farmers problems of cultivation and drainage are as ancient. One possibility is that sols lessivés developed generally in neolithic times, but where illuvial horizons formed rapidly at shallow depth in soils of initially high clay content but low base reserve cultivation was soon abandoned because of water-logging and impossibility of cultivation and was only resumed in much later periods when greater power was available. In some cases, abandonment would have been almost immediate, as the good structure of a forest soil gave way, before much clay movement had taken place. Neolithic cultivators are likely to have tried many soils whose characteristics under forest were good but which did not persist under use. Another possibility is that some areas of brown forest soil remained in good condition through a long period of land use, and only developed into sols lessivés when the cyclic system of use and regeneration of woodland gave way to more intensive agriculture.

The very large amount of soil movement which has taken place under agriculture is apparent in the depths of accumulation frequently found overlying truncated profiles and archaeological features. It is not uncommon to find a metre of plough soil, even in only gently undulating land, and in declivities, particularly in dry valleys, several metres may be present. Dry valleys retain the soil which comes down into them; elsewhere plough soil has reached stream banks and flood-plains. It can be postulated that the entire body of fine alluvium in lowland river valleys has been formed as a result of agriculture. There is some evidence that in mesolithic times lowland rivers had broad, braided, gravelly beds, with areas of peat development in backswamps temporarily ponded up by gravel banks, but no floodplains. The rivers would have shifted the bed load and nibbled at the valley sides and terraces in times of flood, and small quantities of soil would have been carried away, sorted, and redistributed according to particle size, but large quantities of soil were not redeposited as floodloams until land surfaces were cleared of woodland and their soils rendered mobile, run-off from hillslopes increased, and the ratio of load to water volume in the rivers radically altered. Investigation of the history of flood-plains is necessary in order to establish the availability of meadowland for hay production and summer pasture at any period in any area. The idea that the first neolithic herders arrived to find natural riverine hay meadows waiting for them must be discounted. There would have been some patches of grass and herbaceous plants on temporary gravel banks, but the willow and alder carr fringing wide river bed probably merged rapidly into thick forest on the deep brown forest soils of the terraces and slopes. Subsequent dense and prolonged settlement on the river terraces, as demonstrated, for example, by Benson and Miles (1974) for the Upper Thames region, followed not from the presence there of the archaeological myth of "light forest on light soils" but from the fact that the early exploitation of these soils quickly reduced them to a thickness at which base reserves from the gravel were again available, and only as a result of cultivation did the eroded soil begin to accumulate to form a flood plain.

PART III
Soils of Other Regions

9

Soils of The Northern Continental Interiors

The soils of the more oceanic areas of the temperate zone are almost always moist. Under forest they would not dry out and when cultivated or intensively grazed they do not dry to great depth or for long periods. Cultivated soils are bare only in winter, when rainfall is high and temperatures low, and so are not exposed to drying then. As we go into the drier interiors of the continents we come to soils which receive much less rain in summer and which could dry appreciably even under forest and do so for long periods when under grass or cultivated. Harvesting of crops early in the summer leaves soil exposed to the sun during the hottest part of the year.

There are large areas of the northern continents where loess provides a parent material which dominates soil formation. In the more humid areas considered so far the loess is thinner and more patchy, and it has been involved in the solifluction and cryoturbation of the moister conditions which obtained there in late glacial times. It does exert an influence on the mineralogy and texture of the soils but it is only in the drier areas, where it was thicker to start with and has been less disturbed, that the characteristics of the loess itself still dominate the soil.

Loess and the redeposited loess of river terraces and flood plains provide a great depth of porous and easily weatherable materials which are rich in bases. The gain size is sufficiently small to provide a large surface area for attack on primary minerals but not small enough to clog up the drainage. In the areas of the thick loess blankets winters are cold and much of the winter precipitation falls as snow. In spring rapid warming provides an abundance of water from snow melt and a quick start to plant growth. Most of the rain falls in spring and early summer and then the rest of the summer is hot and dry. There may be more rain in autumn before the cold sets in again. Total annual precipitation in the zone of chernozem and brunizem soils is about 300 or 350 mm to about 650 mm.

Chernozem

Chernozem is the black soil of the steppes. In the porous soil grass roots reach much greater depth than they normally do in either drier or wetter regions, and they continue to extract moisture as the dry season advances. Each year a high proportion of the root mass dies, contributing organic matter throughout the depth of the root system. Abundant earthworms bring mineral and organic materials into close association and burrowing animals complete the mixing process. Humification is rapid in spring and early summer, and at the same time alteration of primary minerals is rapid. The profile is wetted throughout its depth but little soil solution drains away to the groundwater, except perhaps in early spring; most of the precipitation is lost again from the profile by evaporation and transpiration. The soil solution concentrates as the summer advances and the newly formed humus substances undergo concentration reactions, forming large, compact molecules, reducing the number of peripheral reactive groups and making them very resistant to microbial attack. There is abundant calcium, released from primary minerals, and the humus substances combine with calcium to form the very stable, dark coloured "calcium humates".

Clay minerals formed under these conditions of high cation concentration are predominantly montmorillonitic, giving the soil a high exchange capacity and a strongly swelling character. The shrinkage on drying brings clay minerals and humus substances into close association, humus molecules being trapped in micropores between crystals and even entering into inter-layer spaces, where they are inaccessible to micro-organisms. These montmorillonite–humus complexes are also very dark in colour.

The chernozem profile has an **A** horizon which is up to one metre deep, very dark brown to black, with a very well developed strong crumb structure. Whereas in a moist soil worms are usually only active in the upper 20 cm or so, as a soil containing decaying roots dries the worms retreat with the falling boundary of moistness while remaining active. The structure of the chernozem **A** horizon is produced by the combination of drying by grass roots, promoting shrinkage of the soil, and the casting of worms within it. Burrowing mammals are common, and their holes, filled with material of contrasting colour as they pass through horizon boundaries, are characteristic of the chernozems and are called *krotowinas*. They illustrate the way in which the burrowers contribute to mixing and to extension of the humic horizon downwards. Below the humic horizon there is sometimes a cambic **B** horizon, which is brown and has a blocky structure. The **A** or the **Bw** horizon merges into relatively unaltered loess of pale brown or grey colour, though it may be coloured yellow to some depth by oxidation of iron where the depth of wetting is greater.

Calcium carbonate accumulation is a constant feature of chernozems, and its presence appears to be essential to the long term stability of the profile. Fresh loess may contain free calcium carbonate in the form of

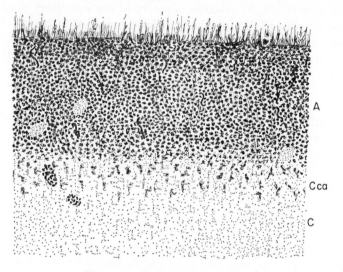

Figure 23. A chernozem profile.

particles of limestone derived from the rocks exposed to glacial and periglacial attrition. Where loess has been redistributed by water, lime taken into solution as the surface waters washed over the fresh glacial deposits was precipitated in the loess-loams as the waters subsided (Lukashev *et al.*, 1970). As soil formation proceeds, the primary and precipitated carbonate undergoes dissolution in the most leached and humic upper part of the soil but in the absence of prolonged leaching secondary carbonate is deposited lower down in the profile. Chernozems have calcareous horizons at greater or less depth depending on rainfall, within the **A** horizon or entirely within the **B** or **C** horizon. In the horizon of greatest concentration the calcium carbonate forms nodules, "loess dolls".

Variations in climate produce soils showing different degrees of development of the calcareous horizon, and at different depths. In drier areas the **A** horizon is paler and thinner, there is less likely to be a **Bw** horizon and the **C** horizon shows less coloration by iron oxides. Wetter areas have soils showing maximum development of the **A** horizon and a strong tendency to develop the **Bw** horizon as the underlying material is wetted more frequently. Meadow chernozems are a variant which occurs in flat country and in valley bottoms where the soil is affected by high groundwater levels and gleying sets in.

Chernozems have a very high agricultural potential. The high content and stable form of humus, the abundance of weatherable minerals and the high exchange capacity of the secondary minerals all combine to form a buffer against depletion of plant nutrients. In spite of climatic aridity, cereal crops can, like grass, extract sufficient moisture for abundant growth, and they ripen rapidly in the high summer temperatures.

Brunizem

The transition to soils of more humid regions and thinner loess deposits is by way of the *brunizems*, soils similar to the chernozems in development of the dark, humic **A** horizon, but more strongly leached. The **B** horizon is well developed, either as a cambic horizon or, in the more fully developed and characteristic profiles, as an illuvial clay horizon, and there may be a calcareous horizon.

The degree of leaching indicated by the **B** horizon, higher than that appropriate to the chernozem-type **A** horizon, indicates a history of change of vegetation and moisture regime of these soils. In America they are the prairie soils and their vegetation is mostly grass. Pollen analysis suggests encroachment of grassland into a forested zone (West, 1961), and remnants of forest on steep valley sides where bison do not graze suggest that the grassland may be the product of grazing, grazing land being extended into the forest zone under the influence of hunters and fire. Apart from the valley sides, the prairie soils are only forested where their parent material is calcareous or particularly base-rich (Soil Survey Staff, 1960), which might suggest that only here could forest regeneration occur, after reduction of bison herds in recent times.

In Europe, Stepanovits (1971) describes soils of brunizem type, though without clay eluviation, which occur on the fringes of the Great Plain of Hungary and in smaller relatively dry basins within the mountains around it, where loess is known to have been forested. He states that these profiles have been produced by forest clearance and cultivation, development of the chernozem type of **A** horizon being a direct result of the change in leaching regime when trees are removed.

There has been much discussion among Russian soil scientists as to the history of the brunzems. In Russia, these soils occur on loess where it is forested or is known to have been so, and it is generally held that deciduous forests on loess are subsequent to a grassland phase. It is reported from one direct observation that downward movement of moisture was not greater under forest than under grass in the same locality (Ivanova *et al.*, 1970). However, Choni (1971) reports a lowering of the level at which calcium carbonate occurs and a reduction in the thickness of the **A** horizon as a result of afforestation of a chernozem and it is generally found that the amount of calcium carbonate in the upper part of the soils is less in the forested soils, and the importance of the maintenance of free carbonate in the profile for the stability of the chernozem is stressed by Russian pedologists. The forests are therefore seen as inducing degradation of chernozems.

The differing interpretations of the brunizems are due to their development in regions in which the transition occurs between the relatively stable but very different systems of chernozem supporting grassland and brown soils supporting deciduous forest. Soil profiles are thus particularly sensitive to changes of climate and to changes of vegetation induced by man. Human exploitation has been intense, under efficient and well organized hunting in America and under a very long period of agriculture, taking over from hunting, in Europe, and such exploitation has resulted in an increase in the stability of the soil profile by the formation of the chernozem type of **A** horizon and reduction in leaching, allowing use of the soils to continue, limited by lack of moisture in summer but not by loss of plant nutrients. Changes in the opposite direction, resulting in degradation of the chernozem profile, may be the result of a slackening of human pressure, or of a change in climate, with reversion to a forested landscape after a period during which grassland was maintained.

Grey Forest Soils

In the northern part of the chernozem zone soils which have been called *grey forest soils* or *degraded chernozems* make the transition from the grassland through the deciduous forests and the mixed forests to the podzols of the coniferous forests. Through this zone the loess gives way to glacial, fluvioglacial and lacustrine deposits, many of which are calcareous, and the grey forest soils are found on any of these materials as well as on the loess. The transition is marked by a reduction in thickness of the humic horizon of the chernozem and an increase in leaching of the upper part of the soil which is not carried right through the profile, so that a grey eluvial horizon and signs of podzolization occur above an illuvial clay horizon and often a calcareous horizon.

In both North America and Eurasia it has been suggested that these soils are degraded chernozems, produced by a spread of forest over the steppe or prairie as the climate became cooler and more humid after the postulated post-glacial climatic optimum. They would be comparable in their genesis to the brunizems, at least in the sense in which the Russian brunizems under forest are interpreted. Some pedologists include them with the brunizems in classification (Fitzpatrick, 1971). However, the brunizems are formed under relatively warm conditions compared with the grey forest soils, with a long growing season and an abundance of plant residues in the forest, a high rate of weathering for much of the year and not a very pronounced dry season. The humic horizon is maintained and the rate of turnover of organic material and the rate of weathering are sufficient to prevent leaching in the upper part of the soil becoming too drastic. The grey forest soils are found in regions where there is a long season of snow cover, during which neither weathering and organic activity nor leaching occur, and a short growing season followed by a dry summer. During the short wet spring and autumn the effects of leaching are severe but it does not persist long enough to clear the profile, and the calcareous horizon persists beneath a soil in which podzolization is active. The grey colour of the soil is produced by the visibility as a white powder of the silt grains on ped faces in the leached horizons, showing that as well as eluviation of the soil colloid, sufficient podzolization has been active to clean the mineral particles of iron oxide coatings. More strongly podzolized soils are sometimes distinguished as *grey wooded soils*, and in these the humic horizon is mor, whereas in the grey forest soils it is still the mull form.

In Russian soil terminology *sod-podzolic soils* are recognized. These are soils showing some degree of mull humus form in a podzolized profile, and the greater content and greater stability of humus in these compared to the normal podzols is attributed to their development under grass or at least under a forest with a grassy forest floor. The sod-podzolic soils are found predominantly in the southern part of the forest zone, and are a further stage in the steppe-forest transition, giving way northwards to the ordinary podzols.

Some of the sod-podzolic soils have a lower humus horizon which has been the subject of considerable discussion among Russian soil scientists. The horizon is darker than the surface mull humus and lies below the eluvial horizon. It was formerly thought that it was an illuvial humus horizon, part of the present profile, but recent work has shown that the humus is of more condensed form than that of the surface humus horizon and more akin to the chernozem humus forms (Kuz'min, 1969), and radiocarbon dating has shown it to be very much older than the surface humus (Tolchel'nikov, 1970; Dobrovol'skiy *et al.*, 1970). It is concluded that these soils are another form of degraded chernozem. The earlier soils were meadow chernozems, and the lower parts of their **A** horizons have survived while the upper parts have lost their humus and given way to podzolization under increased leaching as forest encroached upon grassland.

The evidence for spread of forest into the steppe zone along its northern margins is considerable. The grey forest soils and the lower humus horizon of some sod-podzolic soils show increased leaching associated with a change from grass to forest vegetation. Continuing high levels of leaching are shown in many podzolic and grey forest soils by a tonguing of the eluvial horizons into the illuvial ones, the illuvial horizons being destroyed at their upper boundaries or moved downwards by further destruction and translocation of clay after they have formed. These northern areas have too short a growing season for successful crop production under primitive agricultural systems and until recently hunting has been the main human activity there. The encroachment of forest is usually attributed to the onset of cooler and more humid conditions of the sub-Atlantic period, the length and intensity of the summer dry period being sufficiently reduced to allow tree seedlings to survive where they had previously not been able to. The possibility remains that the spread of forest could be due to the reduction in the herds of wild herbivores throughout the steppes and prairies. The animals are migratory in habit and as the human population increased in the agricultural and herding areas they would increasingly be hunted, partly for their food value but also because of the damage they do to the crops and the need for the grazing land they occupy. The effect of reduction in numbers would have been noticeable in the less populated areas to which the animals migrated as the dry season advanced.

Whereas changes in vegetation zones which are entirely climatically induced may be expected to be consistent, any of which are the result of interference by man may be somewhat variable in distribution and timing. Contrary to the widespread evidence for spread of forest is a report of a buried soil beneath a burial mound in the Kazan steppes from which no significant change in the position of the forest-steppe and steppe zones were found (Ivanova *et al.*, 1970). Some signs of a change in the opposite direction are reported, in the accumulation of humus and calcium in the upper horizons of grey forest soils, which would indicate a reduction in leaching and possibly a change from a forest to a grassland vegetation (Ivanova *et al.*, 1969). The poly-genetic soils of the fringes of the chernozem zone are still only beginning to be understood.

Chestnut Soils

In the drier interior regions of the Northern continents soils are rarely wetted throughout and are dry for the greater part of the year. Periodic leaching of bases to some depth leads to formation of a well developed calcareous horizon, and there may be a cambic **B** horizon, which is brown to reddish-brown depending on the degree of dehydration of iron oxides, becoming redder under higher summer temperatures. The **A** horizon is dark, the humus substances forming the dark coloured complexes with calcium and clay as in the chernozems, but it is much thinner since the soil is not wetted to so great a depth. The Russian name for these soils, *kastanozem*, has been widely adopted.

The chestnut soils occur in North America to the west and south-west of the chernozems, in the rain shadow of the Rocky Mountains. In Eurasia they occur to the south of the chernozems and extend from the Black Sea right across Asia. The predominant parent material is loess. Rainfall of about 200–350 mm annually falls mainly in spring and autumn, and the natural vegetation is grass and scattered spiny shrubs. In much of the region of the chestnut soils the winter is very cold and soil processes are at a standstill. Where winters are rather less cold, plant growth and organic processes in the soil may be inhibited but leaching still active. These soils are less strongly dominated by calcium; in Russia they are recognized as a distinct group, the *grey cinnamon-brown soils*.

In the warmer and drier parts of the chernozem zone and in the zone of chestnut soils the soils become increasingly susceptible to damage by deflation when their stability is reduced by cultivation and by over grazing. The deep and strongly structured humic horizon imparts a very high degree of stability to the soil, but in the drier areas it becomes less deep, and as the dry season lengthens and the spring comes earlier in the more southerly regions the soils are exposed for longer periods after the crop is taken, and in the grazing lands the grass dries up earlier and animals' hooves break up the surface peds. Some of the soils have been severely deflated, leaving shallow sandy soils which cannot hold enough moisture to extend the season of grass growth sufficiently for the humic **A** horizon to be maintained. An example is given by Gayel' and Malan'in (1971). They describe soils on the sandy terraces of the Don steppes which have been so severely deflated that in some places deflation hollows bring the surface down to within the capillary fringe of the water table, allowing birch trees to grow in newly forming soils.

Sierozems

Conditions still drier than those of the chestnut soil zone reduce the profile to a thin greyish-brown **A** horizon resting directly on a calcareous horizon. Plant growth and decay occur very rapidly after the brief and occasional rains, and the rate of organic activity is then so high in the warm soil that a very high proportion of the plant residues is totally mineralized within the same short cycle as they are produced. There is no opportunity for the formation of stable humus forms before any newly formed humus substances are used up. The thick calcareous horizons are probably relict of deeper soils, perhaps chestnut soils, which were exposed to erosion by heavy grazing. Vinogradov *et al.* (1969) report ancient soils in the Kyzyl Kum desert which have suffered drastic deflation since Neolithic times. They postulate a wetter period during the past-glacial for the formation of these deeper and more weathered soils, but they could be relics of Pleistocene environmental conditions, and have remained stable until their protective vegetation was reduced by grazing.

10

Soils of The Mediterranean Region

The mediterranean climate is one of mild winters during which most of the year's rain falls and warm, dry summers. In the Mediterranean Basin rainfall is strongly related to altitude and aspect. The coastal plains are dry and rainfall increases markedly as one goes up the mountains and is particularly abundant in the mountains backing the west and southwest facing coasts which receive the full influence of winter depressions coming in from the Atlantic or originating within the basin. The high mountains receive much of their winter precipitation as snow, but tend to have rain in summer as well. The hinterlands, particularly the plateaux of Spain and Turkey, are markedly arid, and in the south and west of the basin the arid regions of North Africa and Western Asia merge into the deserts.

The region now has few and restricted areas of natural vegetation, and the soils are mostly severely eroded and much modified by prolonged cultivation and heavy grazing. It is difficult to establish what would be the natural distribution of soils and vegetation, and to work out which characteristics of the soils are the result of processes in harmony with the present climate, which are relict, and from which periods of the Pleistocene, and which are the result of interference by man.

Red and Brown Mediterranean Soils

The soils which are regarded as typically mediterranean are certainly relict. They occur throughout the region which now has a warm temperate, sub-humid climate with marked summer drought, and are often so eroded as to remain as only thin stony soils. In many places their profiles have been decapitated, and younger soils are forming in fresher material deposited on top of the remaining horizons. Their materials form a high proportion of the soil sediments which occur on coastal plains and river terraces.

Red mediterranean soils, widely known as *Terra Rossa*, occur on hard limestones and other calcareous rocks. The full, uneroded profile has a reddish-brown **A** horizon and a textural **B** horizon which is reddish-brown to red and has a strong blocky structure with very marked coatings of illuviated clay in pores and on structural faces. *Brown mediterranean soils* occur in similar situations on a very wide range of non-calcareous rocks and have a similar profile but a browner colour. The **B**$_t$ horizon in the brown soils may become a hard pan in its lower part, and this characteristic points up the difference between these soils and the clearly related sols lessivés. Under higher temperatures more mobilization of silica can occur. The silica moves downwards and in the intense dry season may be deposited as a cementing agent in the lower part of the profile.

A degree of leaching leading to intense clay migration in limestone soils suggests a high degree of maturity, established during a long period under a climate which provided an abundance of moisture but also seasonal drying intense enough for the dehydration of iron oxides, which enhances the red colours, to have occurred, though it is probable that ageing alone may cause reddening of some soils. The uniformity of soil type over rocks having a great variety of texture and mineralogy also points to a long period of soil formation, the tendency of climatic factors to eventually predominate over those due to parent material, which is seen too in the other parts of the world in which abundance of young surfaces and immature soils does not obscure it, being well established. These are probably soils formed under interglacial conditions. The vegetation would have been the various types of deciduous forest and evergreen, more or less sclerophylous forest which today occur on the few mountain slopes where they have not succumbed to the depredations of man and goat, but which would then have covered the lowlands too. The present interglacial climate could support forest in most parts of the Mediterranean Basin; in the drier parts the reduced rate of transpiration of evergreen trees and the cover they provide reduces the period for which the soil is dry in summer and enables them to survive periods of extreme aridity, and the canopy protects the soil from the force of rain of high erosive potential which is characteristic of such areas. There is evidence that forest cover was widely established in the early part of the present interglacial (e.g., Greig and Turner, 1974).

Colluvial Deposits and Calcareous Crusts

Much of the stripping of these relict soils from hillslopes and their deposition on flatter land nearby must be attributed to a breaking up

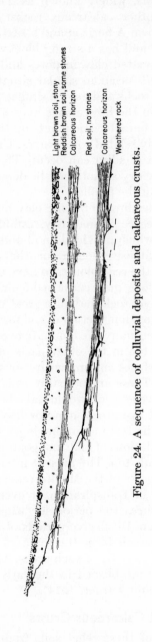

Light brown soil, stony
Reddish brown soil, some stones
Calcareous horizon
Red soil, no stones
Calcareous horizon
Weathered rock

Figure 24. A sequence of colluvial deposits and calcareous crusts.

of the forest cover at certain stages of climatic instability, aridity or coldness during glacial periods. Colluviation alternated with periods of stable soil formation, and successive layers of colluvial deposits contain less of the highly weathered material, as the vulnerable areas of mature soils were reduced and the younger soils in their turn suffered erosion. The common association of calcareous crusts with colluvial deposits is largely attributable to catena effects (Ruellan, 1965), though a number of other encrustation mechanisms have been discussed (Durand, 1959). The lowland soils, and in particular those on the terraces and lower slopes which were the receiving areas for the stripped soils, received base-rich moisture from the soils being leached further uphill during periods of stability. The calcareous horizons formed at a depth in the soil or deposit related to the relative importance of downhill movement of soil solution and direct leaching from rainfall. The rapid increase of rainfall with altitude leads to predominance of direct leaching in higher positions, while lower down as the base content of the percolating water increases and the importance of direct leaching decreases, precipitation of calcium carbonate occurs. In the more humid areas, calcareous horizons occur at some depth and only in a narrow zone near the coast, whereas in more arid parts they formed nearer the surface, perhaps even from trickling of base-rich moisture over the surface, and extend further uphill.

Complex sequences of colluvial deposits and calcareous crusts occur in a large number of places around the Mediterranean Basin and the Atlantic coast of North Africa, and their relationship to coastal features and river terraces provides a means of relating the environmental conditions of which they provide evidence to the geochronology of the Middle and Upper Pleistocene. Where palaeolithic sites or fossil hominid localities have been found associated with the deposits or the coastal features, as in Morocco, much work has been done. The Moroccan deposits, which present sequences of soil and gravel deposits, with calcareous horizons and encrustations, interfingering at the coast with sequences of dunes whose formation and weathering could be related to changes in sea level, were discussed by Gigout (1951), Choubert et al. (1956), Choubert (1962) and Biberson (1962, 1963, 1970). Deposits in Algeria are summerized and discussed by Vita-Finzi (1967) and those in Libya by Hey (1962 and 1963). Butzer interprets a complex sequence in Mallorca (Solé-Sabarís, 1962; Butzer and Cuerda, 1962; Butzer, 1963) and has based much of his subsequent writing on Pleistocene climatic conditions in the region on the conclusions he drew from it. In France, the possibility of more direct correlation with the Pleistocene glacial sequence is introduced by the occurrence of similar types of deposit in sequences in which loess participates or on river terraces directly related to behaviour of the Alpine glaciers (Bonifay, 1962). Deposits in Greece are described in Dakiris *et al.* (1964) and Higgs and Vita-Finzi (1966), the same deposits being differently interpreted by two of the people who have contributed to the description and discussion of such deposits elsewhere. Colluvial deposits and calcareous crusts in the island of Gran Canaria are described in Limbrey (1968).

Interpretation of the stratigraphic sequence and the characteristics of the soils in terms of climatic and vegetational sequences range between the approach of Butzer, who attempts to correlate every feature in a very complex stratigraphy with a universally applicable change of climate, and that of Vita-Finzi (1967), who sets aside much stratigraphic detail and in producing a simple but broadly valid sequence does not exploit the opportunity of investigating variations due to local topographic and climatic differences. The formation of deposits such as these has nowhere been observed in progress in recent times, so interpretation must be somewhat hypothetical. Whatever the processes were, they no longer happened after the end of the last glaciation, but they had contributed greatly to the distribution and characteristics of the soils and soil parent materials available for exploitation by the first farmers of the region.

Soils of More Arid Areas

In the drier parts of the Mediterranean Basin, red and brown mediterranean soils, and the deposits derived from them, give way to reddish-brown and brown soils of arid type and chestnut soils, similar to and forming a continuum with those of the warmer parts of the dry continental interiors and those of the arid sub-tropics. These too are probably largely relict soils. Profiles are often eroded down to a hard calcareous horizon, which in the complete profile formed below a thin **A** horizon, and sometimes a cambic **B** horizon of reddish-brown or brown colour.

Where the arid areas extend towards the deserts of Western Asia loess contributes much material to the soils. Changes in extent of the deserts and in the strength and direction of the prevailing winds during the Pleistocene and Holocene periods are reflected in sequences of buried soils and windblown layers. Those investigated in Israel are discussed in Dan and Yaalon (1971). Correlation between sequences of buried soils in which the degree of development and the characteristics of each member vary from place to place according to the climatic characteristics of the period in which they formed and according to topographic relationships is often difficult, as is their dating. However, the mapping of buried surfaces and investigation of their catena relationships and the changes in their profiles across the areas in which they occur has great potential for area investigation of changing environmental conditions. These accumulative sequences of buried soils are perhaps easier to interpret than the sequences in which erosion and deposition of existing soils of varying maturity adds to the problem the complexities of polygenesis, and there is a greater chance of finding the full expressions of catenas and investigating synchronous surfaces across a region, through a range of climate.

Recent Soils

The soils which are developing under present conditions can be identified by their being on surfaces produced by erosion or deposition in recent times. These young soils on eroded slopes and on sediments have much in common with soils of comparable age in the formerly glaciated and periglacial zones, forming a climatic progression with those of adjacent regions. The mountains, with their abundance of young surfaces and a relatively cool and humid climate have associations of regosols and raw and skeletal soils with rendzinas on calcareous rocks and with rankers on other hard materials, and in more stable situations progressions to brown forest soils on the one hand and to podzolized forms on the other are seen. Podzols are restricted to the coolest and most humid zones, and the balance between the brown forest soils and sols lessivés and the acid brown forest soils is in the favour of the first two, as would be expected in conditions of greater warmth and marked aridity in summer. Calcareous brown forest soils are common, especially in the lowlands. Had the forests not been disturbed, the progression towards red and brown mediterranean soils might have been being achieved by the increasing maturity and eventual leaching of these soils, with the catena relationship between hillslope and plains expressed in the development of the calcareous horizon.

In the areas of denudation, xerophytic scrub grows on the eroded remnants of the former soils, represented by the horizons of greater coherence such as illuvial **B** horizons, calcareous crusts, or the partially weathered rock of the **C** horizon, or on redeposited soils and alluvial deposits. Young soils of arid aspect are often developing in these situations on the partly weathered material and rock debris distributed over the surfaces in the latest stages of erosion. The profiles are therefore polygenetic not only in terms of climate and vegetation but also in parent material.

Vertisols

Distinctive black clay soils have developed on fine grained sediments of basic, often calcareous, material in areas of low relief and in valley bottoms and basins of sedimentation. Some of them may be relict, but many occur on recent fine alluvium and must be recent soils. Similar soils occur in tropical and sub-tropical regions, on basic hard rocks as well as sedimentary material, so they are really *intrazonal* soils, but since they are relatively common in the Mediterranean Basin they will be described here.

The essential climatic factor in the formation of vertisols is a marked dry season. The fine grained basic materials release abundant bases during the period when the soil is moist. The combination of uniform fine grain size and low lying situation reduces percolation, and leaching is further inhibited as the soil develops by the strongly swelling habit of the clay, almost all of it montmorillonite, which forms in such base rich environments. Concentration of the soil solution and contraction of the clay during the summer drying brings clay and humus into close association, forming dark coloured complexes, as in the chernozems. Contraction of the clay on drying opens cracks in the ground and forms a surface layer of loose peds, which tend to fall down the cracks, or be washed into them by the first rains, and as the soil is moistened and swells again these peds are incorporated into a lower level in the soil. Pressure on swelling, increased by the greater amount of material now present in the lower levels, produces a strong blocky structure, and is relieved upwards by sliding movements which produce slickensides. This process results in a continual churning of the soil, introducing humus throughout the depth of the **A** horizon to give a mixed, uniform layer which is limited below rather abruptly at the depth to which drying reaches. There may be a calcareous horizon at or within the top of the **C** horizon.

The habit of breaking into separate peds at the surface, which is essential to the churning process, is called "self mulching" since it produces a loose layer with large pores similar to the mulch which cultivators seek to achieve as a seed bed; it protects the underlying soil from too rapid drying before the crop has grown enough to provide cover by breaking up the fine capillary system by which water rises as it evaporates at the surface.

The name vertisol is based on the root -vert-, indicating the turning action. These soils are known under some systems of nomenclature as *grumusols*, and in Yugoslavia they are called *smonitzas*. In some parts of the world vertisols have been known as chernozems, because of their black colour, which is derived from a similar humus form, but they are distinct in their shallowness and in the high clay content, which inhibits the percolation of water, in strong contrast to the porous and freely draining character of the chernozems.

Exploitation of Soils

Exploitation for pasture and crop growing extends back in some places to a period so close to the end of the Last Glaciation that the establishment of mature soils under forest little disturbed by man and in harmony with a climate similar to the present cannot be assumed. Where forest cover did not survive the climatic conditions imposed by proximity of the ice or by the aridity which may have held sway in areas further away from those which were glaciated, its re-establishment would have been quickly achieved by migration of species from refuges within the basin itself onto soils remaining perhaps only partly eroded from the previous forested phase, and onto soils rapidly forming on sediments containing much soil material or already weathered rock debris. The post-glacial soils had many inherited characteristics, and the early farmers introduced their disturbing activities into a situation in which the equilibrium between climate, vegetation and soils was not established.

The heavily leached profiles of red and brown mediterranean soils have little potential for agriculture. Their clay is predominantly kaolinitic and their cation exchange capacity low, and they have little or no reserve of weatherable minerals in the upper part of the profile. Nutrient reserve is therefore low and once the nutrients made available from the cleared vegetation had been used up, or lost by leaching or by erosion following clearance, cropping potential would have been minimal without heavy manuring. Once the structure of the forest soil was lost, the soils would have been extremely difficult to manage under cultivation and extremely susceptible to erosion. Nowadays, where they remain as soils of any depth on hillslopes they are kept under pasture, truncation to the more coherent illuvial clay horizon often having occurred, perhaps consequent upon clearance and possibly short periods of cultivation at an early stage in their exploitation.

The slopes from which the soils which went to form the colluvial deposits were stripped, and the deposits themselves, in whose upper layers less heavily leached soil material and partly weathered rock debris predominate, provide stony but potentially fertile soils, which are exploited today by terracing on slopes and by irrigation on both slopes and deposits. Availability of water is the limiting factor for utilization of soils throughout the region, and the terracing not only allows control of irrigation water, but also conserves rainfall and prevents erosion of the slope soils, exposed as they are to highly erosive rain when under cultivation. The early history of terracing has been very little investigated; there is evidence for the practice in northern Italy during the Bronze Age, in hilly country whose slopes were too steep to provide cultivable land without terracing (Barfield, 1971).

Much agriculture is now carried on on the lower terraces of river valleys and coasts, whose deposits were laid down in a period subsequent to dissection of the colluvial deposits. There has been much discussion as to the reason for deposition of the material of these lower terraces, which is datable in many places by its burial of buildings of the Roman period, climatic causes and the result of man's agricultural activities being alternative hypotheses (Vita-Finzi, 1969; Bintliff, 1975). Availability of water is better here than on the older, higher terraces, since streams are often nearby, and the water table may remain accessible to plant roots through the dry season. It is worth remembering that at a period when the older terraces were in a similar relation to their dissecting streams there would have been good potential for agriculture on them, utilizing ground water or simple irrigation. The chronology of their dissection, which may have varied considerably from place to place, needs study; some of the settlements which may have been so situated as to be able to utilize their agricultural potential must, of course, have been destroyed by the process of dissection. It has been suggested that early in the history of cereal cultivation crops were grown on naturally watered soils beside streams and many such situations may have been available at the time when agriculture reached the Mediterranean Basin.

Cultivation of hillslopes is possible on rendzinas, whose structural stability gives them better resistance to erosion than the relict soils of kaolinitic nature, and whose base reserves are high. The chronology of the rejuvenation which gave rise to rendzinas must be established in order to determine whether or not these were available in any area to the people living there at any particular time. Some of them are formed on the slopes most severely stripped by the colluviation process and have therefore been available throughout the period of agricultural activity, but others have been created through cultivation of the less eroded profiles of the older soils, and the soils from which they developed must have passed through a period when they were poor in bases and unstable in structure, before erosion reached the fresh base reserves of the C horizon. Some of the material shifted from the slopes which now bear rendzinas is now providing better opportunity for agriculture in the form of the lower terrace soils.

Use of Soil Materials in Building

An interesting corollary of the ubiquity of relict soils and the products of their erosion in the Mediterranean Region is the use of these materials for building. The relative distribution of timber and unfired mud as building materials has been much discussed, and it is often assumed that timber is used where rainy climates make unfired soil materials unsuitable. However, mud has been much used in building up to recent times even in England, in certain areas, and survives a rainy climate well if water is prevented from soaking into tops of walls by suitable roofing and eaves. The mud walls around farmyards in Wiltshire, for instance, are thatched. The critical factor in determining the use of unfired mud is the nature of its clay component. Swelling clays, characteristic of immature and unleached soils, will cause disruption of the dried mud as soon as it is moistened, whereas kaolinite, characteristic of mature, highly weathered, heavily leached soils does not change volume on wetting and drying. In England, mud is used for building where kaolinite of geological origin, or local occurrences of relict soils such as the clay-with-flints, contribute to soil parent materials, or where calcareous soils and muds are available: the crystallization of calcite on drying provides a framework within which shrinking and swelling can be accommodated. Otherwise, building in timber is characteristic of those areas whose mature soils were destroyed by glacial and periglacial processes, and tells, built up by continual rebuilding on the same site with unfired mud, occur in the area in which relict kaolinitic soils and alluvium derived from them are widespread.

11

Tropical and Arid Zone Soils

Throughout the tropics and sub-tropics, old surfaces with polygenetic and relict soils abound, and soils on erosion surfaces of different ages are found in close juxtaposition. Since soil formation is rapid under high temperatures and high humidity, many soils reach an advanced stage of weathering in response to the climatic conditions under which soil formation begins on any particular erosion surface, and little potential remains to be developed under changed conditions. Many soils are therefore relict monogenetic soils rather than polygenetic, though they may have features such as concretionary horizons which owe their origin to later conditions, especially where catena effects come into play.

The concept of the catena was introduced by Milne (1935) in interpreting the distribution of soils in Kenya in relation to topographic position, and it has been applied more often to tropical soils than to others because the intense leaching occurring in the humid tropics brings moisture bearing a large load of soluble substances into soils in lower positions, and changes in base status or concentration of the solution by evaporation lead to the deposition of the less soluble components within the lower lying profiles. However, since different elements of the landscape are often surfaces of different ages, simple monogenetic catenas are less common than complex mixtures of catenas and chronosequences.

Change of climate in the tropical and sub-tropical regions is probably a matter of shifts and changes in widths of the belts under different rainfall regimes, and perhaps changes between greater seasonality of rainfall "displuvial", and more uniform distribution through the year "isopluvial" (Bernard, 1962a and b). On either side of the equatorial belt with rain most of the year, the zone of tropical rain forest, are the forest savannah and tall grass savannah zones which have two rainy periods, as the sun passes overhead in each direction, and considerable aridity in between. Increasing aridity and the loss of first one wet season and then the other through the steppe and desert steppe zones leads to the desert belts where the stream of descending air prevents the normal convectional rain from falling. On the other side of the deserts are the winter rain, Mediterranean, climatic zones.

The concept of tropical pluvials is suffering a welcome decline at present, as the people studying Pleistocene environmental conditions in these regions, particularly in East Africa, are concentrating on detailed studies of the sedimentary deposits in each area rather than making the sort of generalizations for the whole region, based on lake levels in highly unstable situations, and correlations with the high latitude glaciations which have been so misleading in the past. However, while lacustrine deposits should in the end yield a detailed picture of the sequence of past environmental conditions in any one area, evidence from soils can give a broader view of major climatic trends.

Ferralitic Soils

In the perhumid equatorial belt, temperature and rainfall are high all the year. Mineral alteration proceeds very rapidly and the level of organic activity can be very high while residues are available. Under the natural vegetation of tropical rain forest, mineralization of residues is rapid and complete, so soils contain very little humus, and the living vegetation contains a very high proportion of the total nutrient content of the soil/plant system. Bases released from primary minerals and organic residues are rapidly leached from the entire profile, and a lot of silica goes into solution too. Such clay minerals as form are kaolinitic, and the soils have a low cation exchange capacity. Rapid alteration of minerals and the removal of much of their components leaves behind iron, none of which can be accommodated in any of the secondary minerals which can form under these conditions. It remains in the soil as goethite and gives a strong red colour in the absence of humus to mask it.

Under extreme leaching, so much silica goes into solution that more alumina is released than can be accommodated in kaolinite in conjunction with that which remains. Gibbsite crystallizes, and in the end a soil consisting largely of gibbsite and goethite is formed. Such drastic leaching, taking silica even from previously formed kaolinite, may only be effective in the upper part of the soil, and these soils commonly have a lower clay content in the thin humic horizon than below. The transported silica may be deposited lower down, as the ionic concentration of the descending moisture increases, and forms a coating to peds and particles in the lower part of the soil and the upper part of the parent material.

Some of these red kaolinitic or gibbsitic soils have been forming for a very long time, a large part of the Pleistocene or even longer, and can be very deep. Below them the parent material is softened and oxidized to greath depth. The rotten rock, in which only the more resistant minerals remain, embedded in a kaolinitic matrix, is called *saprolite*. If exposed by erosion it has already gone a long way towards soil formation, and soils formed in the next cycle take on the nature of highly weathered ferralitic soils in a much shorter time than if they had started from unaltered rock.

The deposition of silica in part of the profile can create a hardened horizon. Where the silica concentration is not great enough for it to be deposited while leaching continues and the soil remains moist, the hardened layer may not develop unless the soil dries out.

Silcrete

The silica mobilized may be deposited in the same profile, or may be carried away with the groundwater. In some drainage systems it will be lost to the ocean, but in other areas the topography causes the siliceous water to seep into the soils of the lower slopes and depressions. Such areas of internal drainage may also have accumulated soluble salts, and in the resulting conditions of high alkalinity, and with evaporation concentrating the solution, silica may be deposited as a cementing agent or a layer at or near the surface. This is the silcrete of large areas of the interiors of the southern continents. There is some controversy as to the environmental conditions under which it formed, and in particular as to its age. There is no reason to suppose that the process is not going on now in areas where high temperature and effective leaching occur in suitable topographic situations, but the areas in which it is now exposed are often barren deserts, and opinion seems to centre on its formation largely in the Pliocene (Stephens, 1971). Silcrete can be suitable for making stone tools, and has been used for this purpose in southern Africa.

Laterite

The associated process of laterization, forming a *lateritic crust*, can occur in any soil from which much silica has been lost. The iron in such soils is normally immobile, but in conditions of fluctuating groundwater levels, reduction of iron occurs when high levels of organic activity use up the oxygen in a waterlogged soil. Iron goes into solution and is redeposited as the groundwater sinks and the soil is aerated again, by which time it may have moved, either upwards with the rising water, or downwards, in the same profile or into a lower position in the landscape. A ferruginous crust is built up by repeated deposition in a particular horizon, and the migration of iron towards centres of crystallization produces first a mottled and then a reticulate pattern. A vesicular structure is formed as the encrustation proceeds, with the loss in volume of the patches depleted of iron enhanced as increased leaching concentrated through them results in destruction of kaolinite. Silica mobilized by this means can contribute to the encrustation, being deposited in complexes with the iron oxides.

The horizon from which much of the iron has gone is pallid, with mottles of oxidized iron centred on remains of ferromagnesian minerals not yet totally destroyed. The *pallid zone* merges with the zone of iron concentration by way of a mottled zone, where the ferromagnesian relics become centres of crystallization. Lateritic soils show a variety of expression of pallid, mottled and ferruginous horizons, whose relative position depends on direction of movement of groundwater. Rise of water into a previously formed ferralitic soil can be due to change in climate or to a change in drainage pattern, which may be due to erosion following a change in climate or may simply be the result of erosion progressing to a critical point in relation to water movement somewhere in the area. The concentration of iron occurs in the oxidized zone of the capillary fringe above a fluctuating groundwater level, and the pallid zone lies below it. If water seeps from a saturated soil into a lower position in the landscape, where it comes into oxidizing situations closer to the surface, the iron concentration is in the soils of lower slope and bottom positions, and the pallid zone is further upslope.

A ferruginous horizon may reach a high degree of induration while moist, particularly if silica is participating in its formation. It can, however, remain soft until it dries, and then dehydration of the iron oxides is involved in the hardening process. Drying can be intermittent, and due to the same fluctuations in water table as are forming the horizon, but it may not happen until a change in environmental conditions exposes the profile as a whole to drying or results in erosion which exposes the horizon of iron concentration. When exposed at the surface, either already hardened or hardening as it dries, it forms a lateritic crust. It may be exposed in profile, so that as an eroding slope retreats, the lateritic material, already hardened or hardening progressively just ahead of the point of erosion, forms a shelf which, as the softer material below is removed breaks into lumps which lie as gravel on top of the deposits or are left behind as finer particles are transported further.

Exposed lateritic and siliceous crusts, the gravels derived from them and the pallid and mottled materials associated with them have provided parent materials for later cycles of soil formation or sites for further encrustation. In the interior of Australia and Southern Africa stable geological formations have undergone many cycles of soil formation and erosion and the succession of surfaces of different ages carrying crusts and soils of different types hold information which should eventually allow reconstruction of climatic zones at different times during the Pleistocene. The volume of literature, particularly referring to Australia, is enormous. Butler (1959) introduced the term *K cycles* for cycles of soil formation and erosion, and many workers since then have extended recognition and correlation of K cycles in the Australian landscape.

The word laterite was first used in India. Moist, soft material is dug out of the horizon of iron concentration, and cut to shape as building blocks. As it dries in the sun the dark red vesicular material hardens irreversibly. "Laterite", "lateritic soil, or crust", and "latosol" have been used for a large number of different indurated soil horizons and the soils in which they occur, and there has been much confusion as to what is meant by them. The common feature in all soils which may eventually give rise to such ferruginous and siliceous encrustations is the mobilization of silica so as to leave behind a concentration of iron and a ratio of silica to alumina so low that of alumno-silicate minerals only kaolinite can remain. This process has been called *ferralitization*.

The forest cover which protects a soil from desiccation through a short dry period becomes unstable as the dry periods lengthen and transpiration so dries the soil as to deprive the trees of water for long periods. The year by year reliability of the rains also decreases as aridity increases, and further puts the forest at risk. At the margin of the perhumid zone, occasional desiccation leads to soils in which dehydration of goethite occurs. Haematite, which gives the soil a darker red colour with a purplish tinge, may be dispersed through the soil, or at the faces of drying peds, or form pea-sized concentrations, in combination with silica, centred on relics of ferromagnesian minerals.

Shifts of climatic zone, or changes from perhumid to displuvial, or isopluvial to displuvial, bring wider areas which have ferralitic soils formed under conditions of intensive leaching under regimes which allow desiccation. Ferralitic soils containing haematite are common in the present forest savannah and savannah belts.

Desiccation can also be the result of forest clearance. Since most of the nutrient reserve is in the living trees, clearance leaves an infertile soil. The small amount of organic matter in the soil and the residues from clearance are soon mineralized. The coherence of the surface soil, which is poorly structured, particularly if depleted of clay, is further reduced and the soil is exposed to the full force of the tropical rains. Soil stripping following clearance occurs on even low gradients and the saprolitic rock is exposed. However, in these regions, unlike most drier areas, erosion can be beneficial to man. The exposed rock has a higher base content than the soil and the rapid weathering under high humidity and temperature soon makes available to crops an abundance of nutrients. Provided that the newly forming soil is protected from further erosion by terracing, good yields are obtained on valley sides where the uneroded soils of the plateaux are not worth cultivating.

Krasnozems

Soils which are to some degree ferralitic but which, because of youth or a limited availability of moisture, have a relatively high base status and a reserve of weatherable minerals have been called *krasnozems*, a Russian word meaning simply red soil. Desilication has proceeded so far that clay minerals are predominantly kaolinitic, but there may be some 2 : 1 minerals which increase in quantity as the degree of leaching decreases down the profile. On basic rocks the 2 : 1 minerals tend to be montmorillonitic, while on more acid rocks they are micaceous. The 2 : 1 minerals and small amounts of primary minerals which generally remain in these soils give them a weathering reserve which maintains a base status sufficiently high for the drastic desilication of the tropical soils considered so far not to occur. The colloid remains flocculated, and the **B** horizon is a cambic one, of a strong red colour and a blocky structure. The reddish-brown to red **A** horizon may be fairly rich in humus, which is not so rapidly mineralized in regions where dry periods occur as in the perhumid zone. Earthworms may be abundant, and termites are very common in these soils and may so far have modified the profile that its sequence of horizons owes little to purely pedogenetic processes : a surface layer of clay taken by termites from the **B** horizon may supersede a normal **A** horizon.

Vertisols

Black clay soils having the same characteristics as those already described in Mediterranean regions are common in the tropics and sub-tropics. As well as forming on fine grained sediments which either are in themselves base rich or receive base rich moisture from surrounding soils on higher ground, they form on fine grained basic rocks such as basalt even in externally well-drained situations. The climatic criterion for their formation is a dry period sufficiently marked for the self-mulching process and associated churning action to occur.

Vertisols on solid rock are generally shallow soils, since the swelling action of their montmorillonitic clay inhibits percolation of moisture from the soil into the underlying rock. They have a simple profile consisting of a fully mixed black clayey **A** horizon of very strong blocky or columnar structure resting on rock which may be almost unaltered. If drying reaches to the **C** horizon, small rock fragments and finer products of alteration are picked up by the churning action as soon as coherence of the rock is lost. In wetter situations moisture may penetrate further and drying not reach so far, and a cambic **B** horizon or a **(B)/C** horizon develops which may be reddish-brown or red in colour and very rich in montmorillonite. In drier situations, on the other hand, a calcareous horizon can form at the base of the **A** horizon and effectively cut off the parent material from further incorporation into the soil.

The amount of humus in a black clay soil is not necessarily any greater than in a red soil. It has been shown (Singh, 1954, 1956) that the black colour of a montmorillonitic soil is achieved with an amount of humus which scarcely colours a kaolinitic one. The stable form of the humus in combination with montmorillonite, however, gives these soils a longer term reserve which, together with the nutrient reserve maintained by a high cation exchange capacity and the abundance of weatherable minerals, makes them rich and not easily degraded agricultural soils.

The economic importance of these soils in regions where soils on more acid rocks have very poor agricultural potential and are easily lost by erosion, and their distinctive colour and peculiar properties, have led to their acquiring a wealth of local names. Dudal (1963) lists some fourteen names, from all the continents; some of the best known and most important are the *black cotton soils*, or *regurs*, on the basalts of the Deccan in India, and the *tirs* of Morocco.

The strength of the churning movements in vertisols is such that it can disturb or disrupt paving on the surface and posts or pipes within the soil. If the soil is not regularly cultivated, a form of micro-relief or patterned ground develops, in which small mounds, or on sloping ground ridges, alternate with flat areas. The forces set up by the swelling of increasing amounts of clay below the surface are resolved into cells of rising and falling soil currents, within which there is differential movement between particles of different size so that the size distribution is different in the soil of the "puff" and the "shelf".

The size of the cells, giving the spacing of the puffs, is related to the depth to which the soil dries and is generally of the order of a metre or two. These patterns, called *gilgais*, are better developed in soils on sedimentary materials than in those of solid rock, perhaps because the possibility of greater depth of drying gives more room for the cells to develop.

Termites, common in almost all other soils of the tropical and sub-tropical regions, rarely colonize vertisols; maybe their nests would be too subject to disruption in the churning ground, or possibly the tendency of these soils to become temporarily waterlogged after heavy rain makes them unsuitable.

The Relationship between Krasnozems and Vertisols

On basic rocks, particularly basalt, the transition in degree of leaching between the red and black soils is via dark reddish-brown soils, which in Australia have been called *chocolate soils*, and in Africa *eutrophic brown tropical soils*. The base status of these transitional soils is high and there are 2:1 clay minerals present throughout the profile. The dark reddish-brown **A** horizon has a mixed mineralogy, with kaolinite as well as hydrous micas and montmorillonite, a high clay content and well developed blocky structure. The cambic **B** horizon is red, but its clay may be entirely montmorillonitic and strongly swelling in character. The mutual distribution of the red and black soils and the status of the transitional ones has aroused much interest, particularly in Australia, where study of soil types in relation to surfaces of different ages has played a large part in geochronological work. The soils are found in toposequences, whose degree of differentiation and the positions of the transitions in relation to the elements of landscape vary with climate. A prolonged controversy centred on the question of whether the toposequences are simple catenas or in part or in whole chronosequences. Hallsworth (1951) and Hallsworth *et al.* (1952) believe that the catena relationship is in equilibrium with the present climate. The krasnozems, in upper positions, are strongly leached and the bases leached from them influence the pedogenetic environment lower down, so that montmorillonite persists in increasing quantity downslope, producing the reddish-brown soils with montmorillonitic lower horizons in intermediate positions and vertisols in lower positions. In drier regions leaching even in upper positions does not extend through the profile, reddish-brown soils occupy hill tops, merging to vertisols on the slopes. Passing to wetter regions, the top positions become fully leached and krasnozems extend downslope, the reddish-brown soils occur further down and leaching occurs in the upper parts of soils even in bottom positions, until in the wettest areas krasnozems occupy the whole landscape. Simonett (1960) and Simonett and Bauleke (1963) modify Hallsworth's hypothesis, saying that at higher rainfall, leaching is dominant throughout but that as rainfall declines the drainage conditions and the regime of wetting and drying which is essential to the formation of vertisols take over. Corbett (1968) finds that krasnozems of recent formation occur under much lower rainfall than that postulated by Hallsworth, and points out that in this case they are forming under forest rather than grassland.

Other workers, particularly Teakle (1952), Fergusson (1954) and Briner (1963) regard the catenas as chronosequences, the krasnozems, though perhaps in equilibrium with the present climate in wetter areas, having mostly been produced during former wetter periods, and occupying upper positions by virtue of these being the remains of older surfaces. The slopes, being positions of erosion, carry the immature, reddish-brown soils whose leaching is incomplete, and the vertisols form in the bottom situations which receive both soil materials and base-rich moisture from the slopes. Beckmann *et al.* (1974) attribute the development of a toposequence of red and black soils to the effect of erosion, so that slope soils are always kept at an immature stage of weathering as the landscape develops. Even where red soils in upper positions are reduced to too small an area for their leaching to maintain a supply of bases to soils in lower positions black soils continue to develop on slopes, and, having developed, the nature of their clay prevents their ever being changed by leaching. These authors maintain that differences in rainfall only change the rate of development of the pattern, provided the same seasonal distribution of rainfall occurs.

The value of soils on volcanic rocks in the study of geochronology and environmental sequences lies in the fact that vulcanicity produces surfaces for soil formation whose ages, unlike those of erosion surfaces, are independent of the factors of environmental change under investigation. The distinctive characteristics of basaltic soils and their distribution are discussed in Limbrey (1968) as a background to investigation of basaltic soils in Gran Canaria where associations of vertisol and krasnozem occur on volcanic surfaces which can be correlated by means of valley and coastal terraces with Pleistocene geochronology.

Brown Soils

Apart from the basaltic ones there are many tropical and sub-tropical brown soils, less leached than krasnozems and not having formed in the special circumstances leading to domination of a soil by montmorillonite. Some of them are soils on young surfaces, which have not yet suffered sufficient leaching for kaolinite to predominate and which have a high reserve of weatherable primary minerals and sufficient $2:1$ secondary minerals for the darker coloured clay/humus complexes to colour the soil. They may eventually become krasnozems, but meanwhile are not very different from the brown soils of temperate regions.

Under a drier climate leaching may be inadequate to remove sufficient bases for soils to go beyond the brown soil stage. Under these conditions, humus content is low, since plant growth is limited by lack of water and such residues as are formed are rapidly mineralized by the high rate of organic activity in the high temperatures during the limited periods for which the soil is moist. These permanently immature soils therefore have a paler colour than soils in moister regions, and are light brown to light reddish-brown. They are shallow and have a low clay content. Calcareous horizons are common, where the parent materials are sufficiently rich in calcium, since leaching is not carried right through the profile. These horizons have been exposed when erosion and deflation have removed the thin and incoherent soils, and form surface crusts over wide areas which are now too dry for them to form.

Tropical Podzols

As well as the podzols of cool humid mountain regions within the tropics, there are podzols of a different type which develop in the equatorial zone. Deep podzolic profiles form in sandy alluvial deposits such as those of river terraces, where extremely effective vertical leaching is combined with changes in groundwater level within the profile. Depletion of plant nutrients prevents the full development of rain forest and the heathy vegetation contains plants which produce chelating agents. These soils are widely developed in South America, and have been studied in Surinam by Veen et al. (1971) and Veen and Maaskant (1971), and in the Amazon basin by Klinge (1965). Podzols in Costa Rica, reported by Harris (1967) occur on volcanic ash, allowing the time scale of their development to be investigated.

Red-Yellow Podzolic Soils

These soils of humid sub-tropical regions bear the same relation to podzols as do the sols lessivés in the temperate zone, which in earlier American terminology were called grey-brown podzolic soils. The movement of iron is part of the migration of the colloid as a whole and is not due to chelation. The profile shows a brownish-yellow or reddish-yellow eluvial horizon relatively depleted of clay and iron and a textural **B** horizon of yellowish-red to red colour. Red-yellow podzolic soils are not fully developed where termites are active, since the termites bring soil from the illuvial horizon into the upper part of the soil and onto the surface. They are therefore best known where local conditions produce a climate of humid sub-tropical type in higher latitudes, for instance, in the south eastern parts of North America, and eastern Asia, and in southern Russia adjacent to the Black Sea coast. These soils may to some degree be relics of more humid conditions, and some of the red Mediterranean soils may come into this category, particularly those which have lost their coarser textured and more weakly structured **A** horizons by erosion.

Soils of Arid Regions

We have left the soils of each region considered so far at the point at which they take on a desert aspect. In Asia, America and Australia chestnut and prairie soils show a degree of soil formation sufficient to produce a moderate quantity of secondary minerals and enough organic matter to produce, in association with calcium and montmorillonitic clay, a dark coloured **A** horizon, but leaching is insufficient to prevent formation of calcareous horizons. On the desert side of the belts of these soils, the profile becomes shallower, the cambic **B** horizon, where present, is lost, and the dark colour of the **A** horizon no longer occurs. Brown and reddish-brown soils with calcareous horizons occur in the dry grasslands where the period of plant growth is short and the rate of organic activity in the short period of moistness is high enough to mineralize completely such residues as are formed, so that any humus substances formed are likely to be destroyed in the same season.

In the dry regions of the Mediterranean Basin, brown and reddish-brown soils of similar profile occur; there, the loss of organic matter in the **A** horizon may be due to the long periods of heavy grazing.

These brown and reddish-brown soils occur in the steppe belts on either side of the great deserts of the mid-latitudes, merging on the one side with those of either the continental arid regions or the Mediterranean zones, and on the other side with the soils of the sub-tropics. The changes in width and shifts in position of the deserts during the Pleistocene have left many relict soils, on the one hand, and relict desert land forms, such as dunes, on the other. Soils of this type exposed to an increased rainfall will rapidly take on the characteristics of soils in harmony with the climate; left under greater aridity they may persist unchanged, suffer deflation so as to expose the calcareous horizon, be eroded by the flash floods characteristic of deserts, or be overwhelmed by wind blown deposits and preserved as fossil soils. All these conditions are found in the deserts themselves, as well as stony and sandy areas with nothing that can be called a soil, areas of incipient soils and the grey and red desert soils. Grey desert soils are called *sierozems*. They form under conditions rather more arid than those which produce brown soils, and have only a thin grey or light brown **A** horizon, slightly coloured by oxidation of iron and very little humus, and a very low clay content, merging down into a calcareous horizon. Where the parent materials are too acid to provide the material for a calcareous horizon under the very slow rates of mineral alteration prevailing, or where there is not enough moisture to achieve even this amount of weathering, *red desert soils* may form, dominated by a high content of dehydrated iron oxides. However, it is probable that many red desert soils are relict soils, decapitated profiles of soils formed in more humid conditions.

12
Intrazonal and Azonal Soils

Many of the soils considered so far are distributed broadly according to climatic zones, though the relict nature of many soils outside the areas of Pleistocene glaciation tends to blur the zonation. Some soils are so dominated in their development by factors other than climate that their characteristics are much the same through several climatic zones. These are called intrazonal soils and they occur on particular parent materials or in particular environmental situations within areas whose other soils conform to climatic zonation. Some of them are dependent on one aspect of climate which controls soil formation to the exclusion of other aspects, or on a combination of a climatic factor and a factor independent of climate, the balance between these being different in different climatic zones but producing similar soils.

Desert soils are dominated by lack of moisture, whether they form at high latitudes in continental interiors ar at low latitudes in the sub-tropical desert belts. The climatic regimes of these two regions are very different but the soils can be very similar. Vertisols are intrazonal, their occurrence being controlled by a combination of parent material and a climatic factor which is of greater or lesser importance relative to topographic position in different zones. Hydromorphic soils tend to be intrazonal if the degree of waterlogging is such that normal sub-aerial soil processes are markedly reduced.

Young soils are azonal since the initial stages of soil formation are the same under any environment. Climatic zonation takes over sooner or later, depending on a combination of climate, parent material and topography: on steep slopes or on very easily eroded material soils may be permanently young, and under an arid climate a young soil may progress only from an incipient azonal soil to the intrazonal desert soil stage.

Alluvial soils partake of hydromorphic and immature characteristics in varying degrees, and may be intrazonal or azonal from either cause.

The essential feature is the accretion of further material, which may be fresh or already partially weathered, and humic or entirely mineral, according to situation, before that already in place has time to develop soil features characteristic of the climatic zone.

Soils on volcanic ash tend to be azonal. They are called *andosols* and are dominated wherever they occur by the high permeability of ash deposits and by the presence of easily weatherable minerals and glass in finely divided form. Rapid alteration leads to domination of the soil by allophane and, on basic materials, montmorillonite. The soil colloid has a very high exchange capacity, a dark colour, and remains flocculated. Because of the porosity of the ash, andosols in humid regions have a more arid aspect than the zonal soils, whereas in arid regions, where the soils tend to be dry most of the time anyway, the rapid alteration of the volcanic materials makes available abundant plant nutrients and they tend to be more fertile than the zonal soils.

Saline and Alkaline Soils

Circumstances in which soluble salts accumulate in soil occur in those combinations of topographic and climatic situations where rainfall is sufficient for drainage from soil profiles in higher positions to bring moisture into those in lower positions but not high enough for leaching to remove the soluble salts from the low lying soils. Typical situations are closed basins, with or without a saline lake, and areas of low lying undulating land with no surface drainage. Such situations occur in the sub-arid parts of continental and Mediterranean regions and in those parts of the arid regions which receive groundwater originating in the higher rainfall areas of adjacent mountain ranges.

In the river valleys of sub-arid and arid regions the origin of saline soils is slightly different. Salts carried in the river water may have come from distant areas of high rainfall. After spreading over the floodplains during the flood which follows the rainy season in the catchment area, the water drains away either through the soil or via surface channels, or is lost from the soil by evaporation and transpiration. The effect of irrigation in spreading the water over a wider area than it would reach naturally and in preventing it from running back to the river as the flood subsides is to push the balance towards loss by evaporation and transpiration so that the salts are left in the soil.

While groundwater is within about a metre of the surface a supply of moisture to the surface is maintained through the capillary fringe, and as the water evaporates the salts are left as a surface crust. Seasonal fall in groundwater level provides an opportunity for leaching; saline soils occur where rainfall is inadequate to remove from the soil the salts which accumulate during the periods of evaporation from a soil moistened by flood or by a high water table. Soils affected by salts in this way are known as *solonchaks*.

The salts concerned are mainly sulphates and chlorides of sodium, calcium and magnesium. High salt concentrations in the soil moisture reduce the ability of plants to absorb water and nutrients. Chloride and sulphate ions can be toxic to plants in high concentration and the cations involved can also be toxic in excess either directly or by preventing the plant from getting enough of other nutrients. Saline soils support a vegetation reduced in abundance and restricted to salt-tolerant species. Humus content of the soil is low, but since the clay minerals which form in an unleached environment tend to be montmorillonitic the clay/humus complex can be dark coloured.

Formation of saline soils may be a normal response to climatic conditions in a particular landscape or it may be the result of inter-ference by man. Clearance of forest in the more humid uplands marginal to a semi-arid region can cause increased leaching of the soils there and the arrival of soluble salts in low lying areas where leaching is inadequate to carry them away. Salinity can be produced by irrigation from wells, and in river basins by irrigation or by a rise in water table as a result of interference with the flow of the river in some part of its course.

Similarly, leaching of a previously-formed saline soil may be the result of an increase in rainfall or of a natural change in behaviour of a river which lowers the water table in part of its basin, or it may be the deliberate or accidental result of man's efforts. Attempts are made to ameliorate naturally occurring or man-induced saline soils by leaching them with fresh water, or with water that is less saline than the soil solution. The commonest accidental change is probably the lowering of water table brought about by use of water from a river or by modification of its channel.

The effects of leaching a saline soil are complex. The proportions of different cations in the system are critical. If calcium is present in the exchange complex to the extent of about 90% of total cations leaching converts the soil into the normal soil for the region, usually a chestnut soil or brown desert soil. If sodium is present in the exchange complex in excess of about 12% as well as being abundant in the soil solution, as the salt concentration of the solution falls as leaching proceeds the colloid tends to disperse. Structure of the soil is lost and the dark coloured clay/humus complex begins to spread from the thin humic horizon down into the soil. At this stage the soil is called a *solonetz*; it is characterized by a thick, dark **A** horizon which is structureless and very sticky when wet and cracks into columns when dry. Besides being very difficult to cultivate, such soils are poorly aerated, and the high sodium content can induce calcium starvation of plants.

Further leaching of a solonetz produces a strongly alkaline soil. The carbon dioxide released into the soil by respiration of roots and soil organisms normally interacts with water and contributes to soil acidity, or, if there is an excess of calcium present, is immobilized in calcium carbonate. In soils in which sodium is present in abundance in the exchange complex, hydrogen ions tend to displace the sodium, leaving an excess of carbonate and bicarbonate ions in the soil solution which react with the displaced sodium to form sodium carbonate and bicarbonate. Presence of these salts raises pH to 9 or above and under such extremely alkaline conditions drastic dispersion of the colloid occurs. The clay and humus migrate down the profile till they come to the level not yet so leached and are there deposited in pores, forming a dense, dark pan. Silica too is mobilized at high pH and moves down the profile. It is usually immobilized at slightly higher pH than the clay and humus components of the colloid and can be seen as cappings to the columnar structural units of the developing pan and to the rather platy units formed in the paler and coarser leached horizon above the pan. At this stage the soil is a *solodized solonetz*. Its extreme alkalinity is inimical to plants, preventing absorption of nutrients.

Eventually, with continued leaching, the depth of the eluviated horizons increases. The sodium is all displaced from the exchange complex and moves down the profile with the carbonate-dispersed colloid and the upper part of the soil is left with a hydrogen-saturated exchange complex and becomes acid. This soil is known as a *solod*, and only at this stage does it become amenable to agriculture.

The amount of water needed to leach the soils limits the scale on which lands rendered saline by a long history of irrigation can be improved. If water is plentiful but is already itself saline, the amount needed is even greater, and labour or power for pumping it may be the limiting factor. Farmers in saline areas have systems of land use which keep a proportion of the land relatively salt free or achieve small scale improvements. Buringh (1960) describes the methods used in Iraq, where the soils of Mesopotamia are the natural flood deposits of the river channels and the man-made flood deposits of irrigation channels, distributed in an intricate pattern of abandoned and active levées and basins, and accumulated to considerable depth. The relief and variation of texture between levée, where coarse material is dropped, and basin, where fine settles, and the interleaving of these deposits as the channels have shifted or been re-made, provides a variety of situation, soil characteristics and salinity, and land use varies according to the salt tolerance of the crop and the depth of the water table. In the slightly higher areas the water table is below rooting depth and a perched water table is created by irrigation. Here, a fallow system can be used, crops being grown in alternate winters on any one piece of land, and during the intervening period transpiration by salt tolerant weeds helps to dry the soil, lowering the perched water table locally enough for salts to be washed clear of rooting depth when irrigation for the next crop begins. Buringh states that land can be used for 400–500 years on this system before all irrigation has to cease to allow an overall fall in the perched water table to occur.

In lower lying areas the water table is nearer the surface and a system of "isolated salt areas" can be used. An area is left fallow among the irrigated fields and salts accumulate there as water evaporates from it, saline water being pushed from the irrigated towards the non-irrigated areas. The levées of irrigation channels can also be exploited in this way: plants can be grown on a shelf half way up the side of a channel. Evaporation from the bank above draws water up, carrying salts past the level from which the plants are drawing their supplies to form an encrustation on top.

Methods such as these, adapted to local conditions of salinity, relief and soil texture, to the amount of irrigation water available and to crop plants of differing salt tolerance, have evolved as the saline soils developed and extended, and have permitted continued exploitation of the potentially fertile riverine soils of an area of very low rainfall. An attempt to date the spread of salinity by investigating the salt content of mud brick from archaeological sites is reported by Hardan (1971).

PART IV

Soil and The Archaeologist

13

Archaeology and Soil Survey

Association of Archaeological and Environmental Evidence With Soils and Sediments

Introduction

Information about man's activities and about the environment in which he lived lies in the soil itself and in things found in, under and on it. "Intrinsic" information is held in the materials of the soil itself and in its distribution in the landscape. The information concerns the environment, or the pattern of changing environment, under which the soil formed and under which it may have suffered erosion and redistribution; that is, the environmental conditions under which man was living. Since the soil is the substratum on which food supplies depend, it can hold information about the potential of these supplies for exploitation, the degree to which they were exploited at any time, and the way in which the behaviour of human communities developed to exploit it, or the way in which their development was limited or controlled by their inability to exploit it efficiently.

Another sort of information held by the soil is "contained" information, that is, information represented by organic residues and by archaeological entities lying on, contained in or buried under the soil. The contained information is intrinsic to the ecosystem but extrinsic to the soil. Intrinsic information is a palimpsest in which successive images, some instantaneous, others formed slowly over a long period, are superimposed; some of the earlier ones are obliterated or altered by the later ones, others persist in more or less decipherable form. Contained information on the other hand is a stratification or a mixture of information referring to instants of time; it may be selectively preserved, and mixed, sorted or obliterated by some of the processes represented by the intrinsic information. Its absence can itself be part of the intrinsic information.

Oldfield (1972) described a lake as an information sink, a natural rubbish dump, aided by gravity, in which is deposited information integrated over the area of the drainage basin. If an area drains to the sea, much of the information is lost. However, much information never gets to the sink, or gets there after considerable delay. Soils represent some of the pipelines leading to the sink, and the pipes are to varying degrees blocked; information proceeds fitfully along them, preserved for a time in a stable soil or sediment, mixed, redistributed or destroyed during slow, continuous creep or rapid, spasmodic erosion, again static, with new information added, in a new deposit, and perhaps further modified in content and distribution several times before either being destroyed or reaching the information sink. Whereas information in such a sink does have the advantage for some purposes of being integrated over an area, some of that in the pipelines may still be referrable to more restricted source areas, and whereas that in the sink is preserved in a stratification which superimposes time-fixed events or processes on a time-mixed background, the extent and variety of pipeline stiuations may allow a greater degree of time sorting to persist. The information in the sink and in the pipelines is complementary, and each can only be understood in the light of the other.

Information Associated With Sedentary Soil Profiles

Sedentary soils in a stable situation preserve in their materials and profile information which is integrated over the whole period of their development. Whereas a landscape may owe its general form to a process such as glaciation which provides surfaces of uniform age, within it there may be numerous younger surfaces created by erosion and deposition on a local scale, and in areas where there has been no glaciation or other drastic geomorphological process to provide a regional time base, the entire landscape may be made up of surfaces of greatly differing ages. Such surfaces provide a range of starting points for soil formation, though not necessarily providing equivalent environmental situations for soils of differing age. Buried soils provide a series of termination points for profile development, but burial by natural processes almost always involves at least disturbance of the surface, and usually major truncation of the profile, by the agency responsible for transport of the burying material. Soils buried by human agency provide more complete profiles, and often the possibility of more precise dating, but episodes of burial tend to cluster in age according to human activity: periods of building of burial mounds or ramparts, for example. Also, conclusions about the effects of man upon the soil drawn from soils buried under earthworks which might be concentrated in areas of dense settlement, or whose location might be determined either by topographic situation or by availability of unproductive land on which to put them, should not be too widely generalized.

In the absence of an adequate range of soil profiles limited in age at one end or the other, deductions about past environmental conditions can be made from the evidence of developmental history in the profile itself, such as indications of change in weathering relations of minerals or in nature and behaviour of humus substances, occurrence of relict horizons, or lack of equilibrium between soil characteristics and the present environment.

Evidence of environmental conditions and human activity which may be contained in the soil is not usually directly associated with that which is intrinsic to the soil itself. Downwash of pollen or incorporation of snails may occur contemporaneously with soil development but the downward movements involved may proceed at rates unrelated to the wide range of rates of movement of substances in soil processes, and many of the processes of profile development are alterations and differentiations involving no overall movement of materials. However, the presence of a pollen profile is itself intrinsic evidence of chemical conditions in the soil and if the beginning or end of the period of pollen preservation can be dated by reference to the composition of the vegetation it represents, then a fixed point in a period of changing acidity of the soil is obtained. Incorporation of artifacts into a soil by superficial disturbance or by worm action may provide the only evidence obtainable that such disturbance occurred or that conditions were suitable for surface casting worms and for worm sorting to proceed undisturbed, and the artifacts themselves provide a means of dating the relevant episodes or periods.

Direct association of intrinsic and contained evidence does occur in sedentary soil profiles which accumulate mor humus or peat. These are the only materials which can cause a build up of the soil surface without the addition of soil material foreign to the locality, and they provide a stratified accumulation of plant residues, including pollen, which can be directly associated with changes in conditions of soil acidity or moisture regime indicated by the beginning of accumulation and by changes in degree or humification and rate of accumulation. Artifacts contained in the organic material can be directly associated with the intrinsic evidence of soil conditions at the time and with the contained evidence of vegetation and fauna.

Disturbed, Truncated and Redeposited Soils

Because of the rarity of undisturbed soils, many deductions have to be made from profiles truncated by slope movement or more drastic erosion. These negative and reduced features of soil distribution and development are part of the intrinsic evidence relating to soil history, and have to be considered in conjunction with that obtained from the complementary depositional situations. When a soil is eroded and its materials transported, if the energy and distance of transport is low some of the structural units may remain intact. Peds may survive only slightly rounded and reduced, through plough wash or creep, and those of very strongly structured soils may survive the more drastic processes of sheet flow or colluviation and be recognizable macroscopically or in thin section in the resulting deposit (Limbrey, 1968). Some characteristics of the profile of the source soil may also be preserved in the deposit if the process of erosion is sufficiently uniform to strip each horizon in turn and produce a profile fossilized in inverted form. In these cases a high proportion of environmental information intrinsic to the original soil profile may be retrievable, and some of the contained evidence may also be available, but because of the degree of disturbance and mixing inherent in even the slightest transport, its time sequence is likely to be blurred. Evidence recovered from situations such as these must be referred back to its source situation for purposes of interpretation.

Processes of soil disturbance and displacement are discussed in Chapter 5. Information to be derived from the distribution of erosion scars and truncated profiles and from distribution and variations in sub-aerial soil deposits—thickness, proportions of soil and rock from different horizons, degree of dispersal and sorting of materials—may be interpreted in terms of agency of erosion and transport, its spasmodic or continuous activity, variations in its power, disturbance of soil surface or its protective cover of vegetation, stability of soil structure in different horizons. Further information derived from observation of signs of river or coastal erosion and deposition, and archaeological evidence, may be needed before conclusions can be drawn as to whether the observed phenomena are due to progressive activity of the disturbing agencies under constant conditions, creating non-uniform deposits by exhausting supply of particular materials or cutting through them to reach others, to regular occurrence of extreme conditions, to isolated episodes of abnormal conditions, or to a change from one uniform state to another, and whether the agencies are entirely climatic, entirely the result of man's activity, or a combination of the two.

In considering either environmental or archaeology evidence contained in redeposited soils, processes of mixing, sorting, abrasion and selective destruction have to be taken into account. Study of the soil materials of the deposit in relation to possible source areas is necessary to establish the degree to which particle size sorting has occurred and the degree to which similar size grades drawn from different source areas or different horizons of the same profile are mixed, before the contained evidence which has undergone the same processes of mixing and sorting can be interpreted. The environment in which a redeposited soil finds itself is almost always different from that of the source soil, and since deposition occurs preferentially in lower and flatter parts of the landscape, whereas soils are eroded from the higher and steeper parts, it is likely to be very markedly different, particularly in terms of leaching regime and moisture content. In its new situation the material continues to undergo soil processes, which establish upon it a new soil profile, so that differentiation and movement of material is superimposed upon any stratification which may exist in the deposit. Given the common occurrence of deposits in low lying situations and their deposition by water, gleying is a frequent phenomenon of redeposited soils in the temperate zone, and the colour mottling and development of prismatic structure may obliterate those characteristics of the source soil which survived transport. Clay migration may occur and obscure pre-existing textural, structural and stratigraphic arrangements; on the other hand, development of iron pans or podzolic horizons of deposition within a redeposited soil frequently pick out changes of texture and emphasize stratigraphic features. Worm activity in a developing humic horizon mixes and destroys the stratigraphy within the upper part of the deposit, and worm channels pass through the deposit and may introduce material from the upper horizons into the lower parts, perhaps concentrating the small stones of worms' aestivation chambers at a textural boundary. Roots may pick out a layer of high water retention; in passing through stratigraphic boundaries, root channels blur the stratigraphy and their humification contributes to modification of the materials. Thus, soil drawn from and retaining some properties of the various horizons of the source soils undergoes modification towards the condition appropriate to its position in the newly forming profile. Some of these modifications may reverse those of the source soil: mineralization of organic material at depth in the deposit, turning **A** horizon material into that of a **B** horizon, for example. Other characteristics cannot be obliterated and remain as fossil features in the new profile.

In spite of post depositional changes, some deductions must be made about the environment of preservation in the source soils in order to determine whether or not they contained bone, snails, pollen or other organic materials. Since environmental and artifactual evidence relating to the process and environment of transport and to the new situation will be incorporated into and deposited on the sediment it is necessary to know whether they would have existed within or upon the source soils. The problem may be slightly reduced if a distinction can be made between autochthonous and derived material on grounds of surface abrasion during transport or differences in mode of preservation apparent in condition or colour.

Where several stages of redeposition have occurred the problem of re-assembling information from a variety of depositional situations with reference to possible source situations becomes increasingly complex. With increasing distance of transport and number of stages the degree of size sorting and of mixing of equivalent size fractions from different sources increases, and the degree of abrasion and differential destruction of contained materials increases. Eventually one reaches the sink situation, where the value of integration over the area of the drainage basin becomes important. A lake basin receives a predominance of fine sediments, sorted from the coarser material or derived from it by abrasion during transport. Artifacts and macroscopic animal remains are unlikely to reach it. Pollen from source soils and intermediate situations will travel with the appropriate size fraction, taking into account its lower density than that of mineral material, once the humic aggregates with which it is associated in the soil have been dispersed, but is likely to suffer considerable damage on the way. Seeds and remains of arthropods which are sufficiently resistant to decay to survive in the humus of acid soils and in peats will also reach the lake basin. Some quite large organic particles have sufficiently low density to remain afloat in a fast and turbulent stream, and travel with a much finer size grade of mineral particles. In a fairly clear stream they can be seen travelling as a floating layer above the bottom, while the mineral particles roll and bounce along. They may therefore escape some of the mechanical damage suffered by denser particles, and so eventual preservation is of material sorted and selected by size, density and strength. All these organic residues are joined during transport and deposition by those falling into the water from the air and vegetation. Thus, unless it can be asserted that the drainage basin contained no situations in which a certain item of environmental evidence could have been preserved, or unless autochthonous and allochthonous material in a deposit can be distinguished by condition or colour, much of the information obtained from stratified basin deposits will be contaminated and its time sequence blurred by material whose source, period of residence there and in intermediate situations, and time of travel to its ultimate destination are unknown.

However, the appearance and the rate of arrival of derived material in the basin are themselves intrinsic evidence of soil disturbance in the surroundings, and any contained evidence which can be correlated with changes in type of material or rate of deposition, or with changes in the environment of deposition, can be referred back to possible source areas. The studies of Mackereth (1965–6) on the sediments of the English Lake District, referred to in Chapter 8, are an example of this approach.

In arid and semi-arid regions preservation of residues occurs because of insufficient moisture to maintain organic activity, either in the source soil or in redeposited soils. In an active soil in a warm, dry region, the period of moistness adequate for plant growth is generally also adequate for complete mineralization of organic residues, but deposition may occur in a more permanently dry environment than that from which soil or residues were derived. In particular, where wetting, and therefore the entire cycle of organic activity, is confined to a shallow zone at the surface, organic material buried at greater depth may remain permanently dry.

Wind Erosion and Deposition

In arid conditions, wind transport assumes great importance. The winnowing effect upon a soil surface leaves its traces in both deposit and source area as intrinsic evidence of climatic aridity and surface exposure. Artifacts and coarse soil particles left behind as fine particles are removed may become polished by the fine particles in the wind and they are gradually lowered and relatively concentrated, so that artifacts and bones, for example, starting from a number of levels within a soil or deposit may arrive as a concentrated layer at the level to which deflation has reached. Such a layer subsequently buried can be very misleading.

Material deposited from wind has a characteristic grain size distribution, and organic particles have an even greater tendency to travel with smaller grades of mineral particle than during water transport, because of their much lower density when dry. Many seeds have shapes or surface characteristics by which they are adapted particularly well to wind transport, and this should be remembered when considering possible sources of material incorporated into sedentary soils as well as that forming part of wind transported deposits.

Encrusted Materials

Encrustations or indurated horizons within the profiles of many soils of semi-arid regions resist erosion by wind and surface water, and are left as surface crusts in denuded landscapes. When such a crust is attacked by more violent erosive forces than that which exposed it, it may break up and provide gravel to be incorporated into a deposit elsewhere. If it remains as a resistant layer, under changed conditions more soil or potential soil forming material may accumulate upon it and a new soil profile form, whose parent material may be different in mineralogy and particle size from that which gave rise to the encrustation and the relict horizons of the former soil which remain beneath it. Alternatively, changed conditions may initiate dissolution of the crust, releasing the soil materials enclosed within it and allowing rejuvenation of the profile without addition of foreign material. Any items of archaeological or environmental evidence which were preserved within an encrusted horizon may be released, and may remain *in situ*, be destroyed, or be redeposited elsewhere when the encrustation is destroyed by chemical processes within the soil or by physical disruption.

Cave Situations

Dry conditions in caves in otherwise humid areas require consideration in terms similar to those applicable in arid regions. Deposition of material other than that derived from roof and walls or brought in by man is usually mainly by wind or by the occasional passage of water in a situation which otherwise remains dry, and formation of calcareous or siliceous crusts under moister conditions or during moister periods is a common feature of the cave environment.

A cave acts as a trap for anything that is blowing around—fresh organic particles, mineral soil, and organic material already contained in the mineral soil—and the chances of preservation of organic materials are good, either because of dryness or because of the encapsulating effect of calcareous and siliceous encrustation. The sediment may represent the only evidence remaining from certain periods about soil types and environmental conditions in the surrounding area, or it may provide a datable sequence of archaeological and environmental evidence to which the developmental history of soils whose profiles can be studied nearby can be related. As in the case of any soil deposit in which preservation is good, the possibility of contamination of a straightforward stratigraphic sequence with material whose time of residence in source soils is unknown has always to be taken into account.

Since the soil materials in a cave deposit are sorted in the process of transport, and since wind transport predominates and is unlikely to involve much undispersed soil, direct observation of the characteristics of the source soil are rarely possible. Deductions must be made from the proportions of constituent minerals, their particle size and their degree of alteration in comparison to the parent materials of the soils from which they are likely to have been derived. The advantage of a cave deposit over others similarly composed of particles sorted from soil profiles, particularly in humid regions, is that the amount of post-depositional alteration is likely to be low, and its incidence can be estimated by reference to the organic component of the deposit. It cannot be assumed that a dry deposit has always been so, but if in spite of a base rich environment organic materials other than bone are preserved it can be assumed that periods of moistness sufficient for the activity of micro-organisms have been few and brief, and since mineral alteration also requires moisture it can also be assumed that little post-depositional alteration has occurred. The same applies in the case of stalagmite layers, since little organic activity or mineral alteration can take place if the encrustation surrounds particles sufficiently to exclude both moisture and air.

If it is found that post-depositional alteration of minerals is likely to have been slight, provided due regard is given to the effects of particle size sorting during transport, the nature of clay minerals, the degree of dehydration of iron oxides, and the relative proportions of minerals of different susceptibility to alteration, will then give some idea of the nature of the soils of the area at the time of their erosion and deposition in the cave. It is to be hoped that the study of amorphous humus substances which may have remained unaltered in such deposits might provide further information about the soils in which they formed.

Principles of Survey Work

The relationship between people and the landscape in which they live can be extremely complicated. It involves climate, geology and geomorphology as direct determinants of the behaviour of people independently and in various combinations, as well as in their combined effect through the soil in providing food and organic raw materials. Climate and geology combine to determine geomorphology and soil; geomorphology and climate combine to provide water resources and the comfort and security considerations of aspect and shelter. Geology, geomorphology, soil and water provide mineral raw materials, access to them and means to work and transport them. In order to determine what resources were available to people at any particular time, and what aspects of the landscape might have been important to them, it is necessary to work out the time relationships of the various elements of the landscape, and in the process of doing this areas will be identified in which archaeological entities of that period remain *in situ* at or close to the present surface, others where they are likely to be deeply buried, where they are absent altogether because of destruction or removal by erosion or because of deposition or exposure of a sediment of a type unlikely to contain them, and where the eroded and transported remains are to be found incorporated into a sediment. Only when this has been done can the original distribution of the entities be reconstructed. Reconstruction of soil distribution at that period is partly achieved by establishment of the age sequences of surfaces, so that it can be determined which did not exist then, which had greater or lesser extent than now, where there might have been deeper and more complete soil profiles, and where there would have been shallow or incipient soils which have since become more mature. Such changes in area and thickness of soils can be determined from study of land forms and deposits in relation to the time sequence provided by the archaeological evidence. Changes in potential productivity of the soils can be determined by reference to buried soils and the intrinsic evidence of soil characteristics remaining in redeposited soils, but often this type of information will be difficult to find, and without it changes in profile type, which may be evident in the presence of relict horizons and polygenetic profiles, can be difficult to assign to period. In these circumstances evidence of vegetation from pollen or, indirectly, from molluscan analysis, becomes essential.

When archaeological and soil distributions have been established, it may be found that there is no indication of association of human activity with particular soils which cannot be accounted for by choice of locality for some other reason. This might mean that population was so high in relation to resources that all land had to be used regardless of quality, or that potential productivity was relatively uniform, and it should be obvious from the range of variation of the soils as well as from the archaeological evidence of density of settlement which was the case. Where there was a considerable range of variation in soil types available, differential usage is likely to have been practised, and it is likely to have increased the differentiation of the soil types. Thus, referring to Chapter 8, the differences between upland and lowland soils was probably not great before any human activity interfered with the forest-soil relationship. Early agriculture may have initiated or accelerated the degradation of upland soils, so that later people were faced with a differential distribution, and in adapting their system of agriculture and herding to it they extended and intensified the differentiation. Successive occupants of an area encounter the cumulative effects of their predecessors' activities, and have to adapt their habits to make use of the degree of differentiation they find, and to modify them continually as changes occur under their own influence.

Quite apart from the agricultural effects of changing soil resources, it has to be remembered that processes of erosion and deposition, waterlogging and peat formation can cause major changes in accessibility and communications. Routes across upland areas and access to mountain pasture can be blocked by peat growth; valley bottoms become boggy and impassable as soil accumulates and water regime changes; lakes and estuaries silt up and raised bogs grow on them, cutting off short cuts and access by boat.

Many factors and combinations of factors affect differential land use. Hunters and gatherers find more food in open woodland, patchy forest and forest margin areas than in either closed forest or open country, and since soil conditions might either reflect or determine the position of such vegetation, settlement there will appear, when the vegetation itself is no longer available for study, to have been a selection of soil type. If the people maintain or extend the pattern of vegetation they may cause soil changes which influence later people. Agricultural people arriving or spreading into an area not previously cultivated will find some areas easier to clear than others, depending on the type of vegetation and the degree to which it has been disturbed by non-agricultural people who were there before, and having cleared the land will find some soils easier to cultivate than others. Ease of clearance and initial cultivation, however, does not necessarily go hand in hand with either immediate fertility or maintenance of that fertility with use, or with persistence of characteristics which make cultivation easy. After perhaps some experimental attempts on unsuitable soils, particularly if people are moving into an area where parent materials, climate or drainage conditions are different from those of their former lands, so that the range of soil types exceeds their previous experience, use will be concentrated on certain soils.

For cultivation, soils whose texture falls into one of the loamy classes generally provide the best combination of nutrient reserve and cation exchange capacity, moisture capacity and facility of drainage, and, potentially, structural stability. (See Chapter 14 for definitions of soil textural classes.) Some clay soils may provide a higher nutrient content and higher moisture reserve to guard against drying out, but, except for a short period after clearance, they may be very difficult to cultivate unless heavy ploughs and the means to draw them are available, and their high moisture content makes them unworkable with any equipment in wet conditions as all interference tends to cause puddling and damage to structure. In dry conditions too, clay soils can be extremely difficult to work. In the colder parts of the temperate zone, clay soils drain late and warm up late in spring, shortening the growing season, and only where a cool, wet spring is followed by a hot, dry summer might the greater moisture reserve, lengthening the growing season at the other end, be an adequate compensation. Sandy soils, by contrast, are extremely easy to cultivate, but their low nutrient content and low cation exchange capacity provide no protection from loss of productivity with use, and the risks of loss of structure and damage by erosion are very high. Sandy soils can usually be cultivated in wet winter conditions, and they drain and warm up quickly in spring, but except in situations where water is available from the water table, they are very susceptible to drought during the growing season even under relatively rainy climates. Silty soils have some of the advantages of the loams of mixed particle size in their compromise between the extremes of drainage and moisture retention and in the workability and high nutrient reserve, but they can become highly susceptible to wind erosion. Loess soils are the most extensive silty ones, and in their typical expression they have a high organic content which helps to provide a strong and stable structure and prevent erosion, but once this is lost, or in situations in which it was never able to develop, loess goes away on the wind as easily as it came. It is possible that the apparent scarcity of deep loess deposits in the wetter parts of the temperate zone, where their more leached soils would have lacked the high calcium content which helps to stabilize the humus, is in part due to their having blown away rapidly as soon as cultivation and cropping had reduced their organic content. I have recently examined a non-calcareous loess soil beneath a barrow in eastern England, which had only a very thin and unstable humic horizon. During excavation the soil blew about when dry at the slightest draught, and even in the laboratory, examination of and sampling for pollen from a monolith were hampered by the soil being so liable to fly about when breathed upon. Thus, except in those areas where the rainfall regime does not cause decalcification of a loessic soil, they might have been much exploited by early cultivators but soils on the thinner loess deposits would not remain for the use of their successors, though much of the loess might thereafter be present in, and contributing to the texture and nutrient status of, other soils in the region.

For pasture, different criteria apply in selection of soils, and many of the limits which apply to arable soils have greater tolerance under pasture. Soils of more extreme characteristics are therefore likely, in a mixed economy, to be used for pasture, and where cultivation produces declining yields arable land may be let go for pasture, either temporarily, until structure has improved under the influence of grass roots and increasing worm population, or permanently. Since animals can be moved about, use for pasture is adaptable to the condition of the soil at different times of the year. Many wet situations become puddled and unusable if grazed early in the spring, but can be left to grow a rich crop of grass which is either taken for hay, by which time the land is dry enough for animals to graze the aftermath, or grazed as soon as the land drains. At the other extreme, animals can continue to derive nutriment from dried vegetation long after the soil has dried out, and can continue to provide meat or milk even when walking enormous distances to glean enough food. At the very least, they keep the meat supply in good condition by staying alive, even if they are not eating enough actually to grow or produce milk.

Under a cold climate at high altitude, pasture becomes available in abundance where the growing season is too short for any crop production. Here, enough hay must be produced to keep the animals alive in winter, and potential arable soils at lower altitude might be restricted to grass production for winter grazing and the hay crop for this reason, limiting the land available for crop growing.

If there is no archaeological evidence for the type of land use at any period, in the form of distribution of artifacts related to cultivation, herding or exploitation of the natural vegetation, the evidence must be sought in the soils themselves, from pollen or molluscan analysis or from features of the profiles of buried soils and the nature and distribution of eroded surfaces and soil deposits which might be attributable to clearance and cultivation. Variations in land use with time might be the direct result of the erosion or degradation produced by man's activities, or the combined effect of climate and such activities. It might also be a result of advancing technology, such as the introduction of ploughs which could cope with soils too clayey to be tackled by the earlier ones, or which could bite into less impoverished soil at greater depth than before; or improved drainage techniques for wet lands, methods of controlling flood water or of irrigating dry lands. These are all variations related to the soil itself; change in custom as a result of migration or changing economic needs may be unrelated to the soils of the area under study, but could be an effect of soil-related factors elsewhere.

To achieve all the aims of soil survey in conjunction with archaeological work, the survey team should include at least a geomorphologist, a soil scientist and a botanist, and should have adequate back-up facilities for analytical work. Adequate time is also essential, particularly for work in an unfamiliar region. It is a rare fieldworker who can obtain results of much value during a few days of rapid transit through a climatic zone or geological province in which he has not been before. Often, however, assembling a complete team is not possible, because of lack of funds or lack of people, and the archaeologist must be his own soil surveyor, and he needs to make as much use as possible of the information about soils and related environmental factors which might be available in published form, or among the people living and working in the area. He also needs to record accurately the soils, sediments and landforms he encounters so as to be able to enlist the help of specialists in interpreting his data later on. The procedure for recording and sampling soils is given in Chapter 14. Dekker and De Weerd (1973) discuss the importance of soil survey in archaeology and give examples of its use.

Soil Maps

If the archaeologist is working in an area for which soil maps and other soil survey publications are available, he should make full use of them. Soil maps show by means of coloured areas or symbols the distribution of soil types or groups of soil types in the landscape. The maps either have detailed legends or are accompanied by memoirs. These memoirs contain a wealth of information of value to archaeologists. Climate, geology, geomorphology, vegetation and land use are described. Mapping units are defined, with full descriptions and illustrations of typical profiles. The mapping unit used depends on the scale of the map in relation to the scale of the pattern of soil distribution in the areas being mapped. Where two or more types of soil are mapped together as a unit, either because they are too closely intermixed or because their boundaries are too merging to be shown as a line at the scale being used, their relationship to a differentiating factor is given.

It may be parent material or topographic position, so that distribution can be determined with the aid of geological or relief maps of larger scale, or it may be a distinguishing feature such as vegetation, land use, or colour of the surface soil, so that the types can be recognized on the ground.

Soil *phases* are of particular interest to archaeologists. Phases are features such as shallowness of profile, stoniness, waterlogging or salt accumulation which occur in the soils in particular situations. They are indicated on soil maps by suffixes to letter symbols and by neutral symbols overlying coloured areas. Phases are often related to effects of erosion or disturbance of the soil or to irrigation or hydrological control practised by man, and subsequently, whether natural or man-induced, can be limiting factors in use of the soil.

Large scale soil maps are produced largely from field survey. The surveyor defines his soil types by digging pits and establishes the boundaries between them by making frequent borings with a soil auger, and plots his findings on field maps at a scale larger than that at which the map will be published. In the published maps the soil infor-mation is usually superimposed on existing topographic maps; the maps of the Soil Survey of Great Britain are based on the 1:25,000 and 1:63,360 sheets of the Ordnance Survey.

Since it is impossible to establish every point on a map directly from field work, maps have a certain inaccuracy, in the form of imprecision of boundaries and impurities in the units mapped. The amount of field work done is aimed at keeping the impurities below 15%. Aerial photographs are used to fill in the details in areas in which the general pattern of soil distribution has been established on the ground and has been found to produce, or be related to, a visible feature such as surface texture or reflectivity of either the soil itself or the vegetation. Use of geological and relief maps in conjunction with field work enables the surveyor to cut down the amount of work, by following up changes in soil related to a geomorphological feature or a change in parent material. However, soil movement by erosion or creep can blur the boundaries or shift them downhill, and when using a soil map it is necessary always to be sure on which side of a boundary one is and to observe its coincidence or lack of coincidence with other features. Interference by man can be a factor in the disturbance of the natural relation between the soil and the other factors of the environment, and lack of correlation between a soil map and the soils as found on the ground should be investigated by the archaeologist with some care.

The smaller the scale of the map the greater is the ratio of interpola-tion from other evidence to direct observation in the field, and the greater is the uncertainty in the positions of the boundaries. With decreasing scale the complexity of individual mapping units increases and the amount of subsidiary work by the map user, either in the field or by comparison with geological and relief maps, increases. For the purposes of archaeology, soil maps of scale smaller than about 1:500,000 are probably not worth using, except to get a very general idea of the types of soil likely to be encountered in an area one has not yet visited. Any deductions made by the armchair archaeologist in comparing an archaeological distribution map with a small scale soil map would need the most rigorous checking in the field.

Reconnaissance soil maps are available at scales of 1:500,000 or 1:1,000,000, for many parts of the world. They are usually maps of one country or a large region, produced by national soil surveys and based upon small scale aerial photographs or interpolated from geological and relief maps with little field work. They are of very little, use to the archaeologist. Detailed soil mapping is in progress in many countries and maps are being published mainly at scales of 1:10,000 to 1:100,000. No country is yet covered by these large scale maps. A scale commonly used is 1:25,000 and maps at this scale are very useful for archaeological work, sheets of a size suitable for easy handling covering areas suitable for field work and having sufficient precision and detail to be of great value without much subsidiary work with geological and relief maps. An area on the map comfortably taken in by the eye approximates to the area of countryside visible from a vantage point in rolling country. Soil maps of larger scale still are sometimes produced for individual farms by agricultural advisory services, or in some countries for land assessment for taxation purposes, and would be of value in connection with individual excavation sites. These maps are not usually published, but may be available for use by interested inquirers.

The mapping units of large scale soil maps are usually based upon the *soil series*. A soil series is a group of soils having similar profiles, and is named from the place where it was first described. The degree of variation among the soils of a series is defined in the map legend or memoir, and is confined to small differences in depth or degree of development of horizons or variations in texture, due to slight variations in the parent material or to differences of slope or aspect, and to variations in colour or structure in the surface horizons attributable to a different history of land use. Phases are mapped where greater slope, or more disturbance on a gentler slope, has produced a shallowness or stoniness of marked degree while it is still apparent that the soil is, or was, of the same type as the rest of the series, where impeded drainage has induced gleying or where salt accumulation has occurred in profiles otherwise similar to the rest of the series.

Within areas mapped as a series there may be small areas of other soil types. When these occupy less than about 15% of the area they may be regarded as impurities, and their nature and distribution may be defined or left unspecified. A degree of admixture of two or more series too great to be left as an impurity, but too closely intermixed or with too merging a boundary to be mapped separately at the scale being used, is defined and mapped as a *soil complex*. The same map may use the soil series as a first stage unit and the soil complex as a second stage unit in areas of different complexity.

In general, soil surveys are concerned with the soils as they are now, since their purpose is to provide information for people using or handling the soil. In some countries surveyors work in co-operation with pedologists studying soil genesis, and are able to discuss the history of the soil in greater detail. However, they are often hampered, whether they realize it or not, by the lack of a time scale and the lack of knowledge of past environmental conditions and land use. Many soil surveyors include a section on the history of settlement and land use in the memoir, but, not being archaeologists, they tend to draw on rather generalized and often out-of-date publications or to misinterpret the significance of field evidence. The soil scientist needs the results of the archaeologist's work as much as the archaeologist needs his.

When planning work in an area it is worth while to get in touch with the organization responsible for soil survey. In most countries this is a government body, and in some areas there may be local headquarters from which the surveyors work. These organizations and people may be very glad to co-operate with archaeologists, and may lend equipment such as augers and colour charts and explain the procedures for soil description.

Government soil surveys are usually directed towards agricultural or forestry needs. Surveys are also carried out for engineering purposes, and archaeologists working ahead of road building or urban development might be able to obtain useful information from developers or road engineers, though since these surveys are usually carried out by commercial contractors access to information may be difficult to achieve. However, with increasing co-operation between developers and the survey and rescue archaeologists this is an area of common interest which should be exploited.

Still more inaccessible is the wealth of information about soils in military hands. It probably exists in great quantity since it is essential for planning of access and mobility, particularly in the less developed parts of the world, and just those in which the archaeologist is particularly short of information. However, much of it probably lies in the archives of a nation other than that to whose territory it refers, and represents a terrible waste of data, obtained by extremely sensitive aerial survey methods beyond the means of the poor countries who need it for planning and agricultural development.

Survey Method

Survey work is best carried out in the season when there is least obstruction of access and of view of the soil by natural vegetation and growing crops. In temperate regions the winter is best; arable soils are moist and bare, or show only a little growth. Hedges and trees are leafless and allow wider views than in summer, and farm workers are less fully occupied and have more time to talk. In hotter and drier regions survey is also best carried out after crops have been harvested and while the ground is being prepared for the next sowing, but this will be the dry season; later, when the wet season has begun, soil colours may be seen better in the moist condition, but very rapid growth soon restricts visibility and access.

If soil maps are available the procedure of survey is to identify the soil types of the mapping units and observe their range of variation, and in particular the occurrence and significance of phases, to recognize the members of composite mapping units and to understand their mutual distribution, to study the boundaries between units and to observe departures on the ground from the boundaries as mapped, and to try to correlate the archaeological part of the survey with any of these soil features.

If soil maps are not available the archaeologist must start from scratch. His success will depend on the time available and on the complexity of the soils. He can in any case make some attempt to identify and describe the most frequently occurring profile types, and particularly those which seem to have some significance in relation to his archaeological findings, even if the inter-relation of the soils with each other and with factors of the environment does not make sense to him. Once the information is accurately recorded, it can be interpreted later when someone with a greater understanding of soils is available for discussion.

Soil pits dug to examine the profile should be deep enough to reach the C horizon and wide enough to allow small scale lateral variations to be seen. The optimum size is different for different soils, but in practice size may be limited by shortage of time or manpower. Usually a hole about a metre square is both a convenient size to dig and adequate for the purpose. Description of soils, whether surface soils or the whole profile in a pit, should follow the procedure set out in Chapter 14. Pits must be filled in after they have been studied to avoid accidents to people and animals.

Much time can be saved by examining profiles exposed in road cuttings, quarries, drainage ditches, pipe trenches and other ready made holes in the ground, but with the proviso that the location of such holes and exposures might be related to some feature of the site or soil. This limitation can be overcome if there are enough different types of exposure in the area. Quarries are located where solid rock is close to the surface, and this tends to be at the brow of a hill where erosion and soil creep have reduced thickness of superficial deposits or disaggregated rock and the soil they carry. Road cuttings expose the soils of convexities of the landscape, and routes may follow the better drained soils and firmer ground. The same applies to some extent to holes for house foundations, whereas drainage ditches obviously concentrate on lower lying and wetter situations. Large pipe trenches are particularly useful during the brief period for which they are open, as they often take a direct cross country route. Where they do deviate, unless obviously going round some superficial obstruction or avoiding a property boundary, it might be because of hard rock close to the surface, or a boggy area.

Roadside soils carry the further limitation on general applicability that the presence of the road itself may have contributed to their characteristics. Cultivation is rarely practised right up to the edge of a road, and the soils there might be the only example in the area of an uncultivated profile, which may be that of the localized woodland conditions imposed by a hedge or may be a former soil buried beneath a hedge bank, throw out from roadside ditches or the spoil from the road cutting. The vegetation beside a road might give a better idea of the potential productivity of the soils of the area if those of the fields have been intensively grazed or much altered by cultivation, but some caution is needed here, since road dust contributes a great deal of finely divided mineral material, which might provide nutrients in short supply locally if surfacing materials have been brought from elsewhere.

Exposures which have been open for some time will have modified the soil, particularly in its moisture characteristics and oxidation conditions. It may be possible to cut back the section sufficiently to get a better idea of the profile in its normal state, but this should not be done at road cuttings or in drainage ditches, where the stability of the cutting must be maintained. With all these reservations in mind, these sorts of ready made exposures usually have the very great advantage of showing lateral variation of soils continuously over much greater distances than is possible in purpose dug soil pits.

Soil study preparatory to excavation involves a similar procedure to that in conjunction with survey work, but in a more restricted area. The aim is to establish the range of variation in the immediate area of a site, and to observe those aspects of the local soils which may have been significant for the people living there or which may be directly related to their activities. Something more is needed than the two or three soil pits on the fringe of the site, which more often than not produce evidence of a greater extension of the site itself than was first supposed or of the existence of subsidiary or even totally unrelated sites. The soil pits can so easily develop into excavations in their own right, and though total investigation of an area may be the ideal in excavation, it is embarrassing if it happens unplanned. In areas of dense occupation stretching over a long period it will be impossible to find an "undisturbed" soil. The number and siting of soil pits will then aim to reveal the distribution of particular effects of disturbance and the effect that each type of disturbance can be expected to have had in the area of the site to be excavated. The information gained not only helps in both the planning and the interpretation of the excavation but also adds to the total amount of information about the effect of man on the environment in the locality. It is a great aid to planning if the soil pits can be dug well in advance of the main season of work on the site, since an idea can be obtained of the depth of soil to be moved and the ease with which it can be handled; it is best done during a period of local exploration during the preceding winter or off season, before the urgency of the work on the site itself distracts attention from the importance of wider exploration.

Local Terms

Farmers in many areas use dialect words to refer to particular types of soil materials, often the **B/C** or **C** horizon where these have distinctive characteristics which affect drainage and cultivation practices. Examples familiar to archaeologists in Britain include the "rab" of Cornwall and Pembrokeshire, which, in Cornwall, is the product of weathering which may go back in part to Tertiary times, and of periglacial activity on the weathered material, on granite, and "shillet", the softened, oxidized and fragmented shale C horizon of parts of Devon. Compacted and indurated horizons of periglacial origin are particularly likely to have attracted local names, like the "roach" of parts of the northern Pennines. Discussion with farmers as to the meaning of these terms, the limitations the materials they refer to impose on agriculture, and their distribution, provides the field archaeologist with a short cut to a lot of useful information.

14

Study of Soils and Deposits During Excavation

Soil Description

Standard terminologies exist for the description of soils. They differ slightly from country to country but are closely comparable and the difference lies mainly in the emphasis placed upon different characteristics rather than on the terms used. Description of soils is made in the field and includes assessment of properties determined only by visual inspection and manual manipulation and the only accessory needed is a standard colour chart. The terms are precisely defined and their use allows comparison between soils described by different people. It is perhaps unfortunate that the terms are those of ordinary descriptive language, so that a description made with the best of intentions by the untrained person may use all the right words in the wrong sense and be more misleading than no description at all.

The archaeologist needs to understand the terminology in order to read soil maps and obtain information about the soils of the area in which he is working from the publications of soil scientists. He needs to be able to use the terminology himself during survey work so that comparison can be made between soils seen at different times or described by different members of a team, so that he can make comparisons and correlations with work in other areas, and so that he can discuss the soils with soil scientists.

During excavation the practice of making standard descriptions of all soil and sedimentary material encountered has four functions:

1. Within the site itself, to ensure that everyone knows what is meant when a particular material is mentioned, and so that comparison can be made between materials encountered on different parts of the site and at different stages in the excavation. Subjective assessment of

colour and of tactile properties depend very strongly on context: one man's greeny-grey clay is another man's pale brown loam, depending on the colour and texture of the surrounding materials in the area in which each is working.

2.   For communication with a wider public, other excavators who may encounter similar materials in their own sites and who need to be able to get an accurate idea from the excavation report, and soil scientists who may need the information to increase their knowledge of the history of the soils of an area or who may be able to help the archaeologist in interpretation. Soil scientists simply will not read excavation reports if they suspect that terms are being used loosely and inaccurately.

3.   For communication with specialists working on other aspects of the site, who may need to know, for instance, the context of preservation of recovered materials or biological remains.

4.   For interpretation. Full and accurate description greatly aids understanding, at the time and later when the site may no longer be open. The close examination needed to complete a full soil description is itself a stimulus to interpretation: when the description has been made it will be found that some questions have already been answered. This applies particularly to matters concerning the boundaries between layers since a really close examination of the boundary can give an understanding of the processes by which differentiation or superposition occurred.

Descriptions should be made by the people most closely in touch with the soils as they occur during excavation, often the most lowly of site personnel who in their many hours of trowelling can become very sensitive to variations in soil characteristics detectable by hand and eye and should be encouraged to record them as they go along. Then, when the time comes to draw sections and plans showing the materials to which the descriptions refer all the information is available and includes details not necessarily observable in section or present just where the section or plan is placed.

The process of teaching descriptive procedures to the trowellers can give the excavator a closer familiarity with the materials than he sometimes gets in the course of directorial visitation to the various parts of the site. The material being described must be handled by teacher and taught and the correct assessment decided by discussion between them, and the teacher then forms the link between people working in different areas, ensuring that their descriptions are equivalent, and are sufficiently precise to bring out both the similarities and the differences in the materials being encountered. On large excavations it would be useful to appoint one person in a supervisory capacity, present throughout the excavation, to act as co-ordinator of soil descriptions.

The characteristics to be included in a full soil description are: colour, texture, structure, consistence and swelling, and the nature of the boundaries between horizons or layers. For soil surveyors and most users of their surveys description is most important for that moisture condition in which a soil is usually to be found, though a full description always includes characteristics in both moist and dry state. For the farmer it is pointless to describe a soil dry which never dries out in the normal course of events, but in the archaeological context, where the need is to record as much as possible for the purpose of comparison, and where excavation exposes materials to abnormal conditions in any case, differences in characteristics under different conditions do provide valuable information and give increased chance of recognizing similarities and differences. In particular, colour and consistence, which change markedly between moisture conditions in some materials but hardly at all in others can be very sensitive indicators.

Colour

Standard description of colour requires the use of a Munsell Soil Colour Chart, which should be available on every survey and excavation. Its high cost might prevent purchase by many individual field workers but every institution organizing or sponsoring field work should make them available. The charts are in the form of a loose leaf book in a plastic-covered binding, and are very robust. Under normal conditions they survive years of constant use, and the cost per excavation is trifling compared to cost, for example, of photography. Munsell charts are obtainable in England from the address given on page 280. The Japanese Standard Soil Colour Chart is available at about half the price of the Munsell chart. Its colours and notation are equivalent to those of the Munsell chart but the standard colour names which are used in conjunction with the numerical notation are slightly different, and it should always be stated which is being used. The major difference between the two charts is that the Japanese one has no holes in the cards on which the colour flashes are mounted, so the sample must be held in front of the chart. This has two disadvantages: the sample cannot be held flush with the page and so equivalent light incidence and close comparison are difficult to achieve, and whereas with the

Munsell chart only the backs of the cards come into contact with the soil and the colour flashes themselves are kept clean, with the Japanese chart the flashes will very quickly become dirtied and scratched by the soil. It is particularly difficult to work cleanly with moist soil under field conditions, and I suspect that the Japanese charts have a life expectancy so much lower as to cancel out or reverse the initial advantage in cost.

The range of colours covered by the two charts is slightly different, and since the Japanese one was specifically designed for use in Japan it may well be better suited to work in countries having a similar range of climate, and in particular a similar abundance of volcanic soils, than the Munsell chart, which was developed in temperate America. In general, practical differences in handling apart, it is probably best to use the chart which is in use by soil surveyors in the country in which one is working. This will help to avoid confusion when colours are referred to by name alone.

Colour charts are arranged by hue, value and chroma, the three parameters by which any colour can be described. The brown colours of soils can mostly be represented by hues in the red, R, through yellow-red, YR, to yellow, Y, range, of different value (darkness) and chroma (intensity). Each page is a different hue: 10R, 2·5YR, 5YR, 7·5YR, 10YR, 2·5Y, 5Y. On the pages the colours are arranged in columns of equal chroma, with the low chroma column, 0 or 1, on the left and high chroma, 8, on the right. In each column value increases from low, 2·5 (2 in editions of the Munsell chart prior to 1971) at the bottom to high value, 6 or 8, at the top. The range of value and chroma covered varies from page to page, since the charts have been developed in conjunction with soil survey so that the most useful parts of the range are most fully represented. Extra pages are available of hues redder than 10R, for use with tropical and subtropical soils, and a page containing neutral greys, some low chromas selected from the standard 5Y chart and some green-yellow, green, blue-green and blue hues of low chroma, is available for use with gleyed soils.

The notation is written in the form: hue value/chroma, e.g. 10YR 4/3. Estimates can be made between the colours given on the charts when it is impossible to decide between adjacent flashes. When this is done, 0·5 is used to express a point between whole numbers. Estimates can be made beyond the range of the charts on the few occasions when a soil of higher chroma or lower or higher value than those given is encountered.

Full description of colour includes a name in words as well as the numerical notation. The standard names are printed opposite each page of the charts. Since each name refers to more than one numbered colour flash, they are less precise, but names are much more convenient to use. While the full description of a soil gives both name and notation, the name alone can usually be used in continuous text or in speech, and these standard names should always be used in place of the perhaps more evocative but highly subjective colour descriptions in common use.

Light conditions for colour description are important. Though the sample and the colours on the chart may remain in the same relation under different lighting, the sensitivity of the eye to slight colour differences is reduced in poor light and in bright sunlight, and choosing the best match from a group of possibles is more difficult. When working in bright sun samples should be taken to a large area of shade or indoors to a position near a window. Otherwise, even if chart and samples are held in the shade the eye takes time to adjust to the change from sunlit surroundings and work is either inaccurate or very slow. Early and late in the day, though the light may be bright enough, colour contrasts are reduced as the colour of the light itself changes, and in artificial light determination of colour may also be difficult or inaccurate in certain parts of the colour range. In England, for several months during the winter all colour determinations have to be done by early afternoon.

The ease with which colours can be matched varies in different parts of the colour range. In particular, the eye is not very sensitive to slight differences in the region where both chroma and value are low, especially if colours of high chroma are also visible. Masks are supplied with the charts so that all colours on a page can be hidden except for a block of four between which a choice is to be made. A black mask greatly facilitates determination with dark soils and a grey one helps when working with light soils.

Colour should be determined on soils in both moist and dry condition. The technique is to gouge out a lump of soil, break it open and determine the colour of a fresh, untouched surface. If the soil is already moist, a piece is left to dry and the colour again determined on a freshly broken surface. If it is dry to start with, a lump is moistened until it does not darken any more, and left until it is uniformly moist throughout and then broken and the colour determined. Soils pass rather rapidly through the range of moistness during which the colour changes, and it is not difficult to achieve consistent results. Observation of colour changes during drying, and of different rates of drying of the materials in a section can be very useful in showing up layers or horizons of different texture, structure or organic content, but care must be taken to ensure that all colours recorded in the standard description refer to equivalent moisture states, and to note separately any colour contrasts which may be observed as a section dries but which disappear when uniform dryness is reached.

Few soils and archaeological deposits are uniform in colour, and the pattern of variation in relation to changes in texture, structural planes, presence of roots or animal burrows and positions of stones or archaeological features should be fully described. In the normal profile description carried out by a soil surveyor, colour variation is dealt with fairly briefly, since the experienced surveyor will usually understand its significance. In the stratified deposits and disturbed soils of an archaeological site details of mottling and colour variation may give clues as to the origin of a layer or the completeness of a profile, or help to sort out the features caused by post-depositional alteration from those of the original material, and so should be described with care.

Colour variation may take the form of discrete mottles, a continuous mosaic or marbled effect, whose relation to the soil structure should be noted, thin lines or diffuse bands, which may or may not be continuous and may be related to textural boundaries, or a gradual change from one colour to another, vertically or horizontally or in relation to some feature. The description of mottles and bands should be given in terms of sharpness or diffuseness, size and shape, frequency and distribution, and it should always be remembered that one is viewing a three dimensional effect in two dimensions, whose orientation may be arbitrary in relation to the processes which produced the effect. It should also be remembered that trowelling can elongate and orientate a pattern produced by any material that smears, and that relationship of colour variation to soil structure cannot be observed if the structure is obliterated with the trowel.

Texture

The feel of the soil to the fingers, its texture, is determined by the relative proportions of particles of different sizes. Description of texture is made in terms of defined particle size grades and textural classes composed of mixtures of these. Different definitions of the size grades have been adopted in different countries and for different purposes; those in current use are given in Fig. 25. Mixtures of sand, silt and clay can be represented by positions in a triangular diagram such as that given in Fig. 26b which is the form used in conjunction with the U.S.D.A. size grade limits, or Fig. 26a which has recently superseded the other for use by the Soil Survey of England and

Wales (Avery, 1973) and is based on the size grade limits used in engineering.

Accurate placement of a soil in its textural class requires laboratory determination, but the classes have come into use by way of field survey and the chosen limits largely represent those which can be judged in the hand with some confidence. The effect of differences between the size grade definitions are to some extent cancelled out in the process of assigning a soil to its textural class, since assessment of texture is learnt in the field.

The ideal way to learn soil texture is from an experienced soil surveyor. Failing this, a more laborious way is to study the profiles

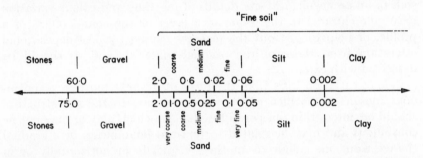

Figure 25. Particle size grades according to the British system (British Standard 1377) above, and the American system below. Logarithmic scale in millimetres.

described in the memoir accompanying a soil map. Sometimes the same exposures can be found, or others can be examined which compare well with those described. If the archaeologist does this in the area in which he is working, he will learn to recognize the textures of soils commonly occurring locally, even if he doesn't cover much of the possible range of textures, and in doing so will learn a great deal about the soils related to his site or survey. The process could easily be combined with soil studies preliminary to excavation. However, this way of learning is only possible in those areas for which a soil survey has been published. In the absence of both survey and surveyor the following guide to texture will be of some help.

Clay. Particles are too small to be seen or felt. When worked with the fingers clay leaves a deposit on them.

Silt. Particles are too small to be seen with the naked eye, but towards the coarser end of the silt grade they can easily be seen with a ×10 hand lens. Silt particles cannot be detected individually by the fingers

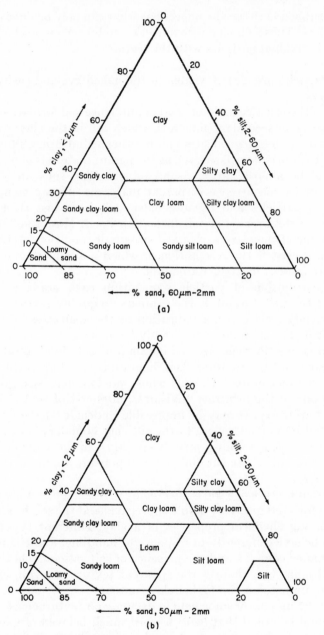

Figure 26. Soil textural classes. (a) British system (B.S. 1377). (b) American system.

but silt rubbed between the fingers has what can only be described as a "silty" feel! Those who have no objection to biting a morsel of soil can feel the individual particles with the teeth.

Sand. Particles are clearly visible to the naked eye and easily felt by the fingers.

Few soils consist entirely of sand or silt, since soil formation implies the presence of secondary minerals, which fall in the clay grade, but deposits on archaeological sites and in other sedimentary situations of interest to archaeologists, as well as the parent materials of many soils, may well be almost entirely sand or silt. Clay soils are common, because of the occurrence of parent materials already composed of clay sized particles and because older soils have lost their coarser particles in the process of soil formation itself. Most soils, however, fall into one of the classes of mixed size grade, and the determination of textural class is the recognition of which grade predominates and which are present in smaller amounts.

The distribution of particle size within each grade affects the apparent texture of a soil. It is customary to qualify textural description of sandy soils with an indication of the coarseness of the sand component. For example, a fine sandy loam is a sandy loam in which the sand is fine. Particle size is a continuum and division into grades is arbitrary, but the textural class diagram is based upon the concept of three distinct grades. In soils whose predominant size grade falls close to one of the arbitrary divisions, judgment of textural class is similarly arbitrary. It may be impossible to decide what to call a soil dominated by particles close to the silt-clay boundary, but it is very different from a soil falling, strictly speaking, in the same position on the diagram but whose silt and clay components are of very different particle size.

Apparent texture of a soil is affected by several factors. Clay minerals may be clustered into very stable aggregates and the soil feel like a silt while having chemical properties of a clay. Different types of clay mineral behave differently in response to handling, and the nature of the exchange complex and the presence of salts also affects the feel of a soil. Humus, too, changes the apparent texture of a soil, in general giving an impression of a higher clay content.

Figure 27 may be of some help in deciding on textural class of a soil.

It should be noted that *loam*, a word much beloved of excavators, is not just the brown humic topsoil and similar materials but is a precisely defined textural class. The term implies nothing as to colour, fertility, or ease of cultivation. Many soils described as loams

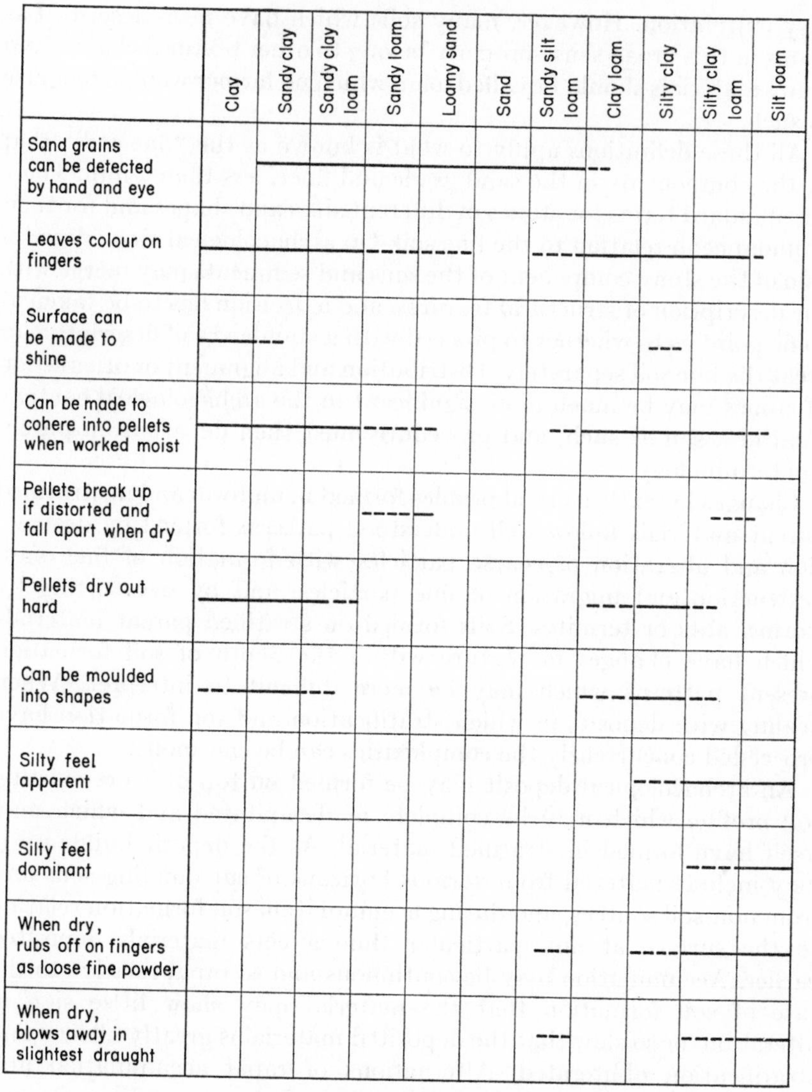

	Clay	Sandy clay	Sandy clay loam	Sandy loam	Loamy sand	Sand	Sandy silt loam	Clay loam	Silty clay	Silty clay loam	Silt loam
Sand grains can be detected by hand and eye		——	——	——	——	——	— —				
Leaves colour on fingers	——	— —	— —	——	——		— —	——	——	——	—
Surface can be made to shine	——	— —									
Can be made to cohere into pellets when worked moist	——	——	——	— —	—		— —	——	——	——	—
Pellets break up if distorted and fall apart when dry				— —	— —						— — —
Pellets dry out hard	——	——	— —					— —	——	— —	
Can be moulded into shapes	— —	——	— —					— —	——	— —	
Silty feel apparent									— —	—	
Silty feel dominant										— —	—
When dry, rubs off on fingers as loose fine powder							— —		— —	—	
When dry, blows away in slightest draught							— —			— —	— —

Figure 27. A guide to assessment of soil textural class.

by excavators may in fact be so, since the term was adopted, perhaps unfortunately, from the one in common use by farmers and gardeners and meaning a fertile soil containing clay and humus to provide and hold nutrients, allow development of a good structure and hold moisture, and sufficient sand and silt to give good drainage and allow easy cultivation. However, many soils which have been described as loams in this broad sense properly belong to other textural classes, and many materials should be called loam which no farmer would recognize as such.

All these definitions apply to what is known as the "fine soil", that is, the components in the sand grade and finer, less than 2 mm. There are standard terms for stones of different sizes and shapes and for their abundance in relation to the fine soil. On archaeological sites, description of the stony component of the soils and sediments may merge with the description of structural features, and a decision has to be taken at some point as to whether to proceed with a standard soil description or treat the fine soil separately. Distribution and alignment or orientation of stones may be much more significant in the archaeological context than in a soil as such, and procedure must then depart from purely soil terminology.

Changes in texture in soil profiles formed in uniform and unstratified parent materials follow well understood patterns formed by destruction and alteration of coarse particles with formation of fine ones, destruction and migration of fine particles, and by size sorting by worms, ants or termites. Soils formed on stratified parent materials which have changes in texture within the depth of soil formation present patterns which may be more difficult to interpret. When dealing with deposits in which stratification and soil formation have proceeded concurrently the complexities can be enormous.

An archaeological deposit may be formed on top of a pre-existing soil profile, which may be complete or decapitated and which may itself have formed in stratified material. As the deposit builds up it may include material from various horizons of surrounding soils and from non-soil sources, and during accumulation soil formation relating to the surface at any particular time affects materials deposited earlier. Accumulation may be continuous and so rapid relative to the rate of soil formation that the material may show little sign of alteration, or so slow that the deposited material is greatly altered and stratification obliterated. Alternations of rapid accumulation and periods of stability may occur. Temporary surfaces may be preserved or may be eroded away before accumulation is resumed. Finally, after accumulation ceases a long period of soil formation may superimpose its effects on all those of the earlier history of the deposit.

In interpretation of such soils and sediments study of texture is the tool which gives the best hope of recognizing changes in material and mode of deposition, provided that the processes by which texture can be changed *in situ* are taken into account. For interpretation of soil processes texture has to be studied in conjunction with observation of colour and structure.

Structure

Structure is the way in which the soil materials are arranged among themselves. In Chapter 4 some attention has been given to the formation of soil structure and to the structures characteristic of different types of soil. Some structural forms are illustrated in Figure 21, and the standard terms referring to them given. Once it is realized that soil has structure, it becomes a most noticeable characteristic, and observation in garden and countryside will soon provide examples of most forms.

Structure may be developed to varying degrees. Some soils, and particularly sediments in which little soil formation has taken place since deposition, are structureless, being either *massive*, that is, holding together as a coherent mass, or *single grain*, incoherent and disintegrating into separate particles when disturbed. *Weak* structure is visible but the peds are indistinct or poorly formed and break easily when disturbed, many of the peds disintegrating into structureless material. *Moderate* structure has mostly well formed peds, but on disturbance some break up or disintegrate entirely. *Strong* structure will stand up to much rough handling, such as that suffered during cultivation, and the peds remain whole.

There may be more than one stage of structure. For instance, a weak coarse granular structure may break into strong fine blocky peds.

Care should be taken to avoid misuse of the word *crumb*. In common parlance a soil is often referred to as having a good crumb structure. What is usually meant is that it has a good structure, the word crumb being unnecessary and often wrong. Crumbs are the irregular, porous peds looking much like bread crumbs which are characteristic of worm soils. Crumb structure can be most readily observed in the section of a lawn at the edge of a flowerbed. Where there are no earth worms, structures are granular or blocky or one of the other forms.

The presence of structure is a useful indicator of the degree to which materials encountered in archaeological deposits have been subject to soil formation, since, apart from swelling clays, sediments which have not been in an appropriate situation long enough for plant cover to be established before they are buried too deeply to partake of soil processes develop no structure.

The development of soil structure involves decrease in bulk density of the soil. In a buried soil, compression may occur, closing up the pores and packing the peds more closely together, but ped boundaries

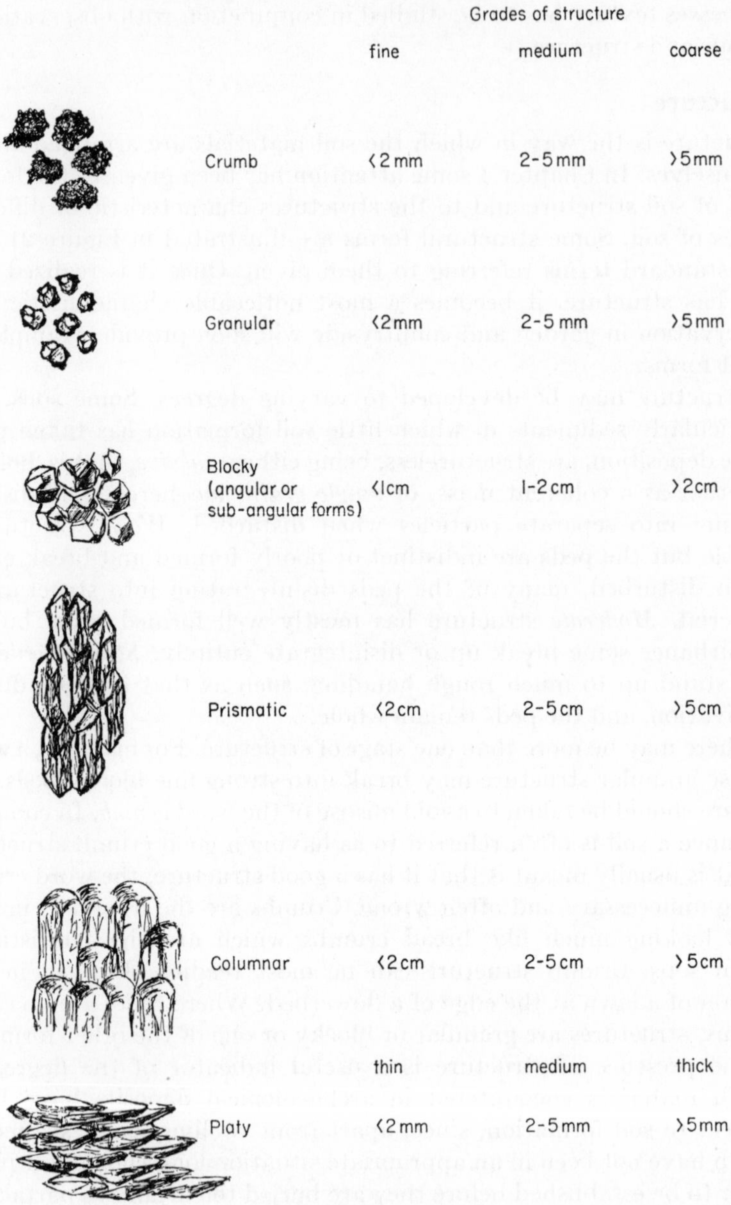

Figure 28. Soil structural forms.

usually remain detectable. The well developed, strong structure of an A horizon of mull form undergoes modification under burial so as to develop a blocky or prismatic structure on which is superimposed a large scale vertical cracking, as seen in Fig. 35. The difference in structure between **A** and **B** horizons of a buried soil can be a useful indication of completeness of the profile when mineralization of organic matter and changes of colour due to reduction and mobility of iron have obscured the colour distinction. In situations of gradual accumulation variations in degree of development of structure indicate periods of greater or lesser stability.

In examining the structure of a soil, presence or absence of *clay skins* should be noted. If developed to any marked degree, clay skins are visible with the help of a hand lens, and can usually be seen by the naked eye, in the form of linings to pores and coatings to ped faces, where the shiny, fluted surfaces called slickensides may also sometimes be seen. Clay skins are often a slightly different colour to the soil in which they occur, since they are composed of the finer component of the colloid of overlying horizons. They may be a darker brown colour if humus substances are present, or a redder colour, since the amount of iron oxide carried by or associated with the fine clay particles is high in proportion to the total content. An argillic horizon often has a reddish tinge imparted by the clay skins.

Presence or absence of clay skins is essential evidence of soil profile type, and can be a valuable indication of completeness of a profile: since clay skins cannot appear at the surface of a soil, a buried soil in which they do so must have been truncated, or so disturbed by cultivation as to bring soil from the illuvial horizon to the surface at a time not long before the surface was buried. Imposition of an illuvial clay horizon on stratified deposits is mentioned on page 302.

Consistence

Consistence is the way in which a soil responds to handling, and is described for the dry, moist and wet conditions. The terms for dry soil are *hard* and *soft* and the corresponding terms for moist soil *firm* and *friable*. In both conditions they describe the resistance of the peds to crushing between finger and thumb. Hard or firm soil is difficult to crush, while soft or friable soil gives way under very little pressure. A soil which is friable when moist is not necessarily soft when dry; it may indeed dry out to become very hard. These terms can be qualified: slightly, moderately, very, extremely, to give a wide range of possible responses. Cemented soils are a special category of hardness, which does not alter on wetting unless the cementing substance is soluble in water. Very strongly cemented soils are *indurated*.

It should be noted that "friable" is another of those words used freely and loosely by excavators in describing soils; it may be difficult to remember that it refers only to moist soil and only to the disintegration of individual peds, not to the way in which a soil breaks into peds.

Consistence when wet, as distinct from moist, is expressed in *stickiness* and *plasticity*. Stickiness is assessed by the degree to which the wet soil pressed between the finger and thumb sticks to them when they are pulled apart, and is given in degrees from non-sticky, which leaves the fingers clean, to very sticky, which forms an extensible bridge between thumb and finger rather than letting go. Plasticity is assessed by the degree to which "worms" can be formed by rolling the wet soil in the fingers, and has degrees from non-plastic, which will not form worms, to very plastic, which readily forms long, thin and coherent worms. This characteristic is very sensitive to moisture content, but is usually adjusts itself as the soil is worked so that a point is reached at which, in a soil that is at all plastic, its shape begins to be amenable to working.

Stickiness and plasticity are clearly of the utmost interest to the archaeologist dealing with pottery using cultures since they relate to use of materials for potting. Like the other aspects of consistence they depend not only on the proportion of clay in the soil but also on the type of clay minerals present, their size distribution and the nature of the cation exchange complex, and on the amount and nature of iron oxides and humus associated with the clay.

Swelling

The last property of the soil itself which should be observed in making a soil description is whether or not the dry peds swell when moistened. Clearly this does not apply to structureless, single grain soils, and is not relevant to very sandy soils since the cause of the swelling is the expansion of the colloidal material. Soils may be non-swelling, slightly swelling, swelling, and strongly swelling. The swelling nature of a soil is strongly related to the development of structure. Blocky, columnar and prismatic structure are to a large extent produced by contraction and expansion of the colloid. Swelling clays are no use for pot making, and the use of unfired soil materials in building is also limited by the property of swelling.

Other Terms Referring to Soil Characteristics

Terms in common use which do not form part of a standard soil description are *tilth* and the adjectives *heavy* and *light*. Tilth is referred to by Russell, E. W. (1971) as that which every farmer can recognize with his boot but no soil scientist can describe. It is a quality compounded of structure, texture and the state of the colloidal material. A soil with a good, or *mellow*, tilth retains its structure against both disruption and compaction, whereas one with a *raw* tilth disintegrates into a structureless mass and compacts into clods when cultivated. The "farmer's boot test" is that when walking over ploughed fields, where the tilth is raw the soil sticks to the boots and where it is mellow it leaves the boots clean. Tilth degenerates as a soil continues in cultivation, as its organic matter content and nutrient status fall, and is restored by a period under grass.

The terms heavy and light are more strongly allied to texture than is tilth, a clay soil generally being heavy and a sandy one light. However, the terms originated from the ease with which the soil could be cultivated, and structure and consistence are very important: there is no regular progression from heavy to light as clay content falls or as sand content rises, and soils change their characteristics under management. A clay soil with a good tilth, such as one recently cleared of forest, with a high organic content and base status and a strong structure, will not be a heavy soil. The concept is important archaeologically, since a light soil could be cultivated by primitive tools, whereas a heavy one could not. A light soil requires one horse to pull the plough, a heavy one needs a team of four, or eight oxen, and its cultivation may imply co-operation between neighbouring holdings in order to assemble large enough teams.

Boundaries

The thorough study of the boundaries between soil horizons or stratigraphic layers is of importance at least equal to that of the materials themselves, since boundaries give information about processes, either the processes of differentiation and migration of materials inherent in soil formation or the processes of superposition of deposits.

Boundaries may be *sharp* or *merging* (or, in American terms, abrupt, clear, or gradual). A merging boundary may be formed by a gradual change in the proportions of different components of a material, or by a gradual change in one or more characteristics of one or more of the constituents, such as texture, colour or structure. Merging can also be by mixing of the materials above and below. In describing a merging boundary the width of the zone of merging should be given. A sharp boundary is simpler, but it may be associated with merging features, such as a colour change which is sharp at the boundary itself but fades out a little way from it, and it may be marked by the presence of a material not forming a discrete layer, such as a scatter of stones, a line of flecks of charcoal or a segregation of some material making a very thin line.

In English terminology boundaries are *even, undulating* or *convoluted*, and the equivalent American terms are smooth, wavy and irregular.

In the study of boundaries methods of excavation are important. Every troweller knows the situation in which he thinks that the material is beginning to change but cannot be sure until he has gone well down into a zone of merging, and needs to get right through this zone before he knows whether or not it is a layer in its own right, or, in the process of cleaning up the surface of a layer he removes the upper part of it and destroys all evidence of the nature of the boundary. Use of sections is very important, and the practice of continually creating and destroying temporary sections during trowelling is to be encouraged. Whether or not major sections are left will be a matter of site policy, and in weighing the advantages of the open area method of excavation and the more traditional methods the needs of soil studies will usually come down on the side of retention of sections. This is because the short sections used in the open area method generally go from boundary to boundary and do not allow examination of the boundary itself, which is always just at the join with the horizontal or is a remnant suffering maltreatment at the top of the section. Where boundaries are sharp and even, the problem is minimal, but merging boundaries are much better studied in a section standing well clear of the present excavation level.

Another argument in favour of high sections is that they allow the study of processes of soil formation superimposed on stratification and expressed through several layers. The archaeologist who also has an understanding of soils will be able to make the decision as to method in the light of the overall needs of the site; where the two specialities are expressed in different people their differing needs can lead to a fight over every section.

Use of the Trowel

The archaeologist's trowel, if sharp and if used in the right way, is an ideal instrument for examining soils. Soil scientists are quick to recognize its superiority over the pocket knives and other assorted ironmongery with which they arrive in the field. However, it has to be handled in such a way as to bring out the characteristics of the soil and not obscure them.

Use in Excavating

Scraping in the traditional manner is not necessarily the best technique, and the use of the trowel should be adapted to the texture and structure of the soil so that it comes away in peds rather than smearing the surface. If the soil is made to part along its structural planes it almost always comes away from any archaeological items which may be present, unless they are very much smaller than the smallest stage of structure. The peds can be broken once they are loose to check for the very minute items such as seeds. Soil also comes away from stone structures and the sides of features such as pits much more cleanly if a picking rather than a scraping action is used. When the soil is not scraped all its characteristics are apparent during the trowelling process, and slight changes are noticeable, so that the boundary problem is reduced.

Use in Cleaning Sections

Only in very sandy materials is scraping a suitable technique for cleaning sections. In clay soils, scraping smears the clay, bringing finer particles to the surface, elongating mottles into streaks and pulling out stones. Experimentation is necessary to find the best technique for each material present which exposes an untouched surface in which soil structure can be seen and boundaries properly studied. A levering or flicking action, a right-handed person working from left to right so that each mark left by the trowel is removed as the next piece comes away, is often the best. It can only be done with a sharp trowel. The worn down, rounded trowel cherished by those who fear that their excavation experience might not otherwise be immediately apparent will not do the job. Trowels, like any other tools, must be replaced when they cease to function well. Very clayey materials may be too wet when excavated for section preparation to be possible. They should be left to dry until the structure begins to develop. A trowelled clay surface can sometimes be prepared after a few hours drying by rolling or flicking off the surface layer with the finger nails.

Examination of Sections

Sections should be examined when newly prepared, and preferably moist. It is always worth while, however, to look at them again after a period of drying, since differential drying and crumbling away or wind etching of less structurally stable materials can bring out features of stratigraphy very clearly. A site abandoned for some months may show differential lichen or algal growth picking out moister layers or those carrying more nutrients. Frost can also play a part in etching a section: sometimes after frost and in suitable conditions of moistness the whole surface is loose and can be flicked off to leave a fresh surface behind. Although features emphasized in these ways should have been detectable in the fresh state, differences between materials may have been underemphasized, or the pattern of variation missed. Also a section which when fresh did not present enough contrast of colour or texture to show well in a photograph may yield a dramatic picture after some weathering.

The best light for looking at a section is given by a bright but overcast day. In sunny weather it is best to work when the section itself is in shadow, but difficulties arise when the surroundings are in bright sunlight and every time you look up it takes some moments for the eyes to adjust again to the shaded section. The difficulty is even greater when drawing or making notes on white paper if this cannot also be held in the shade.

Oblique sunlight, from high overhead or from the side, casts shadows of projecting particles. This can be very useful in exaggerating textural differences: even sand grains throw tiny shadows, giving a strongly textured appearance to sandy layers in a generally clayey section. However, detail is obscured in the shadow of larger stones in an otherwise sunlit section. Direct sunlight from in front can be useful for seeing fine detail, but colour contrast is reduced, particularly as the colour of the light changes towards evening and all colours blend into a beautiful glowing orange.

Thus, repeated looks at a section in all light conditions, and for as long after the site has been abandoned as it still stands, should be combined with detailed examination of the fresh, moist section on a bright but cloudy day!

Really close examination is necessary to pick out some pedologically or stratigraphically very significant features. A hand lens should be used, and lumps of material should be prised out and broken open for careful examination. Sometimes it is easier to work in the comfort of the site hut rather than crouched in restricted positions, perhaps cold or wet, or in bad light. Pieces of the section can be taken away by levering them off with a trowel onto the palm of the hand or a sheet of aluminium foil, giving an untouched back surface, and can be taken to pieces working at a bench and using a lens to study the nature of finely interstratified material, the way a boundary is formed, or the details of colour mottling. A well studied section should have a ragged and well used appearance, not the smart and polished look so beloved of some excavators. Obviously, photographs of the whole rather than of details have to be taken before the beginning of detailed examination.

It must be stressed again that communication with the people who are trowelling the materials is paramount; an advantage of the open area method of excavation is that there is little delay between trowelling a material and recording it in section. The same person should do the two jobs if possible.

Techniques of Section Drawing

A practice deplored by many excavators but still frequently encountered is that of going over a section before drawing it and marking with the point of a trowel the boundaries of the layers which are going to appear on the drawing. This is to be most strongly discouraged. It fixes in the excavator's mind his impression of the boundaries and the relation of layers to features before the close examination needed for drawing begins, and allows little opportunity for revised opinions. It leads to the impression that boundaries are sharp, whereas many, probably most, are merging. It tends to produce even boundaries in place of those which undulate and undulating ones in place of the convoluted. Once these lines are drawn on a section they are very difficult to obliterate and unless the section is cut back and prepared all over again they continue to influence the excavator and everyone else who sees it.

The type of approach which leads to lines drawn on sections also produces drawings consisting of lines of dots joined up to make layers. In progress this type of work involves one person, usually the director or supervisor, with a tape, calling out numbers and a minion with a drawing board producing the dots. Often neither can see what the other is doing: I have seen such a pair working twenty feet apart, the draughtsman being above and behind the section. Apart from any other objections errors of reading, hearing and plotting go undetected. The result is like a child's puzzle: join up the dots and find the mystery layer.

If section drawing is carried out in conjunction with close examination, and if symbols are used to show different materials, boundaries and mixtures of materials can be shown as they really are and doubtful points can be cleared up by a little further trowel and lens work. Explanatory notes and sketches can be made on the drawing to clarify complicated areas and subsidiary sections can be drawn on a larger scale of small areas of great complexity or special importance.

The result of all this is not a beautiful clean section drawing which can be rolled up and put away until the site is written up. It may indeed be something of a mess, but it will carry all the information needed from which simplification can be made later, when a further stage of interpretation has been carried out and it is more apparent which features and variations are significant for the aspect of the site which is being illustrated by any particular drawing. Some excavators draw on site what is really the first stage in interpretation, and this is obviously justified if there are no problems. It does however prejudice the excavator in favour of his opinion at the time of drawing and runs the risk of loss of information whose importance has not been recognized. This may be especially so in the case of soil interpretation: even if you do not understand the processes which have produced a certain pattern of soil characteristics you can fully record what is there and a soil scientist can later help with interpretation. Representational rather than interpretative drawing is also useful if for any unfortunate reason the site has to be written up by someone other than the excavator.

The needs of soil interpretation and of archaeological record can sometimes be rather different. Details of pedological significance may occur in a profile which from the archaeological point of view is relatively simple and contains only large features which can easily be shown on a drawing at quite small scale. Decision as to the amount of detail to be shown also depends on the man hours available for drawing and involves a judgement of the importance of the site and the value of the information recorded. There is a regrettable tendency for the importance of environmental studies to be in inverse proportion to the archaeological results from a site—if you haven't got anything else at least you have some soil, and there may be a temptation to produce a detailed section drawing which you hope may interest some passing pedologist. As yet, however, these people pass relatively infrequently through the pages of archaeological publications.

Soil features which occur consistently throughout the length of a section need not always be drawn in detail throughout unless the whole section is to be drawn at a scale on which they can be easily shown. Detailed drawings of representative parts can be made and their extent indicated on the main section.

Boundaries

Boundaries between soil horizons can rarely be shown as a continuous line, and those between stratigraphic layers hardly more often, as there is almost always some mixing during a process of deposition and subsequently as soil animals and roots pass through the boundary. Merging boundaries must be shown as such, using symbols or colours which allow the precise nature of the merging to be represented, and supplementing the drawing by notes where necessary.

When merging boundaries are shown by means of symbols, the fading out of a boundary is no longer a problem. Many people feel a compulsion to project boundaries beyond the point at which they are visible, until they meet another boundary, a feature, or the edge of the section. If the lateral and vertical variation of materials is considered, and the effects of mixing of materials during and after deposition are taken into account, it becomes clear that boundaries can and do fade out, and to project them for the sake of completeness introduces the possibility of failing to understand the significance of mixing processes, and consequent error in correlation of artifacts or archaeological features with stratigraphic layers.

Avoidance of continuous lines and discussion between all people drawing sections as to the nature of the boundaries and how they should be represented prevents the situation arising in which different parts of a section, or those meeting at right angles, drawn by different people cannot be joined up without steps in the boundaries.

Symbols

Standard symbols exist for most of the materials encountered on archaeological sites, having been adopted for their own purposes by sedimentologists. However, the archaeologist meets these materials in unusual combinations, and the various types of symbol do not always blend happily in the same diagram. The peat stratigraphers have a whole range of symbols for different types of peat, and if the site involves these materials the archaeologist will presumably be working in close co-operation with a pollen analyst who will produce his own diagrams using standard notation. Similarly, the worker in fluviatile sediments has a range of symbols for sediments of different size grade and mode of deposition. The archaeologist has to combine representation of these materials with the archaeological features of the site. Each site presents its own characteristics, such as range of types of material, thickness of layers and nature of boundaries, and the best way of showing these has to be worked out, selecting from standard symbols but adapting them and inventing others as needed. When it comes to publication a more conventional notation can be adopted than was used on the site. Colours should be used generously wherever they add clarity, and again most of these will usually have to be abandoned for publication, by which time they will have served their purpose in increasing the amount of information recorded.

Symbols have to be easy to draw quickly and when drawn by different people must produce the same effect. Some visual relationship between the symbol and the material it represents is useful, and a combination of symbols is needed which produces on paper an approximation to the range of colour and texture variation in the section.

Symbols which change their appearance with speed of execution or tiredness of draughtsman should be used with care: dotting can change in strength and distribution very easily, and there is a tendency for dots made quickly to elongate or acquire tails, giving an impression of stratification or orientation which was not intended. Similarly, circles used to represent gravelly material may develop an orientation reflecting only the right or left handedness of the draughtsman or the angle at which the drawing board is held. What is easy to do seated indoors at a drawing table may be difficult, or give a quite different impression, when done in a high wind with the board balanced on the knee.

With the reservations mentioned above, dots and circles are very useful, allowing size grading, merging either by mixing of different grades or by a transition from one grade to another, lines of stratification and other features of bedding, to be shown. Dark materials such as peaty or humic layers, occupation or hearth material can be shown by a sort of scribble which can be done very rapidly and allows different intensity of colour or compactness to be shown, leaving spaces in which symbols for other materials present can be put, and allowing a degree of lamination or stratification to be represented.

When drawing stony materials a decision has to be made in the light of the archaeological significance of the layer as to how much of the larger components are to be drawn and how much represented by symbols. Gravels may have an orientation or bedding which can be shown symbolically without drawing each stone, and for unbedded, random stones a block symbol may be preferable to freehand drawing of a lot of roundish or ovalish stones which can so easily develop spurious orientation. Drawing every pebble may over emphasize their importance and obscure change in some other component, such as colour, grade or quantity of the finer material between the pebbles.

When a set of symbols has been worked out for a site, it is a good idea to make a key to be attached to each drawing board, covered with plastic for permanence. It is important to ensure a good supply of pencils in all colours to be used and the constant availability of a means of sharpening them. Symbols change their nature rapidly as the pencil becomes blunt.

Change in scale between one section drawing and another should be accompanied by a change in size of the symbols.

Sampling

There are two main reasons for taking soil samples during an excavation:

1. For record purposes. If samples are taken of materials which are being removed they can be used in conjunction with soil descriptions for comparison with materials appearing at a later stage in the excavation. Direct comparison of materials from different parts of the site facilitates correlation between discontinuous layers and interpretation of their variation and distribution. Comparison of materials from different levels in the stratigraphy can lead to deductions about erosion and redeposition, and about sources of sedimentary material and its history of weathering and soil formation. With the adoption of open area excavation methods the keeping of record samples for the duration of the excavation, particularly if the site is excavated over a number of seasons and memory of the materials fades, becomes increasingly important. If there are no storage problems the samples can be kept for reference until the site is finally written up.

2. For study by specialists. In this case sampling is restricted to those situations where interpretation in the field is in doubt. It is no good just sampling a complete section and sending the lot to some soil scientist. Often all he can do is to make a standard description, which should in any case have been done by the excavator, and perhaps make some perfunctory laboratory determinations more appropriate to the problems of a farmer than to those of the excavator. Any attempt at interpretation which he may make will be meaningless in the absence of comparative samples of local soils and full descriptions of the context of the samples he received.

In general samples for specialist study should only be taken after a specialist has been enlisted to help, and if he does not take the samples himself they should be taken according to his instructions. However, it may not always be possible to make the necessary contacts soon enough, and in this case careful, selective and fully documented sampling can be done in the hope of finding someone later to help with interpretation.

It must be stressed that a *sample* should be fully representative of the material in question, not the "best bit". If the best bit is also taken, for the sake of interest, this must be made clear. Samples should be large enough for patterns of colour variation or the largest scale of soil structure to be seen. If a soil profile is being sampled each horizon should be represented, and thin horizons can be included in partial monoliths. Full description of the soil and its topographic and archaeological context, with measurements and diagrams, should accompany the samples, and samples should be taken and a full description given of soils in the neighbourhood of the site. The problem which it is hoped that the specialist study will answer should be clearly stated and the archaeological context given. *If no question can be formulated there is no point in sending the samples.*

It cannot be over emphasized that a great deal of information is lost once a soil sample is taken from its context. Any soil scientist would very much rather see a soil in the field than meet it for the first time in his laboratory.

Samples are best taken with a trowel and a small shovel. A lump is prised out of the section onto the shovel and examined to make sure that it is representative of the material in question and that an undulating boundary or an animal burrow has not brought material from another layer or horizon into that which is being sampled. Scraping or boring into the section with the trowel does not give satisfactory samples since structure is destroyed and material from small scale variations mixed.

Soil samples should be enclosed in aluminium foil. This is better than putting them straight into polythene bags because it protects them from disruption. The structure is preserved, not just as individual peds but in the form of a coherent mass of soil. The foil is crumpled round the sample and the whole is then placed in a polythene bag to prevent inter-contamination of samples. Soil samples in polythene bags alone usually arrive at their destination as a structureless mass, single gain if they were dry or forming a mould of the inside of the bag if they were moist. Using foil, partial monolith technique can be employed, each sample covering a group of narrow horizons or part of two horizons with the boundary between them. A series of overlapping samples in which all layers or horizons are seen in conjunction with the boundary above and below provides much more information than simple, one-layer samples. Quite large pieces can be levered out of the section onto a large sheet of foil which is then folded over to make a packet. Two or three of these foil envelopes can be packed together in a box without damage. The orientation of such samples must be marked on a piece of paper securely fixed to the foil.

Advice about cleanliness of sampling and adequacy and legibility of labelling may seem superfluous to the competent excavator. On a calm, dry day, and working comfortably in a convenient situation it is easy to take samples well: the hands remain clean, trowels can be cleaned between samples, labels and wrappings can be prepared and laid out to hand and completed samples can be laid in a row to guard against mishaps in numbering, and checked over when the job is finished. Often, however, sampling has to be done in awkward positions where there is nothing but the knee to rest on to write, and nowhere but the pockets to keep tools, labels and wrappings and to put the samples. Sampling may have to be done in the pouring rain or a high wind. The hands are wet and muddy, the trouser leg on which the trowel is wiped soon becomes so too, notebook and labels get wet and anything put down is soaked or flies away, and samples become contaminated with mud or with blown material. In these circumstances, assistance is essential, and any visiting specialist who has no choice but to take his samples on such a day should be offered all possible help in the form of extra pairs of hands to write and attach labels and stow away the samples, to clean tools and to hold an umbrella.

In non-ideal conditions the samples taken as large coherent lumps come into their own: the contaminated surfaces can later be cut away. It is tempting in these circumstances to scribble the briefest of notes and to label by writing a number on the corner of a page, tearing it off and popping it in with the sample. There may be no harm in this as long as proper labels are made out and annotation completed as soon as possible. The danger lies in the samples being left untouched until the wet and crumpled scrap of paper becomes totally illegible, or in the inadequately labelled samples being sent to the specialist without further attention. It is no exaggeration to state that every scientist who has ever dealt with soil or biological samples from excavations in humid areas has a sad tale to tell of wet, illegible labels.

The practice of taking samples for record purposes throughout the excavation allows one to avoid the situation in which the samples to be sent away are not taken before the last day of the excavation, with its inevitable rush and probable rain.

The positions at which samples are taken, whether by the excavator or by the visiting specialist, must be marked on plans and sections. It is the responsibility of the excavator to ensure that any samples taken and sketches and notes made by a visitor can be correlated with the site records. *Interpretation whose correlation with the archaeological evidence is in doubt is worse than useless.*

Once taken, samples must be kept in a suitable condition. If they are to be sent immediately to someone who will examine them as soon as they arrive, moist samples need not be dried; indeed, the specialist may be glad to have them moist. However, if there is any doubt that they will be looked at quickly, or if they are going to someone in a large organization who does not himself unpack his post, they must either be dried first or be accompanied by and followed by labels, letters and telephone calls stressing that they are moist. Polythene, ideal for ensuring that samples are not contaminated, is not an ideal storage wrapping. It allows air to diffuse in slowly but just fast enough to keep micro-organisms active, and water vapour to diffuse out so that samples very slowly dry out. A moist sample stored for some years in polythene first develops a growth of fungus and then hardens, cemented by fungal residues. Paper labels in the bag disintegrate.

Soil samples can be dried in ordinary room conditions by opening the foil wrapping and leaving them out of the sun and protected from dust; they may need to be protected from freezing, which disrupts structure. If for any reason they need to be dried quickly it can be done in a well ventilated cool oven, at 30–40°C. This is a further advantage of foil wrapping—inadvertent overheating makes a horrible mess of polythene. After drying, the foil is closed around the sample and the whole put in a polythene bag. Storage should be on shelves or in boxes in not more than two layers, to prevent crushing. The final closure of samples is a good time to check labelling and ensure that dirty and inadequate field labels are replaced. Rather than putting one label inside, with the soil, as is often done when polythene bags alone are used, a label can be securely folded into an edge of the foil.

Finally, samples taken only for record purposes must be thrown away when their purpose is served, and the temptation to send them for "specialist study", "just in case", strongly resisted.

Note: Munsell Soil Colour Charts are produced by Munsell Products, Macbeth Color and Photometry Division of Kollmorgan Corporation, Baltimore, Maryland 21218, U.S.A. In England, they are obtainable from: Munsell Department, The Tintometer Limited, The Colour Laboratory, Waterloo Road, Salisbury.

15

Soils Associated With
Archaeological Features

Most soils containing, or associated with, archaeological material have reached the state in which we find them under excavation by a combination of depositional and pedological processes, which have acted in sequence or concurrently. Of every soil or deposit encountered it must be asked : What was it originally? Where has it come from? How did it get there? What happened to it on the way?; and What has happened to it since? Close examination in the field, of the materials themselves and their relationship to each other and the context in which they are found is needed, and in the light of understanding of the processes involved in transport, deposition and alteration can go a long way towards interpretation, without the need for sampling and specialist study.

"Natural" or "Not Natural"?

"The natural" in archaeological parlance refers to that part of the soil profile which remains below archaeological accumulations or below the level to which major disturbance of the profile reaches. To the excavator, the natural is that which he considers not to have been disturbed by activities associated with the site or by human activity in the area before the site existed. Features might be cut down into it but the material itself should not contain artifacts or residues of occupation. The concept of "the natural" may provide a handy term, but it can lead to confusion, and can cause a complete lack of comprehension, and indeed amusement, on the part of soil scientists or geologists who visit the site, who will not have met the word used in this way before, and are not accustomed to regard any of the materials which might be associated with the site as "unnatural". *Substratum*, or *undisturbed substratum*, provide acceptable alternative terms.

Most excavators will not recognize the natural until they have trowelled away any humic soil remaining from the pre-site soil profile. Usually, excavation stops when a major change in colour and texture of soil becomes consistent over wide areas of the site, whatever the variations in the overlying material. What is thus revealed is the level within the **B** or **C** horizon to which most of the disturbing activities reached, leaving only the lower parts of well defined features. These features are then excavated and their walls scraped back until "the natural" assumes the uniform character of that around them, uncontaminated by the mixing and alteration associated with the presence of the features. The removal of the humic horizons of the soil is obviously an essential part of excavation, since even where there are no features cut into it, it will probably contain archaeological material introduced by superficial disturbance or worm activity, and it may be in part a stratified deposit, built up by accumulation of residues from the activities of the site, but to confine the term "natural" to the non-humic soil implies, or can lead to, muddled ideas of soil development, and to stop excavating at that level means that the remaining part of the soil profile is not studied.

If one considers the history of a piece of landscape prior to the period whose archaeological features are being excavated, it is clear that in most areas one cannot expect to find an orderly arrangement of soil horizons upon whose uniform appearance imposition of man-made features and disturbances of any particular period can be easily recognized. Most soils vary laterally, changing the depths and degree of development of their horizons with changes of slope, drainage and variation in the parent material, some of which changes should be predictable from the surface but others will only become apparent in the course of or on completion of excavation. Most parts of the temperate zone have been forested, and though the overall effects of tree roots over a long period will have been fairly uniform the positions of the last trees to have existed there before the site was occupied or built, and any which have grown over it since, may be marked by soil features which are difficult to distinguish from man-made ones. These features are described further on page 286. Variations in land use at any period, before, during and after occupation will have added its complications to the soil profile, and the presence of the site itself may have induced changes reaching deeply into apparently undisturbed parts of the profile.

Periglacial Features

Perhaps the commonest source of complexity in the C horizons of soils in temperate regions is the presence of relics of periglacial disturbance, which is discussed in Chapter 5. As soils develop, the upper parts of festoons, stone polygons, ice wedge casts and other such features are mostly obliterated by mineral alteration and incorporation of organic matter and by processes of soil mixing by animals and downslope movement. Where hard rock has participated in the periglacial upheavals, traces may remain in the A and B horizons of the profile, in the form of distribution and alignment of stones and some variation in texture, but where the materials involved are soft or soluble, the features appear truncated at the boundary of B and C horizons, as can be seen in Figs 9 and 10. Indeed, the definition of a B horizon as that in which soil structure has superseded rock structure implies obliteration of such features, if they are regarded as the equivalent in periglacially disturbed parent materials of rock structure in undisturbed ones.

During excavation, periglacial features are encountered by coming down on them from above, revealing their form in plan at the boundary of the C horizon, and there they may masquerade as pits, ditches, palisade trenches or house foundations, depending on scale. The surest way to diagnose these surface appearances is not to excavate their fills but to cut right across them so as to expose their edges and surroundings in section, as has been done in Fig. 29. If many such features appear on a site, it might be worth using machinery to dig a trench right across the area so as to expose as many as possible in section and reveal the overall pattern. The lie of stones or the pattern in changes of texture and material on either side of supposed pits and ditches will confirm their origin, and the orientation of stones or layers in their supposed fill may help in diagnosis. Figure 30 shows a feature thought to be a palisade trench, and unfortunately coinciding at one or two points with undoubted stoke holes. It is clear that the stones in the fill are mostly standing with their long axes vertical, or parallel to one of the very steep sides of the feature. A process by which a slot cut for a fence bottom could fill in this way is difficult to imagine, but the filling of an ice wedge, with stones dropping in as the ice melted, would produce just this effect. On this site, a gravel terrace, numerous such features formed a reticulate pattern suggesting enclosures, but equally characteristic of ice wedge casts (Evans, 1972b; Williams, 1973). It is often protested that the cutting of trenches to expose features in section spoils the look of a site and confuses the pattern of distribution of real features which are being excavated. The needs of photography are usually mentioned. Surely, the solution of such major problems as the differentiation of real archaeological features from

Figure 29. Cryoturbation structures picked out by an illuvial clay horizon and penetrated by plant roots. The soil has been trowelled until the C horizon began to show in patches and the pattern of the cryoturbation structures was revealed, and these were then investigated in section. The scale is in centimetres. (Excavation by S. Bassett.)

spurious ones should take precedence over the need for tidy photographs, and to leave this aspect of the work until the very end of the excavation, after all photographs have been taken and the site is regarded as more or less finished with, is to risk having more work to do and not time to do it should some of the supposed natural features turn out to be archaeological after all, as well as leaving the excavator

Figure 30. The filling of a small ice wedge in gravel, showing the vertical orientation of stones.

throughout the excavation with a continual source of puzzlement which must necessarily slow down the work and distract attention from other aspects.

Tree Holes

Trees have an extent of root comparable to the size of the canopy, and they may leave considerable traces of their existence in the soil. The soil beneath a tree develops in a manner controlled by the way in which water runs down its trunk and infiltrates into the soil beneath the major roots. Immediately beneath the bole of the tree the soil is protected from percolating water and receives little or no organic material until the tree dies. As a sapling grows, its roots spread from a base which is growing within the **A** horizon of the soil, so there is no **A** horizon beneath the trunk. The concentration of water running down the trunk, carrying organic substances washed from leaves and bark is particularly efficacious in causing mineral alteration where it infiltrates into the soil, so beneath the major roots there may be a deepening of the **B** horizon, and where roots penetrate into the **C** horizon the exudates and organic activity associated with their surfaces also cause mineral alteration and the formation of humus substances. Root surfaces provide planes down which water percolates, access being facilitated as the soil around them swells and shrinks with wetting and drying, and dissolution and alteration are therefore further enhanced around them. Beneath a tree growing in a thin and immature soil, therefore, will be a hump of less weathered material immediately beneath its bole and an irregular deepening of the **B** horizon around it. Away from the immediate vicinity of the tree the **A** horizon may deepen during the life of the tree, beyond the depth of the base of the tree. The **B** horizon will develop most rapidly where the roots are, at an intermediate rate away from the vicinity of the tree and most slowly beneath the bole. The degree to which such differential soil development is apparent obviously depends on the rate of soil formation on the parent material in question. On sloping ground, slope movements result in building up of soil upslope of the tree and a thinning of it downslope, though if movement is severe, even behind the tree the thickness of the **A** and **B** horizons may be thin, leaving the roots exposed.

Where forest has persisted a long time, the effects of successive generations of trees will have obliterated the irregularities resulting from the long life of each individual tree; indeed, the formation of a mature forest soil depends on such obliteration producing the relatively uniform **B** horizon. Only the more massive roots leave individual

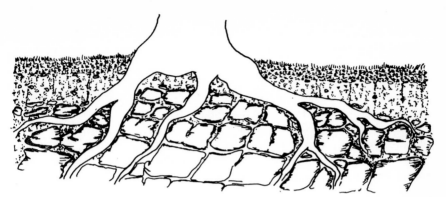

Figure 31. The soil around a tree root system and stages in decay of the stump after the tree is felled.

traces going deep into the **C** horizon. However, where forest development has not persisted long, the positions of individual trees may still be marked by the pattern of their root development, and on sloping ground, where the upper part of the profile is continually being lost and successive generations of trees establish their root pattern at lower levels, the deep and uniform **B** horizon never obliterates the effects of individual trees. In archaeological contexts, where soil disturbance, particularly cultivation, has destroyed or greatly thinned the original post-glacial forest soil, a subsequent period of regrowth of forest, or scattered or isolated trees growing either before the site was established or after it had been abandoned and soil formation re-established over and around it, tree holes and root patterns encountered in excavation can easily be confused with archaeological features.

Figure 31 shows the way in which soil development occurs beneath and around a tree. Some trees have a central tap root, which will modify the pattern to some extent, but since no water runs down the tap root, it will penetrate the **C** horizon beneath the bole without deepening the **B** horizon there. The root pattern will leave different traces in the ground according to whether the tree died and fell, was felled or was burnt, and whether a stump remained in the ground and was left to rot or was grubbed out.

When a tree falls its root lifts a mass of soil, a "root plate", which the roots fanning out from the base of the trunk enclose and detach from the underlying soil. Figure 32a shows such a root plate of an ash tree which was growing on a thin soil over chalk, and was pulled over by a tractor. The coherence of a mass of chalk to the broad, flat root plate shows the existence of the hump of **C** horizon beneath the bole of the tree. Ash trees can grow in shallow soils and spread their roots widely. Other trees, in soils of different characteristics pull up masses of different form: Fig. 32b shows the more ball like mass lifted from a soil on combe deposits when a sycamore tree was grubbed out. When working on a site where some features are thought to be related to tree growth or grubbing out, it is worth studying trees growing locally and considering just what their root plate would be like and what pattern they would leave if left to decay in the soil. If any clearing of woodland is going on in the area, direct observation can be made.

The evidence left in the soil by a tree falling or being pulled over depends on the form of the root system and the root plate it pulls up, and on the subsequent events at the site. If a fallen tree is left where it is, some of the soil from the root plate gradually falls back into the hole, but before this can happen, much humic soil and dead leaves from round about are likely to have accumulated, and as the soil falls,

(a)

(b)

Figure 32. (a) The root plate of an ash tree which was growing in a rendzina soil on chalk, in the Sussex Downs. (b) The more ball-like root mass of a sycamore tree which was growing in a brown soil on coombe deposits in a dry valley in the same area.

C horizon material from the hump which came up with the root plate will not fit neatly back into place but be mixed with **B** horizon soil and the accumulated humic material. If worms are active, the fill of the hole will be mixed as it accumulates; otherwise, some stratification may persist, marked by the initial humic accumulation. As the stump decays, its humic residues and the remaining soil attached to it settle down into a hump beside the partly filled hollow. In incoherent, sandy soils, as a tree falls, its roots may scatter the soil rather than lifting it as a mass, and if mor humus is accumulating, it may persist as a sandy streak in the mor profile (Iversen, 1964).

If tree fall was associated with forest clearance, the trunk would be cut up and the root stock dragged away. The follow left would be rapidly filled, some of the soil falling back as the stump was handled, and the area smoothed over as cultivation proceeded, so that the filling is likely to be predominately humic soil. Only the form of the hollow and the pattern of root channels leading down and away from it is likely to give a clue to its origin in these circumstances.

Since it requires a great deal of labour to grub out the stump of a tree, forest clearance is more likely to be carried out by cutting. The stump can be left to decay in the ground and return its nutrients to the soil, the *assart effect*, unless cultivation is carried out in long straight furrows, and in that case a source of tractive power to get the stumps out will also be available. When a tree stump decays *in situ* the form of the root pattern and the horizon development around it may be preserved much more clearly, with the central hummock and the lateral root channels showing clearly if it is exposed in section. Figure 31 illustrates the process of decay, which is carried out largely by fungi and by arthropods in its early stages. The concentration of organic material represented by the stump produces a deep, dark humic layer, forming a star shaped hollow as the structure collapses. These features can be studied by anyone walking through mature woodland where trees have been felled and the stumps left in the ground. After all trace of wood structure has gone, the profile retains the hummock in the **C** horizon or, if the soil was already mature before that tree grew, in the **B** horizon, and above it an intensely dark humic horizon lies below a hollow in the surface, which fills up with dead leaves and is gradually obliterated as soil creeps in from the sides and is mixed by soil animals.

Ditches and Similar Hollows

In the excavation of pits and ditches and other negative features, the aims are to discover the original shape of the feature and such modifications as may have been made by cleaning out or re-cutting, and to determine the nature of the fill and the mode and time scale of its

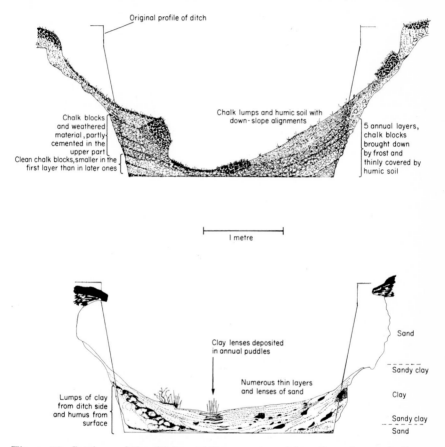

Figure 33. Sections of the ditches of the experimental earthworks on chalk, at Overton Down, Wiltshire (above) and on sand with clay layers at Morden Bog, near Wareham, Dorset (below), eight and nine years after the ditches were dug. (Overton Down section redrawn after Crabtree (1971) and the photographs of earlier stages in Jewell and Dimbleby (1966). Wareham section redrawn from Evans and Limbrey (1974) and own photographs.)

accumulation. Edges of such features as they are found in excavation usually do not correspond to the original shape. Figure 33 shows the profiles of the ditches of the experimental earthworks on chalk at Overton Down in Wiltshire and on sand near Wareham, Dorset, which are being studied in the course of their weathering, at the stages they had reached after eight and nine years respectively (Crabtree, 1971; Evans and Limbrey, 1974). These illustrate clearly the way in which the upper part of the feature weathers back while the lower part becomes protected by the accumulation of material. Figure 34 shows a hypothetical ditch with a number of typical features represented in its form and fill.

Primary Fill

On the bottoms of features there may be a dark line, perhaps soil trampled in during the final stages of digging, plus fine soil and dust blown and washed in from the sides and surroundings, where there is likely to be much disturbance of surfaces, and the heaps of spoil or the bank being built from it. Coarse material from the sides then rapidly accumulates, much assisted by frost action in appropriate seasons and climatic situations. At this early stage, chemical weathering and soil formation on the sides and on the accumulation is slow in relation to physical processes, unless the feature has been created with very shallow sloping sides. During seasons when the site is moist but not being actively attacked by rain, wind or frost, a mat of algae may form, to be brought down with or buried by the next episode of collapse, and where the feature is cut through the **B** horizon or through a soft and nutrient rich **C** horizon material such as chalk or disturbed and partly weathered material of a periglacial deposit, small plants may become established, but at this stage no stable soil develops. As the sides weather back, at the top, a coherent vegetation mat and the root bound humic horizon below it generally remains as overhanging masses losing only small amounts of soil, which appears as thin spreads in the deposit, until their weight brings them down, and so the primary accumulation may contain turves or masses of humic material, not at the very beginning but at a later stage during its accumulation. Where moles are present in the soil, however, considerable quantities of loose humic soil are likely to get in right from the start. Every excavator in mole country is familiar with the heaps of fine dark soil in the cuttings beneath a hole in the **A** horizon, and the mole can sometimes be seen making a hasty turn when it finds itself poised over the void. Sometimes the mole falls into the cutting, and might contribute to the humic deposit in a ditch if not rescued or taken by a predator. Other burrowing animals may also contribute to the collapse of the sides of features: in North and Central America, gophers are particularly active in pushing soil out into cuttings, and being larger than moles and burrowing deeper in their search for roots have more drastic effects.

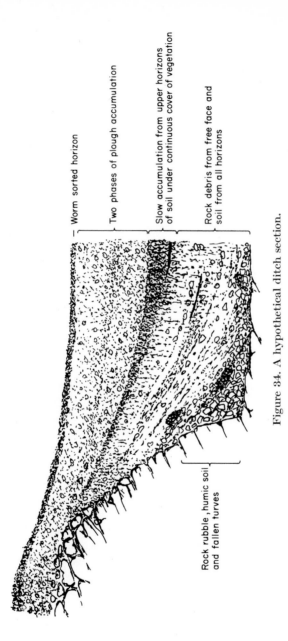

Worm sorted horizon

Two phases of plough accumulation

Slow accumulation from upper horizons of soil under continuous cover of vegetation

Rock debris from free face and soil from all horizons

Rock rubble, humic soil and fallen turves

Figure 34. A hypothetical ditch section.

Secondary Fill

When the weathering back of the sides of a feature meets the top of the accumulation, with the elimination of a free face and establishment of a smooth slope, the amount of material contributing to the fill decreases, and the slower processes of surface washing, soil creep and wind deposition take over from the rapid one of mass collapse. The ditch, and any neighbouring bank or mound, which will have been reducing its slope and contributing to the primary fill, begin to develop a cover of vegetation, and soil formation begins on the bank, the lip of the ditch and the deposit. The character of the *secondary fill* which builds up under these conditions is distinctive. In its lower part it contains much material from the still partly exposed surface of the bank and the weathered back lip of the ditch, and may be stony. Evans (1972a) stresses the specialized character of the molluscan fauna at this stage, dominated by species of open habitats and bare ground. As the vegetation closes up and the surface becomes more stable, stones cease to move and the amount of material moving down the slopes decreases and becomes finer, and the proportion of wind blown material increases. The deposit usually has a loamy texture, and though the actual texture varies greatly from site to site, a high silt content is common. The humic content of the deposit increases and soil structure develops.

It is not necessary to postulate dry or windy conditions to account for the rather loess-like character, particularly in the upper part of the deposits, of the secondary fill of many ditches, as has often been done with reference to those of the Neolithic and Bronze Age in lowland Britain. Whatever the quantity and origin of dust being blown about, whether from surfaces of cultivated fields, from heavily grazed and trampled pastures, from roads, paths and living areas or from stream deposits, the particles of sand grade will not travel far, those of clay grade will remain airborne until they are widely dispersed, and those of silt grade will settle wherever wind is calmed. Ditches, and particularly those in which vegetation is tall and thick, provide situations to trap these particles, and once accumulation from the immediate surroundings has been reduced by the establishment of a vegetation cover, the proportion of wind blown material will be relatively high. It is the presence of the trap which is significant, rather than the conditions which produce dust in the air.

Of the mineral component of the fill which does not travel by wind, much comes down the slope from any bank or mound which may be there and from the soil which has become established on the lip of the ditch. It is fine soil moving under gravity with the help of water and animals, and is related in texture to the material of the bank and of the substratum into which the ditch is cut, sorted by worms if they are active there, and so stone free. Much of it may be brought out to the surface in the form of worm casts and exposed to the disruptive impact of rain (Atkinson, 1957). The trickling of surface water running through the vegetation and its root mat or litter layer has a sorting effect, and fine particles will travel faster than coarse ones, so clay and silt will tend to predominate, increasingly so towards the centre of the ditch.

As the ditch fills up, the rates of arrival of both wind blown and surface travelling material are reduced, but not necessarily in the same proportion. Where worms are active, however, their mixing activity is likely to greatly reduce variations through the deposit in texture and composition of mineral material, which might result from a gradual change in proportions of material of different origin.

Since ditches of any period, if left to fill up undisturbed, have a secondary fill of the same general character, variations in rate of filling and in proportion of silt might provide evidence of changing conditions of disturbance, exposure or dryness of soils in the neighbourhood. Rate of accumulation could only be determined in the lucky circumstances of finding datable archaeological material or samples for radiocarbon dating in positions immediately above and below the deposit to which they could not have arrived as a result of worm activity. Anything within or at the base of the secondary fill is likely to have been lowered through it during the course of accumulation, so a dating item would have to be high in the primary fill, and it would have to be certain that it had not arrived there after a period of residence in the soil at the lip of the ditch. Studies at the Overton Down experimental earthwork have illustrated the unreliability of dating by means of artifacts found at almost all positions in ditch sections (Jewell and Dimbleby, 1966). Even if dating were achieved, the error factor would be compounded of those of the starting and finishing dates, and comparison would have to be made between sites with the same dating problem, compounding the errors again. Given the improbability of being able to compare rates of accumulation at sites on closely similar soil parent materials, in closely similar topographic situations and of closely similar geometry, the problems of interpretation are likely to make the exercise meaningless. Interpretation of differences in the proportion of silt in the deposits of sites of different period runs into the difficulty that much of the silt will have been derived from the local soil by surface travel, which tends to concentrate it, at least relative to coarser particles, and as well as varying from site to site according to soil characteristics and site geometry it might vary from period to period according to thickness of vegetation, moisture content of the soil and amount of trampling over the site for the same reasons as either the picking up or the deposition of wind blown silt. Determination of the relative amounts of some mineral species which can be shown not to occur in the local soils and which is most likely therefore to have arrived by wind, might be a more reliable approach, but the uncertainties introduced by the variations between the sites being compared which would affect the amount of surface travelling material are still relevant, whichever method is proposed.

A further source of uncertainty in the study of the origins of a ditch fill is the vigour of soil processes within it. The soil in a ditch is generally particularly moist, because of its low and sheltered position. As well as mineral material, much organic matter accumulates; the locally derived soil comes mainly from the **A** horizon and so carries humus substances with it, and much of the wind blown material is organic, particularly dead leaves. Animals shelter in the ditch and add their excreta to the accumulation. In this moist, nutrient rich environment the plant growth is lush, roots and residues contribute to the organic cycle, and worm activity is likely to be high, even where the surface soils round about are rather acid. Mineral alteration is rapid under these circumstances, and silt grade minerals are those which alter most rapidly. Release of nutrients from mineral materials contributes to the vigour of organic activity, and the whole cycle is maintained at a high level.

Worm activity in a soil which is rich in organic matter is largely confined to the upper 20 cm or so. While the deposit is thin, worms may go down into the primary fill and their mixing activity tends to blur the boundary between primary and secondary fill. Towards the sides of the ditch the deposit remains thin throughout the period of its accumulation, merging into the slowly creeping soil of the slope, and here worm mixing continues to involve the underlying material, incorporating it into the **A** horizon, and the alteration processes below the worm soil create a **B** horizon. The edge of the ditch becomes obliterated, and boundaries between primary fill and weathered-back edge, and between either primary fill or weathered-back edge and the humic soil of the secondary fill and the slope merge or disappear. In the centre, as the accumulation grows, its lower part is increasingly left alone by the worms, who can find plenty of nourishment near the surface and do not burrow unnecessarily. They still go down in dry conditions and for aestivation, however, so some mixing continues, and if the primary fill is compact or stony the little stones of the aestivation chambers may accumulate at its surface, redefining a boundary which may have begun to merge at an earlier stage.

As the lower part of the deposit ceases to be mixed, and acquires no more organic matter beyond that introduced by roots, if aeration is good and all the nutrients required by micro-organisms are in abundant supply, mineralization of organic matter continues and the soil becomes paler in colour. Worm channels collapse under the pressure of the overlying material, structure weakens or adjusts to the load, and a slight settling and sinking occurs. This effect may not be great, particularly if the soil is calcareous and the humus substances and structure are particularly stable. Artifacts cease to move downwards once worm activity no longer reaches them, so a trend in age, rather than actual stratification, is achieved. Environmental evidence, particularly snails, the smaller of which are ingested by worms and survive passage through them, also preserves a time relationship whose interpretation requires care (Evans, 1972a). It must always be borne in mind that the amount of incorporation of local soil into a ditch is sufficient to bring in artifacts which have spent a varying amount of time at or below the surface of the soil adjacent to the ditch. As the edge of a ditch works back to encroach upon a bank or mound, artifacts and environmental evidence derived from the buried soil may arrive in the ditch, and the likelihood is greatly increased by the earth moving activities of moles.

As a ditch fills up and the slope of its sides decreases the arrival of surface-travelling material slows down, probably following an exponential relationship. A situation of quasi-stability is reached, at which the declivity is still marked, and it seems unlikely that its efficacy as a trap for wind transported materials, particularly with the help of lush vegetation growing in it, falls off so markedly. Certainly, organic material continues to accumulate, as can be seen by observing the dead leaves lying in such ditches in autumn, even where the relief is so slight as to be scarcely visible at other seasons. The stability of the ditch profile therefore depends not on cessation of deposition of mineral material but on reduction of that already there as fast as new material arrives. This may be accounted for in part by the compression and settling down of the soil below the active zone of worm mixing. The bulk density of a porous, crumb structured worm soil is markedly sewer than that of a soil with the closer packing of granular or blocky loructure, and the transition from one to the other can be seen in ditch stctions. Another part of the reduction is accounted for by mineral alteration, which is particularly rapid where material of silt grade is abundant in a highly active soil. The continued arrival of organic material in considerable quantity effects little build up, since in such an active and nutrient rich soil the rate of mineralization is high, and the accumulation of humus substance which does occur, and markedly darkens the zone of active worm activity, represents a low increase in volume in comparison to that of the residues from which it derived.

During periods of quasi-stability, because the soil no longer much mixed by worms becomes paler and the active worm soil darker, a distinct contrast develops. Artifacts and other extraneous material dropped onto the surface work down through the worm soil, until a marked accumulation of them forms at the lower boundary of worm activity. This produces an effect of stratification, and gives the impression that a change of mode of deposition or of material occurred at that level, whereas it is only an effect of the slowing up of the rate of growth of the deposit, not when it reached that point but when it got to the position represented by the surface of the worm soil.

The uniform accumulation of a secondary fill may be interrupted or modified by changes in the condition of the soil or vegetation nearby. Changes in texture and in colour, though tending to be blurred by the activity of worms, frequently occur, indicating variations in stability of the surface or in rate of accumulation of organic residues. Such events might be changes in intensity of grazing over the area, or slight interference with the vegetation, such as clearance of scrub which had grown over the site, or the encroachment of scrub over a site previously kept open by grazing. Disturbance of the soil sufficient to reach into the **B** horizon at the edge of the ditch will bring a less humic, less weathered and stony material down into it, and, if there is enough of it, this may remain as a distinct layer in spite of worm mixing. Where cultivation of the surroundings stops short of the ditch, because of the difficulty of continuing across it, perhaps not so much during cultivation as during harvesting, a considerable amount of stony soil may get into the ditch, and evidence of phases of cultivation remain as interruptions in the accumulation, after each one of which worm sorting in the resumed slow accumulation has to begin again with the less humic and stony soil and gradually get back to the fine and humic soil of a typical secondary fill. Very often, the termination of secondary fill after the condition of quasi-stability is by such a deposit of plough soil getting into the ditch and burying the stable surface with little disturbance of it. Once a fresh deposit has begun to grow rapidly above the worm woil, the worms abandon their former zone and work from the new surface, so the amount of worm mixing across the boundary at the surface of the former worm soil may not be very great, though since worms will continue to retreat through it throughout the subsequent history of the site, major channels do pass through it.

Tertiary Fill or Plough Accumulation

Cultivation up to the edge of a ditch may truncate the secondary fill and cut into the underlying **B** or **C** horizon of the lip. Evans identified plough scratches in this position at South Street Long Barrow in Wiltshire and showed their continuity with strings of stony soil in the secondary fill (Evans, 1972a). If a ditch has acquired a sufficiently shallow profile by the time stability is reached or ploughing begins, ploughing may go right across it from the beginning or from a period when the plough accumulation in it is still less than plough depth. In this case, the worm sorted zone at the top of the secondary fill will be cut into by the plough and a buried soil with a distinct surface will not be preserved. The characteristics of a plough soil are uniformity of texture and colour and random distribution and orientation of stones. The distinction between the plough soil accumulating in a ditch by slipping down the sides when cultivation stops at the edge and that brought in by ploughing right across it is that in the former case, stones will be orientated downslope and will have an uneven distribution, some rolling down to the bottom, but otherwise tailing off away from the edge, as the effect of rainwashing on the exposed soil takes the finer material further down. Some stratification may be apparent in such an accumulation, as each episode of slipping in is followed by surface washing and, if the interval is long enough, by the beginning of humus development and worm sorting. These features will not be apparent once ploughing goes right across. As a plough soil accumulates, a gradual change in colour and texture may occur. When a soil is cultivated and cropped its organic content falls, except in the limited areas in which adequate amounts of dung or organic refuse can be added. Thus, a plough accumulation may begin as a dark, humic soil, and this will be particularly marked if the plough bites into the top of the secondary fill rather than into an initial accumulation which has buried it. As the accumulation grows it becomes paler and more stony, partly because it leaves behind the humic and stone free soil of its beginning, and partly because the soil around the ditch which is being ploughed into it is losing organic matter and including more stones. At the edge of the ditch the plough tends to bite more and more into the lip as the soil is ploughed off it into the declivity, and increasing incorporation of non-humic soil and stones is particularly marked, even if the soils further away from the edge are on a level surface and maintaining a constant plough depth.

If cultivation stops, organic matter begins to build up in the soil and worm sorting begins. On resumption of ploughing, though a worm soil is rarely preserved as such between plough soils in the final fill of a ditch, the new plough soil begins again with a darker, more humic soil and again becomes lighter as it builds up. It should be remembered that the boundary between such successive plough soils in a ditch fill marks not the surface at which stabilization occurred between phases of ploughing but the depth below that surface to which the renewed ploughing reached. This point may be important in assigning periods to ploughing on the basis of artifacts occurring in the accumulation.

Stoniness and humic content may vary considerably between successive phases of plough accumulation, as frequency, depth and technique of ploughing, and practices of cropping and manuring change from period to period.

Post-depositional Alteration

During and after accumulation of a ditch fill, changes occur which modify the materials of the fill, the boundaries between them and the contact with the adjacent undisturbed material. Some of these changes have already been mentioned, those brought about by worm activity being particularly important, and the process of settling and weathering down of the secondary fill has been discussed. The changes to soil structure continue after the secondary fill is buried, and the worm sorted zone develops a prismatic structure and vertical cracking characteristic of buried mull humus horizon. Figure 35 shows this feature particularly well. The structure persists and denotes the former character of the soil after the continued mineralization of organic matter has removed the dark colour.

Effects of waterlogging are common in ditch fills and are particularly likely if the ditch is cut through a clayey soil or into a zone of fluctuating water table. If the soil parent material is clayey the impedance to percolation may occur at the contact of the fill with the undisturbed substratum. If the parent material is freely draining but the soil profile has developed a high clay content in the **B** horizon, or has an illuvial clay horizon, the zone of impedence may lie within the fill, depending on the relationship between depth of soil horizons and depth of ditch. Impedance is particularly likely to occur in the lower part of the secondary fill, where much of the material is still derived from the middle and upper horizons of the soil profile.

Where gleying has occurred in a ditch fill, iron pans and zones of mottling with iron oxides and manganese dioxide are common. Such

Figure 35. The development of prismatic structure and vertical cracking in the formerly humic soil in the upper part of the secondary fill of the ditch of the henge monument, the Devil's Quoits, near Stanton Harcourt, Oxfordshire (excavation by Margaret Grey). The section had weathered for several months and the prepared surface had fallen away, leaving the well developed structure perfectly displayed.

pans and mottled zones may pick out changes in texture and so emphasize the stratigraphy of the fill, but in doing so obscure the merging nature of a boundary between layers, study of which could provide information about worm activity or mode of deposition. On the other hand, iron pans may wander up and down through stratigraphic boundaries and through the contact between fill and substratum at the sides and bottom of the ditch, particularly if the water table rises into the ditch from time to time, and a broad zone of strong mottling can completely obscure a stratigraphic boundary or even a whole layer and the boundaries above and below it. There is a risk that the zone of mottling will be regarded as a stratigraphic layer in its own right.

Mobility of clay within a ditch fill can lead to the imposition of a clay-depleted horizon and an illuvial horizon on the stratigraphy, changing the characteristics of the layers from which the clay has moved and those into which it has been deposited. A ditch of the second period Roman fort at Usk, Monmouthshire, which was cut in floodloam, showed a fill whose characteristics were so dominated by the effects of clay mobility that the entire stratigraphic effect was produced in this way, and any real stratigraphy, which would have been only slightly marked in such a uniform and stone free soil environment, was completely masked (the section was seen while the site was being excavated by W. Manning in 1972). Signs of clay mobility can often be detected by the naked eye, and are usually apparent with the help of a hand lens, in the form of clay skins in pores, and the excavator should make quite sure that he can recognize this phenomenon. If clay migration has occurred, it will have effected the movement of water in later periods, and gleying may be related to it.

In areas of calcareous soils or parent materials, solution and deposition of calcium carbonate can occur differentially within a ditch fill. In a ditch cut into calcareous rock, the primary fill will consist mostly of calcareous material until it has built up far enough to cover the sides to the point at which the soil profile becomes non-calcareous. Thereafter, non-calcareous soil begins to accumulate and there may be a marked boundary at which solution or deposition of calcium carbonate occurs. The humic soil of a secondary fill may be non-calcareous from the start, may become so as it builds up, or may remain calcareous until the rate of build up slows down and the rate of decalcification can keep up with the rate of accumulation. In such a fill, recrystallization of calcium carbonate in the lower part may be the counterpart of decalcification in the upper part, so that pores and structural planes become lined with secondary carbonate. The boundary between non-calcareous and calcareous soil can be very sharp, marked by a zone of heavy secondary crystallization. As in the case of iron pans calcium carbonate may pick out a change of texture and emphasize stratigraphic boundaries, or it may be deposited in the middle of a layer and give the appearance of a stratigraphic boundary where none exists. If the fill includes a gleyed zone, secondary deposition of calcium carbonate may be concentrated there, crystallization occurring in the crevices of the prismatic structure developed by the gleyed soil, producing a white veining following the marbling of colour of the gley. This effect becomes pronounced as an excavated section is exposed and crystallization from calcium-rich moisture in the gleyed soil occurs.

Where a ditch fill maintains a stable profile for a long period and decalcification has occurred, and then a calcareous plough soil is deposited on top of the decalcified soil, the buried soil may be recalcified by moisture percolating from the plough soil. If the plough soil was produced by the mixing of a non-calcareous humic soil with calcareous rock fragments derived from the C horizon, once the lower part of the deposit, containing a high proportion of the humic soil, ceases to be mixed by the plough, solution occurs around all the rock fragments, at the contact of rock and humic soil, leaving voids around them, and the calcareous solution descends through the fill to build up the concentration of secondary calcium carbonate lower down.

While all these alteration and migration processes are going on within a ditch fill, the adjacent undisturbed soil continues to undergo soil forming processes, particularly where it is in contact with a humic layer of fill, which acts as a soil **A** horizon and influences the adjacent soil in the direction of a **B** horizon. Where the ditch is cut into hard and insoluble rock, the changes may be slight, only a little oxidation of iron and a little softening close to the edge, with deeper penetration along cracks and structural planes. Where the rock is soft or calcareous, the influence of the ditch may be very marked, and the position of the side of the ditch can be difficult to establish during excavation: only by over cutting and examining the side in section can its position be estimated, among the worm and root holes which penetrate it and the gradual obliteration of rock structure as the side is approached. At the lip of the ditch, the persistence of worm activity throughout the history of the site at a point where soil remains thin because of its slope, may cause a particularly deep zone of merging between humic soil and substratum.

A ditch cut into or through a periglacially mixed substratum may develop a shape related to the variations in soil composition and resistance to chemical and physical weathering. In the first place, such a ditch may have been easier to dig through the disturbed and partly weathered material, and be abruptly terminated or markedly narrowed when hard rock was reached. Then, during the early stages of weathering back, while the primary fill was forming, the softer material would succumb more rapidly to disruption, so a distinct break in slope may occur at its base. During accumulation of the secondary fill, the soft and partly weathered material would contribute much fine material to the deposit, and its slope work back rapidly, and finally, during alteration after the ditch had filled up its further alteration would be particularly rapid, producing a more merging zone of transition than where hard rock occurred. Figure 10 shows the face of a ditch cut through such a periglacially disturbed rock, and the change of slope is particularly marked.

Pits

Many of the considerations applying to the study of ditches during excavation apply equally to pits. If pits are left open to fill naturally many of the same processes of weathering back of the sides and accumulation of primary and secondary fill, and final obliteration of the declivity by ploughing takes place, though relatively more rapidly because of the smaller ratio of volume to sides, and the processes of alteration in the fill and in the substratum into which the feature is cut are the same. However, in excavating a ditch, problems which arise in one area can often be solved by excavating a bit more, so as to eliminate apparent variations in depth or form which may be due to local factors not connected with the ditch itself, and to identify irregularities caused by tree roots or variations in the substratum. Pits are so variable in form and fill that comparison may not serve the same purpose.

Pits usually have a greater component of deliberately deposited material in them than do ditches; a secondary fill of ditch type is uncommon, though layers of fine soil accumulated by the same means may represent intervals between active dumping. The rubbish dumped in pits has two major components: ash from fires, which after a long period in a soil environment during which the mineral part of it is dissolved is represented by charcoal and other charred residues; and organic rubbish, which, unless the pit becomes waterlogged, undergoes humification. The nature of these types of deposit is discussed on pages 322–327. In pits, they are usually mixed and interstratified with soil from the sides of the pit and the surroundings. The deposits undergo the same type of alteration process as those of ditches, but those which depend on a high organic content tend to be more strongly marked in a pit which has been used as a dump for organic rubbish. Formation of iron pans within the deposit and around its walls may be particularly marked if the fill or the substratum are clayey. The alteration of the walls by the passage of worms and roots and by the proximity of humic soil is also likely to be intense. To determine the original form, much the best method is to reveal the whole pit in section by cutting the surroundings well back.

Variations in the normal pattern of fill are characteristic of the bell shaped type of storage pit. Anything which falls in through the mouth falls to the centre of the floor. Though the sides are not directly attacked by wind and rain, the overhang and particularly the partly unsupported soil at the edge of the mouth are susceptible to damage and to material falling away, disintegration being promoted by wetting and drying, the overhang being particularly likely to dry out, and by frost. This material too falls near the centre in greater quantity than against the walls. A dome of primary fill builds up, in which material from high in the soil profile, where it is most undercut, predominates. On a site which had a deep soil profile at the time when the pits were open but has since been ploughed off to a thin and stony one, the origin of the soil forming the dome of primary fill may not be immediately apparent. As the dome rises, some soil, particularly large particles, may roll down into the corners, but the doming persists as long as some overhang is maintained, flattening as more material rolls to the sides and as the edges of the overhang break away, until it finally collapses or is truncated by ploughing. The occurrence of a domed fill in a truncated pit attests its original shape, and the material of the dome provides evidence of a soil which may be lost.

Shallow pits may have their entire fill within the zone of worm mixing throughout the subsequent history of the site, so that the fill is identical to, and continuous with, the surface soil. On sloping ground, where the entire level has been reduced by ploughing and features truncated, fills formerly below the zone of worm mixing may come back into it and their stratification be destroyed.

Drainage Ditches and Gullies

Shallow features which come under this heading have more in common with pits than with the large ditches which fill up mainly by natural processes. They tend to have fills which represent an accumulation of rubbish mixed and interstratified with the soil of the surroundings rather than the regular sequence of fill types of the large ditches. Humification of organic residues is likely to have been intense if a drainage ditch fulfilled its function, since this implies either sufficient slope to keep water flowing, or a substratum which allowed infiltration, and even when full of rubbish and soil, percolation along it or downwards should have been adequate to prevent stagnation and inhibition of organic processes for lack of oxygen. If a drainage ditch filled up only after it had become blocked the fill tends to be much like that of a pit of similar dimensions. If the ditch silted up during or at the end of its active life, while water still flowed, it may have a distinctive fill, a sorted deposit in which silt grade particles predominate, and if the water carried much organic material humification of residues deposited with the silt produces an intense black colour.

Seepage through the bottom of a drainage ditch during its functional life may produce intense alteration in the soil beneath, and deposition of humus substances and iron oxide in crevices in the substratum may occur. If percolation in the substratum is free, such deposition can extend to considerable depth. Otherwise, an impervious iron pan developing beneath a drainage ditch which depends in part for its functioning on seepage may be the cause of its malfunction, leading to its being left to silt up, or to its being recut. Since iron pans, once formed, are exceedingly difficult to disrupt, it is possible to envisage abandonment either of a habitation site or of a cultivation area because of this problem.

Stake Holes and Post Holes

Stake holes and post holes generally pass through the **A** horizon of the soil, or through pre-existing occupation deposits, into the **B** or **C** horizon. The depth must be adequate to give stability, and the greater coherence and compactness of the lower horizons gives a better grip than the more open structure and less firm consistence of a humic soil. Stakes are rammed in, posts inserted in previously dug holes and wedged with stones, packed earth or subsidiary stakes. The trace left in the soil depends on the type of soil and on whether the wood decays in the soil, is burnt, or is pulled out. In excavation, it is even more essential than in the case of pits and ditches to section right across stake and post holes rather than trying to determine their form by emptying or sectioning their fill. The shape of a stake can be seen, and if it was forced into stony soil or rock, the pushing apart or tilting of stones and rock fragments can be seen. The packing of a post hole can be studied much more easily in a section right across than by attempting to section within it, among the packing stones.

The wood of a stake or post decaying in the ground undergoes humification by the processes appropriate to any organic residue in the prevailing soil conditions, and it acts as a tongue of concentrated A horizon going down into the lower part of the profile. A succession of soil animals and micro-organisms carries out its degradation, initiated by fungi. The fungi provide a source of nutrients for arthropods, particularly ants and woodlice, which chew up the wood to get the nutriment made available by the enzymes of the fungi and in their cells.

If wood is examined in the course of decay in well aerated soil, it is often found that at a fairly early stage much very loose, soft, fine granular material, very dark reddish-brown and black, partly fills the space formerly occupied by the wood, and there may be an irregular core or fragments of quite hard wood remaining among it. As soon as fungal attack has softened the wood, it is chewed up by arthropods, and the dark material is entirely composed of their fecal pellets. The two colours appear to be distinct stages in degradation, perhaps before and after the fecal material has been affected by a further phase of microbial invasion, the black form representing a more humified product. The degradation products occupy much less space than the original wood, much carbon having been lost as carbon dioxide, evolved by animal and microbial respiration. If there are no worms present, in an acid soil environment, the space is filled by the collapsing in and infiltration of soil from the sides of the hole and from overlying layers, but the position of the wood and its compressed and distorted outline may remain visible as a dark stain. Where worms participate in the degradation, they ingest mineral soil and the humified wood, and excrete in the voids, to produce a fine dark mull humus filling the original space of the wood. The worms passing through the sides of the hole and mixing its contents with the surrounding soil tend to spoil the clarity of its ouline.

Participation of ants, as one of the principal arthropods which live on the fungi attacking wood, may be important in many cases. Ants build their nests close to their food supply and may bring soil from the surroundings into the space left as the volume of organic material is reduced, and build their passages and chambers there. The soil may later be reworked by worms, but the fine granular nature of many wood replacement soils has a close similarity to the soil of ant nests, and it may be that this structure survives longer than the larger scale structure of worm casts.

Roots often pass down stake and post holes and through the bottom into the substratum, greatly confusing the excavator. Roots will grow preferentially in any water-retentive environment, and so appear to seek out the humic fill of pits and post holes, and if the hole is narrow the residue of root decay and the alteration of the surrounding soil in the root environment may quite obliterate the form of the hole and the original nature of its fill. Since roots decay in the same way as any other wood it may be difficult to distinguish between natural root holes and those which have re-used a stake hole and extended it downwards. Criteria of regularity and verticallity are all that can be used.

During the process of decay, the vigour of animal and micro-organism respiration may use up oxygen faster than it can arrive, and iron in the mineral soil in and around the hole may be reduced. Re-oxidation at the edge of the hole or a little way into the substratum then produces an iron pan or thin line of staining, lining or outlining the hole. Whether any complexing of iron with the organic substances produced in the process of humification also contributes to the mobility of iron, as a localized podzolization effect, is uncertain; it is not necessary to invoke this process to explain the phenomenon of iron pans forming around stake holes, though it may participate, at least in already podzolic soil environments. Brongers (1962–63) gives a procedure for emphasizing local podzolization effects around post holes by spraying on a chemical reagent. This will not, of course, distinguish between a post and a root.

Chemical investigation of the organic products of wood decay does not as yet allow us to tell whether there was in fact wood in a supposed stake or post hole, or what that wood was. The amount of re-working of the fill by successive generations of animals and micro-organisms may be such that no diagnostic organic molecules would persist, but work needs to be done to find out whether any characteristic patterns could be picked up by chromatographic studies.

When fillings of dark soil are not seen in a stake or post hole it is often suggested that the wood did not decay *in situ* but was pulled out. It is difficult to extract even a small stake from the ground by a simple upward pull if it has been well enough inserted and remained firm enough to serve any useful purpose. To get a stake out it is usually necessary to rock it backwards and forwards, producing a funnel shaped hole, and posts too big to be got out in this way must be dug out. Either process would leave traces in the shape of the hole and the disorder of the backfill. If the shape is well preserved and packing stones apparently in functional positions, absence of a well defined ghost of the wood is likely to be because mineralization of humus substances has proceeded far enough for remaining traces of organic matter to be undetectable against the background of infiltrated and mixed in soil, or because worm mixing has achieved complete homogeneity between the fill of the hole and the humic soil of the **A** horizon through which it is cut.

If a stake or post is part of a structure which is burnt, the burning may extend some way down it into the ground. Since access of oxygen will be restricted, production of charcoal is likely. The parts of the wood against the sides of the hole and cooled by contact with the soil, and the lower part of the wood, are unlikely to burn, which explains why charred remains of posts are not usually associated with signs of extreme heating of the soil surrounding the hole. The core of the wood is converted to charcoal, and the rest later undergoes humification in the normal way. The charcoal may be broken up later in the course of the settling and infiltration into the spaces left by the wood which decays, and disturbed and mixed up by the activities of worms. If only the centre of the uppermost part of the wood within the hole chars, the amount of charcoal will be small, and it is impossible in these circumstances to tell whether the tiny fragments in the fill are in fact derived from a charred stake or post or are a chance mixing in by worms of fragments which happen to be in the soil round about. Charcoal will also remain in a stake hole if the point was charred to harden it before it was rammed in.

Buried Soils

The amount and mode of the alteration which goes on in a soil after it is buried depends on the soil type and nutrient status and on the thickness and nature of the burying material. Factors causing alteration are compression, modification of water movement through the soil, and continuing soil processes within it. The physical effects of compression and behaviour of water have not been studied in any detail in the archaeological context, but the data needed for such study must exist in great quantity in the field of soil mechanics associated with engineering and mining activities.

Continuing soil processes may go on in isolation or within the reach of the active soil at the present surface. If the initial depth of burial is slight or if erosion has reduced the cover, penetration of roots and worms, and eventually total involvement in the active soil profile, may obliterate the buried surface and the characteristics of the profile. In less extreme cases, occasional larger roots, and the deeper channels of worms, may produce some mixing of boundaries and introduction of extraneous matter into the profile without altogether disguising its former character.

Processes in Soils After Burial

Formation of Iron Pans. Where depth of burial is sufficient to isolate the soil from the active soil processes associated with growth of vegetation, no fresh residues are added, or introduced in the form of roots. Organic activity continues so long as there is adequate oxygen or oxidizing power in the form of iron oxides and other reducible compounds, and as long as all necessary nutrients are available. If the soil before burial was close to limiting values for any essential nutrients, activity soon ceases. In many cultivated soils, and in strongly acid soils, limiting availability of one or more nutrients is the normal situation, the micro-organism population being low and largely dormant except for short periods after the addition of fresh residues, and so little further humification is likely to occur when a soil of this type is buried. When a podzolic soil with a mor humus horizon is buried, activity may continue until fresh residues, that is, the plants living at the time of burial, and fibrous material in the litter layer are converted to amorphous humus, and thereafter the entire organic horizon remains as a compact black mass, of very characteristic waxy consistence when moist, which shrinks on drying to produce a strong blocky structure. A moder humus horizon similarly undergoes little further modification, but having a mineral skeleton is not so compressed and concentrated and may retain a loose structure and friable consistence. The progress of degradation in a buried podzol is being studied in the experimental earthwork near Wareham, Dorset. So far, after nine years, the remains of living plants are still little degraded, though blackened by initial micro-organism colonization. The activity of arthropods and fungi has continued with some vigour in the organic materials which were buried on the old ground surface, since unlike the local vegetation they are themselves rich in nutrients and apart from the leather samples, do not have the protection of a high tannin content. Parts of the bodies of arthropods have been found among the degradation products, and fungal mycelium and spore bodies are abundant (Evans and Limbrey, 1974).

When a mull humus soil is buried, the surface mat of roots, stems and fresh residues may be converted to a similar black waxy layer, but below it in the nutrient rich mull humus of the worm sorted layer or in a still humic soil of a ploughed mull profile, activity continues, showing that the nutrient supply in the soil solution and exchange complex is essential to the process, which proceeds further in the humic soil than in the purely organic layer at the surface. Unless the overburden is extremely porous, so that oxygen-bearing moisture or air continues to reach the buried soil, depletion of oxygen results from respiration in the soil, and reduction of iron oxides takes place. As the organic content falls and the iron oxides are reduced and move away, the buried humic horizon takes on a grey colour. The iron is re-oxidized when it meets a less oxygen-depleted environment, and an iron pan forms. Where the pan forms depends upon whether there is movement of moisture through the soil or whether all movement of iron takes place by diffusion away from the zone in which it is released into solution. In the former case, the pan forms within, at the base of, or a little way below the former humic horizon, depending on the activity at different levels in the soil, and on the rate of movement of the moisture. Where the iron moves by diffusion alone, a double pan forms, above and below the zone of maximum activity. Diffusion of oxygen from the less depleted soil above and below the humic horizon, can result in formation of a pan still within it. Otherwise, it forms at or below the zone of worm or plough mixing, and at the surface of the buried soil or a little way above the surface in the lower part of the overburden. Associated with an iron pan, is often a zone of black mottling of manganese dioxide. Manganese participates in the same processes of reduction, mobility and re-oxidation as iron, but seems to move further, so that the manganese mottling is usually beyond the iron pan, and it seems to be deposited as discrete mottles on the faces of soil peds and stones, rather than passing right through them as does the iron pan.

Iron pans may be localized, depending on lateral variations in the buried soil, in the overburden, or in the movement of water through them. Patches of pan may be due to the location of individual plants, or concentrations of organic material in the soil or on its surface, such as cow pats, or to large stones in the overburden. It may be that paths or other local variations in the vegetation and compaction of a soil before burial can effect the development of iron panning, and so if it is variable it is worth plotting its distribution to see if it makes any sensible pattern. However, the very common occurrence of a concentric zone of iron pan beneath round barrows does not necessarily imply a lively round dance as part of the burial ritual, as suggested by Fox (1959), since it is clearly related to the depth of burial and the movement of water through the mound. Often the pan covers the whole central area, fading out round the edges where the exclusion of oxygen was less complete, and sometimes being less marked towards the centre, where the soil may have been so protected by the mound as to remain too dry for organic activity to proceed with vigour. If the central structure of the mound is a turf stack, iron pan beneath and within this and not beneath the more porous soil of the outer part of the mound is a very common occurrence.

It is important to establish whether an iron pan in a buried soil is the result of burial or if it was a feature of the profile before burial. If there is no trace of a pan in soils of the neighbourhood, it can safely be assumed that a pan in a buried soil is an effect of burial, since the conditions leading to development of iron pans have acted progressively, and if they have not developed since, anywhere in the area, they are unlikely to have existed at the time of the burial. In a region where development of pans has been a feature of the history of soil degradation, that history can only be worked out if every buried soil is correctly diagnosed as to whether it did or did not show signs of iron pan formation and the associated effects of waterlogging before burial. The things to look for are: continuity of the pan through the boundary between buried soil and overburden; its continuity around, or interruption by, features associated with the overburden, such as stake holes and stone settings; the relationship between the pan in the buried soil and that in the surroundings, in terms of thickness and position in the profile and continuity between the buried soil and the outside, and the occurrence of a pan in soil derived from the surroundings and forming part of the overburden, such as turves. With careful search, it should be possible to find some detail which shows whether the pan was present and to what degree it was developed before the soil was buried.

Rarely, in my observation, does an iron pan form at the surface only of a buried soil, and it can be very misleading to take the position of a single pan as indicating the buried surface. The upper pan of a double one may pick out the surface, particularly if there is a marked difference in texture or organic content between buried soil and overburden, but it often forms a little way above it, and may wind up and down, beneath or through the bases of stones resting on the surface, and up into the soil between them. A single pan usually forms within or below the former humic horizon, and the buried surface is then the top of the grey layer above it. Only where the overburden is more humic than the buried soil or where groundwater may rise through the profile might a single pan form at the surface itself.

Humic Horizons Without Associated Iron Pans. Iron pans and the associated grey layers are most strongly developed in soil of high organic content and high clay content or where the overburden is clayey. In porous and well aerated soils, and especially in those with thin or weakly developed humic horizons and no surface mat of humus or vegetation, mineralization of organic matter may go to completion, or so nearly so that the remaining traces give no visible colour. The former humic horizon has then to be identified on grounds of structure or texture or by the very slight greyness, or thin whitish layer, representing a slight depletion of clay and iron which takes place in the humic horizons of impoverished soils. Thin or weakly developed humic horizons are usually those of soils impoverished by cropping or reduced by erosion, and in these there will be little or no development of a worm sorted layer and of soil structure whose appearance will aid identification. The slight colour change may be very difficult to see, but may be more apparent in section when the soil is dry than in a freshly exposed, moist condition. With exposure, any traces of structure will appear, and the horizon may be more resistant to wind and rain erosion and so stand out in relief. A slightly higher organic content may allow greater development of a film of algae over that part of the section if it is left exposed for a long period. Alternatively, the humic horizon may have been somewhat depleted of clay as well of iron, and have a poor structure, so that with loss of organic matter the difference in texture can be determined, and it may be more easily eroded than the soil below and the overburden.

Where a better developed humic horizon, and one which has lost neither clay nor iron during its active life, loses all visible traces of humus after burial, it must be identified by textural relationships and structure. A worm sorted layer is identified by its stone-free nature and the horizon of stones accumulated below it; a plough soil has a characteristic uniform distribution and random orientation of stones. It may be deeper than normal plough depth if it is in an accumulating situation, but is unlikely to be less, unless a worm sorted layer is developing in the top of it. The vertical cracking developed by buried humic horizons is a very useful criterion for identification (see Fig. 35). A newly trowelled section needs time to dry out slightly before the structure becomes visible, and preparing the section by picking at it rather than by trowelling brings out the structure much better (see page 271).

As a soil profile is excavated by trowelling down through it, the features of merging boundaries between horizons are revealed in plan, and unless the processes of differentiation of the horizons are understood, the patterns produced can be puzzling. Undulating iron pans are an obvious example, and a feature which may be associated with an iron pan or may be encountered in other profiles is the polygonal network which begins to appear as trowelling reveals the upper boundary of a gleyed horizon. The prismatic structure developed in a gleyed soil appears as a network of cracks from which greyness may spread into brown or yellow soil, or vice versa, or which may have been filled or stained by materials infiltrating from the horizon above: infiltration of a humus sol down the interstices of a prismatic structure in a gleyed podzolic profile produces black or brown veining against a grey background. As trowelling proceeds through such an horizon, the colours and the scale of the pattern change, and a constant appearance at which excavation may be stopped is not achieved. The importance of digging a deep cutting to reveal the profile in section is apparent, in spite of the determination of many excavators to maintain a consistent level of excavation.

In studying a buried soil, a cutting large enough to reveal lateral variations and the full depth of profile development is required. Having found the buried surface and studied the profile, the features likely to have occurred since burial can be identified, and comparison made with the range of profile types round about the site today. Conclusions can then be drawn as to the condition of the soil at the time of burial. The presence or absence of a turf line can be seen, the presence and depth of a worm-sorted layer, the results of disturbance, such as cultivation, re-establishment of worm-sorting after cultivation, and other localized or general effects of human or animal activity can be studied. A certain terminological confusion can be sorted out here. It is customary among some excavators to refer to any dark and obviously humic soil as a "turf line". Only a very dark, usually thin, and predominantly organic layer at the surface of a buried soil should be called a turf line, and even this restricted use of the term can be misleading, since it need not have been turf as one is accustomed to see it in a pasture or lawn. A thick litter layer under woodland or heath vegetation will give the same appearance when converted to amorphous humus as does the mat of grass roots, stolons and stems forming a turf. Presence of a distinct turf line implies an undisturbed soil surface, but the nature of the vegetation can only be established by pollen analysis. However, if a mull soil with a good worm sorted layer does have a turf line, it is likely to have been under grass, or at least fairly open conditions, since if worms were so active they would have prevented the accumulation of a litter layer under woodland thick enough to prevent the growth of a ground flora. If the soil is of mor or moder humus type, the litter layer may be thick under any vegetation.

Artifacts in Buried Soils

The position of artifacts in relation to the buried surface needs careful consideration. If the soil has a mor humus, there will have been little or no activity by surface casting worms and artifacts dropped on the surface will stay within the organic horizon unless the soil is cultivated or otherwise disturbed. Whereas stratification of artifacts may exist in a humus accumulation, the degree of compression of the organic material in a buried soil may make it difficult to relate bulky items to positions at the base of, within, or at the surface of the layer. Thin, flat items such as flint flakes and small implements, may still be clearly in one or other of these positions, and this may be very important for the history of soil degradation at the site. In a mull soil, artifacts are lowered through the worm sorted layer and accumulate at its base. Any artifact below the base of worm sorting is likely to have been introduced during an earlier phase of disturbance. Any at the surface or within the turf line may have been quite recently dropped; it has to be remembered that small items can be pushed quite a long way down into the soil by a foot or a hoof, and local poking around with a stick or rootling with a snout, or the digging of holes by a dog, can disturb the worm sorted layer enough to introduce things into it, and the damage will very soon be obliterated by worm activity, so artifacts might get well down into the worm sorted layer soon after arriving on the surface, and older material may come to the surface. Moles are usually active where there are worms, and, as any archaeologist knows who is accustomed to scatter mole heaps to look for artifacts, things deposited several thousand years earlier frequently lie on the surface of an "undisturbed" soil. Moles tunnel where the worms are, that is, within the worm sorted layer. However, during the period when worm sorting is being re-established in a cultivated soil, both worms and moles may be active to the depth of cultivation, or throughout a greater depth if there has been a rapid build up of plough soil in a declivity or against a lynchet, and recent artifacts may get down mole holes into quite deep positions.

Very small artifacts, and environmental evidence in the form of fragments of charcoal, seeds and snails, can slip down worm holes and get well down into a soil profile, but this is unlikely to apply to things larger than about half a centimetre in diameter. Seeds and fragments of shell and charcoal are very likely to be taken down by worms (see page 30). Root channels may also provide a route into lower horizons, but the period during which a decaying root leaves a sufficient gap for anything to slip down is not long, at least in wormy soil.

Mounds and Earthworks

Structures built partly or entirely of soil materials undergo soil formation over the top and alteration within them. Depending on position within the structure, any material undergoes alteration by the processes appropriate to active soils or those appropriate to buried soils, or both, at the same time, or one after the other.

A newly built earthwork suffers erosion at the surface and settling within it, and these two processes combine to reduce the height and slope until a stable cover of vegetation is achieved, after which the slower processes of surface washing and soil creep continue to work upon it, and a soil profile develops. The rates at which these processes work depend very much on whether the structure is built of soil from the humic horizon of the area, from lower down in the profile, or from rock, and on the degree of weathering of the materials from the various parts of the profile. In areas of periglacial disturbance, earthworks often include much material which although it may have been quarried from a deep ditch is already partly weathered and provides a parent material for rapid soil formation. Timber lacing and revetment of an earthwork succumb to the same decay processes as posts in the soil, and under the pressure of overlying soil or rock their decaying remains and the spaces left collapse, contributing to the rate of settling.

As an earthwork undergoes erosion and soil creep, its lower slopes become protected by the skirt of material coming down from above, and the soil around its base becomes buried. The soil profile which is eventually established is therefore formed on material whose degree of weathering changes from top to bottom of the slope. That at the top, or if the top is wide and becomes relatively flat, on the shoulder, may remain a thin profile, developing little horizon depth and continually losing material by creep, while that at the foot accumulates humic and partly weathered material, rapidly at first but at a low rate which continues so long as the elevation persists.

Within the earthwork, humic soil undergoes the same processes of alteration as those of the buried soil. If turves are present, they may remain as clear outlines in which the sequence of turf line, A horizon, and any B horizon material which was lifted with the turf, remains visible, and it can be seen whether they were placed the right way up or upside down, or they may remain only as faint ghosts, with slight greying accompanied by mottling or slight development of iron pan, or slight differences in texture or structure being all that indicates their position. Since the higher they are in the earthwork the less will have been the exclusion of air, the distinctness of the turves often becomes less upwards, and only the lower layers may show the same degree of alteration as the soil of the buried surface. In a region of iron pan formation, study of the turves within an earthwork is probably the best way of determining the state of the soil before the earthwork was built, since the formation of iron pans subsequently will tend to follow lines of drainage through the structure rather than maintaining a consistent level in relation to soil horizons. It can usually be quite quickly established whether or not the turves were taken from a soil which had an iron pan, whereas in the case of the buried soil a detailed search may fail to yield proof one way or another.

Where mineralization of organic matter has been extreme, and where the former humic soil had not lost clay or iron, or where "turves" might have been taken from a deep worm-sorted layer, perhaps with little or no litter layer or turf line, and without lifting **B** horizon material adhering to the base, there may be no way of telling whether the earthwork was built of turves or not. In any case, if one considers the procedure of building a mound, there is obviously continuous gradation from cutting and lifting turves, through lifting shovelfuls of top soil which may or may not retain some coherence, to gathering up loose soil, and the nature of the soil profile and its vegetation and the efficacy of the tools available clearly determine the process.

The question of weathering horizons within an earthwork often arises. Dark lines may represent the position of a stable surface which persisted long enough to acquire a cover of vegetation before being covered, or they may represent a stage in building at which people with muddy boots trampled over it, or they may represent a stage at which a new source of soil was opened up and humic soil spread over the structure. Either the trampling or the spreading of humic soil may have been done to stabilize a structure whose slopes were becoming unstable before going on to build it higher. If the dark line does in fact represent an interval during which weathering and soil formation took place, the initial stages of erosion before stabilization occurred may be detectable in the stratigraphy of the skirt of redeposited soil around the base of the structure. If soil formation has proceeded far enough for a distinct humic horizon to develop, it should also show signs of **B** horizon formation, though if the earthwork is built of soil taken from the **B** horizon such changes would not be apparent. If the interval of soil formation was long enough for worm sorting to become established, this should be detectable, and one would expect a worm-sorted horizon to vary in thickness according to its position on the slope. If it were not a soil in which worms were active, a merging of humic soil into the **B** horizon may be detectable, which would not be so marked if humic soil were spread over the structure and quickly buried again.

The effect of development of a soil profile over the top of an earthwork is to impose upon any stratification, and on the effects of burial on the materials within it, the changes appropriate to the position in the new profile: worm activity, introduction of organic matter in the form of roots, and the effect of oxidation, hydration and solution on materials which in their former position in the source soil had not undergone these processes to so marked a degree. Development of podzolic horizons or of illuvial clay horizons may be superimposed upon the stratification of the earthwork, either obscuring it or picking it out, depending on textural relationships. If the soil profile over the top of the earthwork remains complete, only reduced in horizon depth by the continuation of slow erosion, the effects of it should be apparent. If the earthwork has suffered further drastic erosion, such as by ploughing, after the establishment of the soil profile, the origin of the remaining horizons of the truncated profile may not be so clear.

If an earthwork is low, or has been much reduced, the effects of soil formation over the top of it are imposed upon the buried soil. Penetration of roots and worms through the buried soil or deposition within it of mobile substances may so alter or obscure it as to prevent effective study of it. The extreme case is of worm activity removing the humic horizon of a buried soil altogether. This is discussed by Atkinson (1957). As the overburden becomes so thin that the worms pass right through it in the normal course of their activities, they may find in it a rich source of nutriment: if the soil over the top has been impoverished by cultivation or thinned by erosion, a buried soil may provide a better food supply. The worms ingest the humic soil and cast it out at the surface, and in doing so lower the overburden onto the **B** horizon of the buried profile. A distinct boundary may remain, but if the soil and overburden are of similar material a complete merging may be achieved.

Formation of a soil over a structure such as a cairn, built of stones with no soil in between can be a little puzzling. Often, the interior remains free of soil, or nearly so, with large voids among the stones, but the surface acquires a continuous cover of soil and vegetation. If the rock is hard, the formation of soil from it alone may be very slow, and it would tend to fall down through the spaces. It is probable that much of the mineral soil is deposited by wind and animals, and the blocking up of the spaces is achieved by dead leaves and similar detritus. Following the normal path of soil formation on a bare rock surface, lichen and then moss will become established, and cushions of moss dislodged from the surface will help to block the spaces in the upper layer of stones. Only when a continuous layer of soil is formed does all the mineral material which becomes detached from the surface of the stones in the course of weathering and soil formation remain *in situ*, and until then, much material falls through the structure to form a layer on the buried surface and among the lower layers of stones. Some soil probably continues to fall from the under side of the soil layer at the surface, throughout the history of the structure, and if it is not very high, or if the stones are fairly close packed, the layer building up from the bottom may meet that developing at the surface.

If there are worms in the soil, they will continue to be active below the structure, since there continues to be space for them to void into. They may much reduce the buried soil or remove its humic horizon completely, lowering the stones and putting the soil among them, mixed with that which filters down from above. It is often found that the lower layers of a stone structure appear to be embedded in the humic horizon and resting on the surface of the **B** horizon. If there is no evidence of a cutting or bedding trench having been made, this may be entirely the work of worms, or of ants. Ants often build their nests within a stone structure, taking soil from below and carrying it up to make their passages and chambers among the lower layers of stones or among those higher up; if they construct the nest below the former ground surface, they may deposit the soil they bring out among the stones. A nest built high within a structure may have passages connecting it with the soil beneath and with the surface above, to which the ants go to forage, but after it ceases to be used, these tenuous traces disappear, and a mass of soil remains isolated among the stones, its origin being then obscure. Ants' nests are often used after the ants have gone by woodlice, spiders, worms and other large members of the soil fauna. Those which do not ingest soil live in the passages and chambers and may introduce organic residues. Worms excrete in the spaces and rework the ant soil, so that its original nature may not be apparent.

Edge Effects

Near to the edge of a structure or earthwork the buried soil becomes involved in processes associated with the presence of the overburden and with soil over the top of it and odjacent to it. Drainage through the overburden depends on the materials and structures it contains, but any elevation tends to deflect water down its slopes, so that more infiltrates towards the edges than either in the centre or outside; impermeable layers or stones within it, particularly if they lie in sloping attitudes, increase the difference. The soil in a zone at or close to the edge may therefore become much more deeply developed, perhaps with solution features penetrating so deeply and in such a regular manner as to give the impression of a ditch where none exists.

An intensification of soil formation at the edge may also be apparent where the processes belonging to the active soil of the overburden are superimposed upon those of the buried soil. The material there has undergone two distinct periods of soil formation, so that the profile may be deeper and more highly differentiated than that of the unaffected buried soil away from the edge, and whereas the soil of the area outside would have had a continuous history of soil formation, its profile may have been reduced by erosion or cultivation, while that of the edge zone remained protected by the presence of the structure. As a structure such as a bank or mound is eroded, the edge effects creep in towards the centre, and the buried soil comes within reach of roots. While the roots are making use of both active soil and the buried soil, they have access to a greater supply of nutrients, and extra lush growth may occur. This will be particularly noticeable where surface soils have suffered nutrient loss and the buried soils still retain the nutrient status of an earlier stage in soil history. With intensification of soil processes under the more vigorous growth, the deepening of the soil profile in the edge zone may be particularly marked. The feature may be so reduced as to bring the entire buried soil area within reach of roots; otherwise the effect may remain as a restricted zone which would appear as a crop or soil mark and lead to mistaken interpretation during survey as well as under excavation.

Worms continue to be active in a buried soil so long as there is nutriment for them and so long as there are spaces into which to excrete. Under the pressure of the overburden, existing spaces and their own burrows may close up, but where there are spaces in the overburden, as in the case of cairns, they can continue to live, and the better aeration there may maintain suitable conditions for them until the supply of food runs out. Where there are no spaces in the overburden, only the zone close to the edge may provide less compressed and more aerated soil, and the opportunity to excrete either within the oids of the soil adjacent to the structure or at the surface outside it.

Thus, whereas the buried soil may remain intact under the centre of an earthwork or a floor, at the edges it will be gradually removed, leading to sagging, tilting of stones towards the edge and the impression of a bedding trench, marking out gully or other interference with the soil there.

Floors and Buildings

Any humic soil remaining intact after the disturbance associated with building activity may be buried beneath the floor. In the case of substantial buildings it seems unlikely that the integrity of the soil surface would survive the trampling, placement of scaffolding, assembling of building materials and work upon them which go on during construction, but lighter structures may be imposed on a profile which remains fairly intact.

During the life of a building, changes occur in the buried soil depending on the degree of impermeability of the floor to air and moisture, just as they do in any other buried soil, and the same sort of edge effects apply. Outside the building, detritus from its walls, in the form of chunks of daub, fragments of stone, and finer residues may accumulate. A building undergoes physical and chemical weathering and incipient soil formation appropriate to the material of which it is made and the slope of its walls and roof. Inside it also undergoes slight weathering and perhaps abrasion as people, animals and furniture rub against its walls, and the detritus either builds up to contribute to an occupation deposit or is swept out as dust.

Once a building is abandoned and allowed to collapse, soil formation, and accumulation of material moving by surface washing and soil creep and deposited by wind, continue. Rubble provides spaces in which wind blown materials are trapped, and into which worms can excrete. Worms bring soil up from below a floor and deposit it on top, getting through small spaces in the flooring materials which may not be apparent. As a humic soil accumulates, materials such as lime mortar undergo dissolution, and more passages for worms are created. The contribution of mortar to occupation deposits, fills of features and the soil around a building may greatly alter the chemical conditions there, raising the pH and providing sufficient calcium for an abundant worm population in an area whose soils are generally too acid. The use of building stone foreign to an area may similarly change conditions of the local soils quite markedly.

The weathering down and burial of a building takes place by a combination of accumulation of soil and organic residues from the surroundings against and among its ruins, weathering and soil formation acting on its materials, and the activity of worms and ants in taking soil from below and putting it on top. While it still has enough elevation, or hollows within it, to provide shelter, people and animals contribute their rubbish and excreta to the accumulation. While it provides a climable structure, children and animals, particularly goats, sheep and young cattle participate in its erosion, and, once it forms a mound, like any other elevation it tends to be walked up and stood upon by people and animals, increasing the rate of soil creep.

Components of Archaeological Deposits

Burnt Materials

When organic materials burn, the carbon in them goes off into the atmosphere as carbon dioxide, carbon monoxide and volatile hydrocarbons, the water initially present in the tissues and produced during combustion goes off as water vapour, nitrogen is given off, and the mineral components of the tissues are left as ash, which consists of the oxides of alkalies, alkaline earths and metals. In practice, complete combustion is rarely achieved. In the heart of a fire and in the interior of large pieces of fuel the supply of oxygen is inadequate, and carbon remains in elemental form as charcoal. Volatile hydrocarbons produced in intermediate stages of combustion may be given off in the smoke, escaping from the heat before they are burnt, less volatile ones may remain associated with the charcoal and ash.

Unburnt material remains in the bed of the fire as ash or flies up in the smoke. Smoke contains fine particles of carbon, mineral ash, nitrogenous gases, and hydrocarbons. Some of the hydrocarbons condense on cool surfaces which the smoke passes, others continue to burn as they rise but combustion ceases as the temperature drops and more carbon is produced in very fine particle size which flies off as smuts or adheres to surfaces and accumulates in chimneys. This is the main component of soot. Carbon has strong powers of adsorption, and soot provides sites for accumulation of the hydrocarbons. The charred remains of fire are thus scattered in the ash, and concentrated in the chimney or spread over walls, ceilings, furnishings, people and animals, as ash and soot.

Once carbon is reduced to elemental form it is unavailable to life processes and unless subsequently burnt remains forever in any situation in which it is deposited; it is not destroyed by solution or oxidation in the soil. The carbon is in the crystalline form of graphite and is essentially an inert material. The crystals are extremely fine and form a mesh which reproduces the structure of the material burnt, provided it does not pass through a liquid of vapour stage in the process of combustion. Whether it does so depends on the chemical nature of the material: so far as archaeological materials are concerned, plant materials tend to retain a detectable structure while that of animals tissues is destroyed or obscured by the liquification of proteins and the bubbling which takes place as gases are given off during combustion. Much of the carbon left in the ash of a fire is too finely divided for structure to remain detectable because only certain components of the cells may remain as charcoal, the structure as a whole falling apart, or is embedded in a mass of distorted and bubbled material. It is to be hoped that scanning electron microscope studies may reveal structure not detectable by light microscope, or that characteristic textures of the bubbled materials may give a clue to their origin, though the conditions of combustion rather than the nature of the original material are likely to control the final effect when there has been an intermediate liquid stage. It is also possible that the study of charred residues through the hydrocarbons which remain adsorbed on the carbon may produce useful results. The crystal structure of graphite provides a large number of incompletely satisfied valencies at the edges of the crystals, and these are available for bonding with the hydrocarbons. The amount and nature of the hydrocarbons would depend on the oxidation conditions and temperature of the fire as well as on the original nature of the material, and they would also tend to become mixed from materials in close proximity. Once in a soil environment, the hydrocarbons become available to utilization and modification by micro-organisms. However, within the interior of large charred residues there may be molecules which were sufficiently characteristic of the material which burnt and have been sufficiently protected by being totally enclosed in the graphite structure or lodged in pores too fine for entry by micro-organisms, to provide some hope of identification. A great deal of experimental work needs to be done, to determine, for example, chromatograph patterns characteristic of different materials burnt under different conditions, before there will be any hope of identifying materials which have lain in the soil for long periods. Mixed materials will provide a problem: the identification of charred food residues adhering to a cooking pot is not likely to be achieved easily.

I would like to stress here the value of using the term *charred* to refer to residues of burnt material, in preference to *carbonized*. Carbonization is too general a term, and can refer to any processes by which carbon content is increased relative to other components. There persists a belief among some archaeologists that carbonization can occur in the context of soils and archaeological deposits so as to produce carbonized wood, grain, or other materials without the agency of fire. Without fire, reduction to carbon takes place under high pressure on a geological time scale, producing, for example, coal. In the soil, carbonization in the form of humification does not continue beyond the formation of humus substances, which may give a very dark appearance to humified materials but does not approach the condition of charcoal, producing no elemental carbon. The term *charred* so clearly implies burning that its use would abolish this source of misunderstanding. *Charcoal* can be reserved for macroscopically recognizable fragments of wood, and other substances can be referred to as charred this and charred that, with *charred residue* for materials whose structure is not identifiable.

In discussion of this matter with archaeologists, the objection has been raised that to many people the word charred implies partial burning, in the sense that part of the object remained unburnt. Careful wording should avoid this difficulty.

The mineral component of ash contains the mineral plant and animal nutrients in the form of oxides. Grasses and cereals accumulate silica which is deposited in cells as silica bodies, *phytoliths*, which form an important part of the ash from these plants. Other structures which may remain recognizable after burning are bones, shells and the siliceous and calcareous tests of any aquatic animals and microorganisms which may have been among the fuel of the fire, if reeds or seaweed are burnt, for example. The rest of the ash is an amorphous powder; since the structural components of plants and the non-skeletal parts of animals are the predominantly carbonaceous and nitrogenous materials, and the mineral nutrients mainly in the cell contents, any trace of structure in the ash is likely to be only a negative, or ghost of the tissue, with no structural integrity of its own, and so disrupted by the slightest disturbance. The oxides which form the greater part of the mineral ash are those of the macro-nutrients, and these are all more or less soluble in rain water and soil moisture, and so rapidly disappear. The insoluble oxides of some of the trace elements must accumulate in situations in which much ash is deposited, and their relative quantities might provide information about the nature of the materials that had been burnt, but, so far as I know, no work has been done on this.

When soil is burnt, its organic component either burns to completion or is left in charred form, depending on amount of oxygen available in the fire in relation to the amount of organic matter being burnt. Charred organic matter imparts blackness to the soil, and in large quantities completely masks the colour of the mineral soil. The main visible effects of heating on mineral soil are those which result from dehydration and oxidation of iron compounds. If there is so much organic matter being burned in the soil itself or in the burning material around it that oxygen supply is inadequate, iron compounds in reduced condition will remain so, and impart no redness to the soil. If oxygen is abundant, iron compounds are oxidized, and the hydrous oxides already present are dehydrated, and the strong red colour of haematite dominates the soil colour. Soils which already have a high haematite content may change their colour little on heating: this applies to some tropical and relict Mediterranean soils, and to soils in temperate regions formed on parent materials such as the red sandstones whose iron compounds acquired their colour under hot desert conditions.

The conditions in a fire and in the soil beneath it usually fluctuate between more oxidizing and more reducing states as the fire flares up and dies down, and it is probably partly as a result of this that the mineral maghemite is found in soils. Maghemite is a magnetic iron oxide of the same chemical composition as haematite but a different crystal structure, and the high quantities of it in soils associated with archaeological sites increases their magnetic susceptibility and provides the basis for archaeological prospecting by magnetometer. The way in which maghemite forms is not fully understood, and an alternative to the hypothesis that it is due to fire is that it forms as oxidation conditions fluctuate in the soil under normal temperatures, in the course of cycles of organic activity and moisture content. It is symptomatic of a regrettable state of affairs that the two possibilities were discussed in almost total isolation in papers given within a few days of each other recently, one to archaeological scientists and one to soil scientists. In the first, a physicist working in the field of archaeological prospecting had been unaware that fluctuations in the oxidation state of iron were normal in a humic soil, and in the other the soil mineralogist dismissed the fire hypothesis in a few words as not applicable in many cases, apparently being unaware of the ubiquity of the effects of fire on soils wherever man has been. It seems probable that maghemite forms by both processes in soils, their importance varying between soils of different humic content and different amounts and types of iron compounds.

Organic Residues

Humification of organic materials is impeded under extremely acid conditions, under waterlogged conditions, and where they have become encapsulated in a mineral deposit. Otherwise, the activity of soil animals and micro-organisms destroys their structure.

In acid soils, pollen grains, spores of lower plants, and some resting bodies of micro-organisms survive after all other plant and animal residues have been humified. Under continuously wet conditions, if stagnation or compaction is sufficient to prevent access of oxygen either in air or in water, humification stops when any alternative sources of oxidizing power available to anaerobic modes of respiration have been exhausted. In many deposits of organic materials, conditions vary from place to place, according to degree of compaction, proximity to the surroundings or the surface, and nature of the materials themselves, so degree of preservation is not necessarily uniform. The products of degradation and substances leached from each material contribute to the environment of the other materials; nutrients essential to the organisms attacking one substance may be derived from a neighbouring one, substances which inhibit decay, such as tannins, may be leached from a plant material, or from tanned leather, and protect other proteins nearby.

The organic material which degrades most rapidly in a waterlogged deposit, and which may have been completely humified before organic activity ceases, and before other materials have suffered much alteration, is the soft parts of animals. Flesh rarely survives, but skin, hair, horn and feather may persist much longer, particularly if tanning processes are involved. The chitinous parts of insects, particularly the wing cases of beetles, are also likely to be preserved, having some natural protection from soil processes which fits them for life in the soil. Plant tissue may be preserved in waterlogged deposits in great quantity, but unless preservation conditions are very good, it may be in the form of small fragments. Seeds and bud scales are both relatively resistant to decay and identifiable, but small fragments of leaf and stem may be recognizable as such but not assignable to any particular plant. This is not so much because such parts have no identifiable characters as because the work of assembling reference data has not been done, except in the case of plants of forensic interest such as *Canabis sativa* (Jackson and Snowdon, 1974).

If the conditions of oxygen exclusion have not been maintained, however dark and obviously rich in organic matter a deposit may look, there are unlikely to be any surviving structural remains of organic materials other than pollen and spores. It is usually quite easy to tell whether or not seeds, insect parts and other identifiable residues are present. If the quantity of humified material obscures recognizable residues, it can be dispersed in water with the help of ordinary domestic water softening powder. A solution is made up and a small amount of the deposit put in it and left overnight if necessary. Seeds and beetle remains will often then float to the top, and the sludge can be sieved to check for non-floating remains. It is not feasible to treat large quantities in this way, as the amount of water softener needed is too great, but it is a useful check to see whether there is any point in starting more laborious separation techniques, and can easily be done on site during excavation. If uncharred seeds are found in situations in which their preservation is unlikely, a careful search for worm holes and aestivation chambers must be made.

The question of preservation of the eggs of gut parasites in deposits in guardrobe pits frequently arises, since eggs of such parasites which may be from human hosts have been found (Pike and Biddle, 1966). These eggs do not appear to be particularly resistant to decay, so they are not likely to be found in a deposit in which there are no macroscopic organic residues. If it seems possible that they might have survived, they can be looked for by making up a solution water softener or sodium hydroxide to the specific gravity at which these eggs have been found to float, 1·3 (for sodium hydroxide this is approximately a 3 N solution). The water softener or alkali will disperse the humus substances and release surviving residues, and any floating material will collect at the surface around the edge of the container. The scum which collects there can be skimmed off with a fine paint brush and transferred to a microscope slide. This is a simpler technique than that used by Pike, and has the advantage that the soil does not have to be ground up to release the organic residues, so they are not damaged, and the dispersal of humus substances leaves them in a clearer and more easily recognizable state.

Pollen analysis of organic accumulations associated with archaeological sites, as distinct from that of buried soils or of naturally occurring peat or lake deposits, is in an experimental stage (Greig, personal communication; Buckland et al., 1974). In situations where materials bearing a pollen load have come from a wide range of ecological contexts, and could have been imported from great distances, and where the mineral component of the deposit is derived from soils or deposits which may already have contained pollen, problems of interpretation are very great indeed. Since pollen analysis is so time-consuming it is unlikely that it will ever be generally applicable in these circumstances.

Occupation Deposits

The term "occupation deposit" tends to be applied to any particularly dark coloured layer which does not form part of a soil profile or a pit fill. The assumption is that it is formed by accumulation of domestic rubbish, dust, ash and soot in situations where the floor or living area is not swept. Atkinson (1957) points out the danger of interpreting as an occupation layer the soil which is deposited on top of a floor by worms after the building ceases to be in use. Where a dark layer does appear to have accumulated during occupation of a building over a laid floor, it may be assumed that the floor was covered by matting, rushes, bracken or some other material which gathered rubbish and dust and itself became humified. Where there is no laid floor, or where the building was used for some purpose in which the hardness, regularity, cleanliness or decorative value of a floor was of no signifi-cance, it is not difficult to envisage an accumulation persisting. One only has to look at the floor of a barn or outhouse, whether or not any laid floor does exist, to see just such an occupation deposit forming. If the floor was of trampled earth in the first place, even regular sweeping may fail to prevent an occupation deposit growing, since compaction of newly dropped material onto the surface occurs so easily, particularly in damp conditions.

The organic component of an occupation deposit is usually com-pletely humified. If accumulation is slow, residues at the surface become humified before they are buried in the compacted mass where oxygen may be in short supply. The deposit therefore consists of a very high concentration of amorphous humus, which gives it a black, waxy character like that of a buried turf line, together with charred residues and mineral material. The high level of organic activity, giving rapid humification, is maintained with the help of mineral nutrients derived from ash and from the fine mineral particles of dust. Mineral dust may be derived from the fresh rock surfaces represented by walls, and from activities such as sharpening of implements, often using a stone whose minerals are fresh and of high base content, as well as that which is derived from surrounding soils and blown in, trampled in and carried in on clothing and the hair of animals. Fresh mineral material and ash, and calcium carbonate from mortar or plaster used in construction, maintain a high pH, helping to prevent acid conditions developing in decaying organic residues.

Study of residues in an occupation layer is limited by the poor chance of preservation of organic materials, unless the situation has become waterlogged soon after deposition ceased. As in the case of other humic deposits, the humus may be dispersed with the help of alkali or water softener to see if there are any recognizable residues, and to see if charred grain and other small items such as bone fragments or rodent bones are present. This procedure will also indicate, by the colour of the solution and the blackness of the residue, how much of the dark colour is due to humus substances and how much to fine carbon derived from soot, ash, and broken up charcoal. Occupation layers occurring on different parts of a site or at different levels may be compared by this method, which needs no equipment beyond a number of test tubes, glass beakers, or ordinary domestic tumblers, and a spoon to measure approximately equal quantities of the deposits. A word of warning, though: the humic solution looks very much like black coffee, so care is needed if domestic glassware is used, particularly if sodium hydroxide is used rather than water softener. Once the humus is removed the mineral component of the deposit also becomes clearly visible, and this can be examined for colour, grain size, and identifiable components, giving clues as to its sources and a further means of comparing different deposits.

Pollen and spores may survive in occupation deposits, though not if the pH has been kept high by a large quantity of ash or residues from mortar. As in the case of accumulations in pits and ditches, pollen analysis may not provide results which can be interpreted in any meaningful way. However, if it is thought that the deposit may have incorporated large quantities of a material such as bracken, used for bedding or strewn on the floor, a high concentration of spores might confirm this. Bracken spores are more resistant to destruction than most pollen grains, and may survive in deposits where all pollen has gone. The high polyphenol content of bracken has been suggested as a reason for exceptional conditions of preservation in deposits which have not been waterlogged, the polyphenols tanning proteins and the tanned protein protecting other components of the tissues with which it is associated.

Iron pans frequently form within and below occupation deposits. Their location depends on variations in proportions of mineral and organic materials, the nature of the materials, the degree of compaction and the movement of water through the deposit, and a deposit may be interleaved with iron pans which give the impression of a series of definite floor levels. While a pan may form at a particularly compacted layer or a level at which change in material or rate of deposition occurred, it does not necessarily imply that that level was maintained as a surface. In materials such as beaten earth, clay or mortar forming the floor of a living area or a bedding for a laid floor, a series of iron pans or zones of iron staining may give the impression that the material was put down as a series of layers, whereas it may only be the effect of a number of changes in condition of wetness and mobility of iron, bringing it down to different depths. Again, layers of different texture or boundaries between layers may be picked out in this way, but the phenomenon also occurs in perfectly uniform materials.

Much of this discussion of occupation deposits applies equally to humic layers in pit fills and deposits in gullies and drainage channels associated with living areas. The components are derived from the same domestic and external sources and the conditions of preservation are similar, though with a greater chance of better preservation in pits because of waterlogging.

Chemical characterization of humic deposits is not usually very informative. High phosphate concentrations are to be expected in situations in which organic matter has decayed, and do not necessarily indicate a high contribution from excreta. The variations in mobility of phosphate according to conditions of acidity and presence of complexing substances makes interpretation of phosphate levels impossible without other detailed chemical data and a survey of the general level of phosphate content in the surrounding soils. The nitrogenous compounds and mineral salts in urine are all so soluble that one would not expect to be able to characterize the contents of a pit by this means. High concentrations of phosphate, nitrogen and mineral salts contribute to the environment of decay of the excreta and of other rubbish deposited in byres, dung heaps and guardrobe pits or latrines, but do not necessarily remain at high levels in the resulting humified deposit. Provan (1971) gives a very quick and simple procedure by which the phosphate content of a soil or sediment can be estimated, and a rough phosphate survey of a site or comparison of samples from different deposits is easily done. However, the results usually do not add much to the information which can be obtained by visual inspection.

Soil Accumulations

It often seems difficult to account for the great depth of soil overlying archaeological levels, particularly in town sites, even on hill tops. The soil is built up by the collapse and weathering of the materials of the buildings themselves plus all the organic and mineral rubbish which accumulates at any living site and the soil which arrives by way of wind, muddy feet, clothing and implements, and anything which comes home dirty and goes out cleaner. If one considers the constant stream of building materials, goods and chattels, food supplies and raw materials for manufacture coming into a settlement and the small amount which goes away again, particularly if dung and rubbish remain within the settlement, in pits or heaps or spread over gardens, the accumulation is less surprising. In hill top situations, so long as there are any walls and obstructions limiting the amount of material being moved down hill by surface washing, creep and erosion, much of the material which arrives stays there.

Plough Accumulations

Plough accumulations forming the tertiary fill of ditches have already been described. Soil accumulates in hollows and against obstructions as a result of ploughing, and layers of plough soil are a common feature of archaeological stratification, forming the immediate substratum of sites, burying them, formed by ploughing the soil which formed over them and from their materials and deposits, and in intermediate positions indicating cultivation between periods of use or occupation.

A characteristic of plough soil is the random orientation and uniform distribution of stones of all sizes, provided, of course, that the parent materials or archaeological features are stony or yield stones to soil formation or attack by the plough. Accumulation by downslope movement under the plough implies thinning of the soil upslope, and the plough will eventually bite into the **B/C** or **C** horizon. When this happens the plough soil will become increasingly stony.

Unless manure or other organic residues are regularly applied in large quantities or long fallows allowed between short periods of ploughing the soil will become less humic. Resumed ploughing will destroy the worm sorting achieved under fallow or pasture and incorporate the humic soil into the lower part of the new accumulation. Where it is manured, the soil may continue to be dark in colour as the accumulation grows. As in any buried soil, the activity of micro-organisms can result in redistribution of iron. An effect that I have recently observed, in Orkney, in a soil which appears to be an accumulation resulting from ploughing a soil so poor that it would not be worth cultivating without manuring, is a distinctive fine mottling in a dark matrix. If a poor, acid soil were manured, the manure ploughed in, and the soil buried by further accumulation before humification of the residues was complete, intense micro-organism activity in an otherwise inactive soil would be centred on the particles of raw humus provided by the manure, and just such a mottling might result (observations in association with excavation by P. S. Gelling, at Skaill, Deerness, Orkney).

Lynchets

Any obstruction to soil moving downhill under the plough causes an accumulation on its uphill side and a cutting down on its downhill side. A field boundary running across slope achieves this effect, even if it is only an unploughed headland. Bowen (1961) and Fowler and Evans (1967) discuss the formation of lynchets in detail. Evans, in the above work and in Evans (1972a), discusses the way in which information from the soils buried beneath lynchets and accumulating as they form can be amplified by study of their molluscan fauna.

Plough Marks

Plough marks may be observed where the depth of ploughing is greater than the depth of the **A** horizon, so that the plough bites into an **E, B, B/C** or **C** horizon of sufficiently different colour for the humic soil in the furrows to stand out in contrast to the ridges of undisturbed soil between them. In excavation the pattern of plough marks appears first as discontinuous bands of paler soil or soft rock, and as trowelling proceeds this "negative" effect is gradually developed into a "positive" of dark soil filling furrows. When plough marks are suspected, it is a good idea to investigate them in a section which cuts across several before trowelling down any further, so that their depth can be seen and the best level at which to stop trowelling so as to reveal as many in plan at the same time as is possible is established. In section it may be observed that the grooves produced by the plough are asymmetrical, indicating that the share was set, or the plough held, at an angle. With an ard type of plough this would help to loosen and mix the soil and achieve some undercutting and turning action, effects produced by the share and mould board of more advanced ploughs. Sections do not always reveal a clean cut, and mixing in the bottom of a ragged furrow can sometimes be seen. Mixing of **A** horizon and underlying soil is unlikely to be fully achieved in any case: Fig. 36 shows how mixed the soil lying in the furrows is.

Sections below plough marks may reveal effects of ploughing in the physically undisturbed soil below. In Fig. 36 can be seen lines of clay in the sandy eluvial horizon of the soil. It is thought that these were produced by downwards movement of clay dispersed in the plough

Figure 36. Plough marks in a podzol profile. This is a soil buried by plough accumulations near to the site shown in Figure 19. An area of plough marks was exposed and then sectioned, showing the shape of the furrows and the mixing of humic and leached soil in them. Below the plough soil, lines of clay can be seen, showing movement of clay out of the plough soil into the previously formed podzol eluvial horizon.

soil under conditions of low base concentration and heavy leaching of the exposed soil. Romans reports clay eluviation detectable microscopically, in thin section studies in a buried cultivated soil of Neolithic period (Romans *et al.*, 1973), and the development of a form of illuvial clay horizon known as a "plough pan" is widely known under present-day conditions. Iron pans may form either as linings to the plough furrows, or at a little depth below them, and these too will only be detected if the area of plough marks is sectioned.

It is often asked why plough marks are detectable at all, since repeated ploughing of the same area should result in establishment of an ungrooved lower boundary to the plough soil. This will only be so when the land that is ploughed is flat. Where there is any slope at all, ploughing will move soil down it, and against a lower field boundary or in a hollow the plough marks will be those of only the earliest episodes of ploughing, before the accumulation exceeded plough depth. On convexities and towards the upper boundary of a field the marks will be those of the latest episodes, the plough cutting further into the **E, B, B/C** or **C** horizon each time. Faint traces of plough marks just appearing or just disappearing are probably often overlooked, since it is not apparent that this is what the patches appearing in the trowelled surface are until a regula rstriation is established, and the common occurrence is of two or three sets only. Two sets at right angles or slightly oblique are generally taken to be the result of one season of cross ploughing.

16

Reclaimed and Man Made Soils

Many techniques have been developed for the modification and improvement of existing soils and for the creation of soils from materials which are unproductive so far as man is concerned, because of their situation or their chemical or biological characteristics. These techniques merge with those for maintenance of fertility, drainage or stability against erosion of less greatly modified soils, so it is impossible to define a boundary between the natural and the man made, but certain of them have been so important in the provision of areas of productive land where there was little or none that the archaeologist needs to be aware of them in considering the economic potential of a landscape in earlier periods.

Plaggen Soils

Soils whose depth of humic horizon and productivity have been greatly increased by the importation of material are known as *plaggen soils*. The best known, and eponymous, examples are those of North Germany, Holland and Belgium, where such depth of soil has been built up that some mediaeval towns appear to stand on small hills. The soils occur in areas of severely podzolized lowland heath. The thick mor humus was stripped from the heathland soils and carted into the town or village to be used as litter for stalled animals. The mixture of litter and dung was then used to manure the gardens and nearby fields. Whereas the humus in its natural position was part of a very infertile soil, once the rich source of nitrogen and phosphate in the dung was added to it, micro-organisms could break it down and release the nutrients contained in it.

The creation of these productive soils involves the concentration into the area in which crops are to be grown of nutrients derived from a much wider area, by carting the litter and by feeding the animals on carted fodder in winter and driving them in at night from widely dispersed grazing grounds in summer. Creation of plaggen soils depends on there being some source of nutrients available in transportable form at a distance which makes such transport economically feasible. The most abundant source of nutrients is the sea coast, and plaggen soils have been made using seaweed or shell sand in many coastal areas. Conry (1971) describes those of relatively recent origin in Ireland. Where sand has been used, the insoluble, non-shell component can be identified in the soil to attest its use, but where seaweed has been used there may be no identifiable residue. Seaweed has been used in the creation of soils on bare rock surfaces in the Aran Islands; no one who has seen the classic documentary film "Man of Aran" will forget the spectacle of the exhausting labour of grovelling in the grykes of the dissected rock pavement for handfuls of mineral dust and carting seaweed up from the shore in baskets, to make a soil to grow potatoes in.

Reclamation of Heathland and Moorland Soils

A counterpart of the use of mor humus either for litter, and thus for the creation of plaggen soils, or for fuel, is the reclamation of the stripped soil. Once the thickness of the organic layer is sufficiently reduced for ploughing to reach into the mineral soil, ploughing in the remaining humus together with added base-rich material produces a soil which can either continue in cultivation or be used for a better pasture than that supported by the former soil. Often, the organic layer is stripped right down to its base, but the little-humified litter of the surface layer is first taken off and it is this which is ploughed in. Close enough for carting to the lowland heaths, distributed as they are according to characteristics of parent materials rather than a regional climatic factor, there are materials which can be used to provide the bases, clay, and weatherable minerals which were lost as the soils became degraded. Liming with chalk, ground limestone, or lime made by burning these to produce a more rapidly soluble product, or marling with calcareous clays have long been practised. Around the fringes of the lowland heaths are soils whose profiles give evidence that they are ploughed, reclaimed podzols, and chalkpits, marlpits and lime kilns are common features of the landscape. Evidence from settlement patterns and place names can help to establish the former extent of heathland.

Moorlands of the highland zone present more difficult problems. They cover vast areas, prohibiting the carting of base-rich materials to more than a small proportion, and the amount of materials needed, for the initial reclamation and for maintenance of the improved soil under use, increases as rainfall increases: the higher the mountain, the further away will usually be the source of supply, the more will be needed and the more difficult will be the carting. Gleyed profiles need disruption of a clayey layer or an iron pan and extensive drainage to render them cultivable, and the commonly very clayey texture of the gleyed horizon makes cultivation difficult. Near to the coast, seaweed and shell sand can be used to provide bases, and sand helps to lighten the texture of clayey soils. Improved podzols in coastal areas form a continuum with plaggen soils.

The islands of Orkney have large areas of moorland near enough to the coast and low enough in elevation for carting to have made reclamation possible. While hand labour and the pony and pannier, were the means, seaweed was used; since mechanical loaders and lorries came into use it has been easier to transport shell sand from the fairly extensive areas of fixed dunes, the lings, and when much of this has been used, from the beaches. The Orkney landscape demonstrates the history of recent reclamation of podzols which have existed at least since the Bronze Age, presenting a patchwork of green pasture and dark moorland sharply separated along boundaries established when the rough grazing held in common was divided in the eighteenth century. The soil profiles are as sharply differentiated, the pastures having deep, brown, well structured mull humus, with plenty of earthworms, in place of the black mor humus of the moorland podzol. Reclamation in the first instance would in some cases have been for the growing of barley during periods when its price was high, and some of these enclosures have been abandoned and gradually reverted to moorland.

A precondition of reclamation of moorland is that the thickness of the peaty mor layer should have been so reduced by cutting for fuel or for litter that ploughing can reach into the mineral soil. The cycle whose downswing involved loss of plant nutrients to groundwater and eventually the sea, partly as a result of destruction of forest, can be started on its upswing when lack of wood for fuel leads people to strip off the peat which the poverty of nutrients caused to form; return of nutrients from the sea serves to complete the cycle, in which fire has played a prominent part, and in which much labour has been expended simply in moving material from place to place.

Soil regeneration without the labour input can be achieved by natural processes, but a long time is needed. Dimbleby (1952, 1953) has shown now natural recolonization of moorland by birch trees, if it is allowed to take place without the interference of grazing animals and fire, can bring about soil improvement, even to the extent of dissolution of iron pan, giving access to nutrients in the **B** horizon which have been cut off from the vegetation since, in some areas, Neolithic times. Allowing this to happen requires the sacrifice of even the very low grade grazing that the moorlands provide, and, of course, cessation of deliberate burning for purposes of game "preservation" as well as protection from accidental fire; some labour input for provision and maintenance of fencing and vigilance against fire, is therefore required. The improved soils could be used either for pasture, under careful control of grazing pressure, or, on a very long term basis, for the production of hard wood timber. The use of moorland for the much more rapid production of softwood timber and pulp does nothing for its soils. Growth of conifers is made more productive by using very powerful machines to disrupt iron pans and gleyed layers and cut drainage channels, and artificial fertilizers are used at the nursery stage, but no reversal of podzolization is achieved : the nutrients of the **B** horizon are exposed to further loss once the iron pan or gleyed layer is penetrated, by leaching and by removal in the crop, since the conifers do not achieve so deep and so vigorous a base cycle as do many hard wood trees, and do not induce formation of mull humus, with its stable structure and high base retaining capacity. Many of the conifers used in forestry have a shallow, spreading root system, adapted to utilizing the nutrients available above a water-logged or frozen layer, and they suffer badly from wind rock. When trees fall, the disruption of the surface exposes the soil to erosion.

Reclamation of Bog

Cutting peat for fuel is often the first process in the creation of productive soils on raised bog and blanket peat. The drainage of part of the bog surface brought about by making peat cuttings lets air in and allows micro-organisms to break down the peat, a process usually referred to as oxidation, obscuring the fact that it is an organic process, not chemical oxidation carried out by the air alone. The process is the same as that which produced the darker, more humified peat during periods when changes in the water regime of the bog, from climatic or other causes, allowed a greater degree of humification during accumulation, and at times perhaps periods of complete mineralization of already formed peat, after which fast accumulation of less humified peat was resumed, creating a *recurrence horizon*.

Colonization of the surface by plants which need the greater aeration around their roots and the nutrients released during humification is the first stage in a process of upgrading, producing low grade moorland grazing in place of a bog surface onto which animals could only venture in very dry summer periods, and which even then only provided plants of very low nutrient content. The availability of grazing on bog surfaces during the periods when the highly humified bands in the peat were forming may be an important factor in assessing the economic potential of a landscape at those times.

Blanket peat is usually cut in strips and in a number of stages, depending on its depth and on the amount of water which may still be held above an iron pan or clayey layer in the mineral soil if the cuttings themselves do not lead it off or if drainage channels are not cut. Often, the tracks giving access to the peat cutting areas serve as drainage channels. These tracks can become badly eroded, and tenants' agreements for peat cutting often include provision for their maintenance. Once the cutting is established the top layer of peat, little humified and thus not compact enough for most fuel needs, and uneconomic to transport, is generally put into the bottom of the finished cutting, and thus provides some protection against erosion of the mineral soil and an organic component for the new soil which, provided drainage is maintained, can begin to form there. Once an area is completely stripped, this soil can be improved and maintained by ploughing and importation of base-rich materials, as in the case of the moorland podzols. However, where the peat grew on an area which was already stony, because of a previous history of soil erosion, or if erosion is allowed to take place before soil improvement begins, it may be impossible to create a usable soil without a great deal of stone shifting. It can be easier to shift the soil instead. For example, in Shetland, until recently crops were grown only in small, securely walled plots, and people can still point to areas of extremely stony moorland, cut over for peat long ago, from which soil was gathered and taken to the "plantie krüs" within living memory.

Raised bog growing on a former estuarine area or lake basin can be reclaimed by drainage, in conjunction with peat cutting, but since the situation is likely to be inherently wet, drainage works are needed on a large scale. A small basin bog might be tackled by a single farmer, but larger bogs need a programme of work involving a community occupying a large area and need continuity of the necessary organization for as long as the land is to continue in use. Once drainage is achieved, if peat cutting has reached underlying mineral sediments fine-grained, base-rich material is already present and agricultural potential is high, once soil processes have overcome effects of gleying and soil structure has developed. Where there is still a depth of peat too great for the plough to penetrate and for plant roots to retrieve nutrients from below, marling with base-rich sedimentary material, which is likely to be available nearby, can be carried out to form a new soil.

Fen peat is less of a problem than raised bog peat since it is already rich in bases, and drainage alone may be sufficient to realize its agricultural potential. Again, small, valley bottom fens might be amenable to individual enterprise, but large areas, such as the fenland of East Anglia, require a very large labour input and continuity of organization in space and time to create and maintain the embankment and drainage system. Where the fens cannot be drained to a lower base level, pumping is required. Large areas of flat land whose history has not allowed the growth of hedges suffer badly from wind erosion when cultivated. In the fens of East Anglia, the dark peaty layer forming the plough soil is sometimes stripped off by wind, leaving the mineral sediments underneath, which are rich in bases but have not developed their own soil structure to stabilize them against further erosion. Continual mineralization of the peat once land is drained and aerated, accelerated as it is used and cropped, lowers the land level of reclaimed fen, adding increased danger of flooding to that of wind erosion. The very high value of reclaimed fen land, however, clearly justifies the initial work and the continuing battle with the problems inherent in its maintenance.

Reclamation of Estuarine Land

Agricultural use of estuarine marshes and the floodland in the tidal reaches of rivers also requires large scale works of drainage and embankment, and their continued upkeep. Reclamation of saltmarsh is a way of speeding up the natural process by which continual deposition of sediment aided by the growth of salt tolerant plants which serve to trap it, raises the level of an area until it is eventually only rarely overwhelmed, plants which need dryer conditions can become established, and as rain leaches away the salt, plants of lower salt tolerance and greater palatability create an area of pasture. Instead of using only small areas as they become available, large areas are embanked to keep out the high tide, and leaching and plant colonization can rapidly follow the same course. The development of a good soil can be accelerated by liming, the calcium displacing sodium from the exchange system and causing flocculation of the clay, so that soil structure can develop. Unless irrigation works are installed to bring salt-free water from above the tidal limit of the river, salt marsh can only be reclaimed where rainfall is high enough for natural leaching to be adequate. Where this is so, once the soils are well established occasional flooding can be deliberately allowed, to bring in a fresh load of sediment with its abundant plant nutrients. Utilization of reclaimed estuarine land is yet another example of the exploitation in another part of the landscape, and with the input of much labour, of the materials bearing plant nutrients which have come down from higher land partly as a direct result of agriculture up there. In mountainous areas, the higher temperatures and lower severity of leaching in the depositional areas may allow the materials to be more productive there than they would have been in their original situation, but the loss of nutrients to the sea can only be made good by utilization of food and fertilizing materials taken from the sea itself.

References

Aristovskaya, T. V. and Zavarin, G. A. (1971). Biochemistry of iron in soil. *In* "Soil Biochemistry" Vol. II. (A. D. McLaren and J. Skujiņš, eds.), Dekker, New York.

Atkinson, R. J. C. (1957). Worms and weathering. *Antiquity* 31, 219–233.

Avery, B. W. (1964). "The Soils and Land Use of the District Around Aylesbury and Hemel Hempstead," *Mem. Soil Surv. Gt. Br.*, Agricultural Research Council, London.

Avery, B. W. (1972). Remarks made in presentation of Avery (1973) to the British Society of Soil Science.

Avery, B. W. (1973). Soil classification in the soil survey of England and Wales. *J. Soil Sci.* 24, 324–338.

Avery, B. W., Stephen, I., Brown, G. and Yaalon, D. H. (1959). The origin and development of brown earths on clay-with-flints and coombe deposits. *J. Soil Sci.* 10, 177–195.

Ball, D. F. (1966). Brown podzolic soils and their status in Britain. *J. Soil Sci.* 17, 148–158.

Ball, D. F. and Williams, W. M. (1968). Variability of soil chemical properties in two uncultivated brown earths. *J. Soil Sci.* 19, 379–399.

Barfield, L. H. (1971). "Northern Italy Before Rome." Ancient Peoples and Places, No. 76. Thames and Hudson, London.

Bartelli, L. J. and Odell, R. T. (1960). Field studies, laboratory studies and genesis of a clay-enriched horizon in the lowest part of the solum of some brurizem and gray-brown podzolic soils in Illinois. *Proc. Soil Sci. Soc. Am.* 24, 388–395.

Beckmann, G. G., Thompson, C. H. and Hubble, G. D. (1974). Genesis of red and black soils on basalt on the Darling Downs, Queensland, Australia. *J. Soil Sci.* 25, 265–281.

Benson, D. G. and Miles, D. (1974). "The Upper Thames. An Archaeological Survey of the River Gravels." Oxford Archaeological Unit.

Bernard, E. A. (1962a). Interpretation astronomique des pluviaux et inter-pluviaux du Quaternaire africain. *Actes, 4-ème Congr. de prehist. et de l'etude de Quaternaire*, 1959, 67–95.

Bernard, E. A. (1962b). Théorie astronomique des pluviaux et interpluviaux du Quaternaire africain. *Mem. Acad. r. Sci. d'outre-mer, Cl. Sci. nat et med.*, Nouv. ser. 12, no. 1.

Biberson, P. (1962). L'evolution du paléolithique Marocain dans le cadre du Pleistocène Atlantique. *Quaternaria* 6, 177–205.

Biberson, P. (1963). Human evolution in Morocco in the framework of the palaeoclimatic variations of the Atlantic Pleistocene. *In* "African Ecology and Human Evolution" (F. C. Howell and F. Bourliere, eds.), Viking Fund Publ. in Anthrop. No. 36.

Biberson, P. (1970). Index cards on the marine and continental cycles of the Moroccan Quaternary. *Quaternaria* 13, 1–76.

Bintliff, J. L. (1975). Sediments and settlement in Southern Greece. *In* "Geoarchaeology: Earth Science and the Past" (D. A. Davidson and M. L. Shackley, eds.), Duckworth, London.

Bloomfield, C. (1953). A study of podzolisation. Part I: The mobilisation of iron and aluminium by Scots Pine needles. Part II: the mobilisation of iron and aluminium by leaves and bark of *Agathis australis* (Kauri). *J. Soil Sci.* 4, 5–23.

Bloomfield, C. (1954). A study of Podzolisation. Part III: The mobilisation of iron and aluminium by Rimu (*Dacrydium cupressinum*). Part IV: The mobilisation of iron and aluminium by picked and fallen Larch needles. Part V: The mobilisation of iron and aluminium by Aspen and Ash leaves. *J. Soil Sci.* 5, 39–56.

Bloomfield, C. (1957). The possible significance of polyphenols in soil formation. *J. Sci. Fd Agric.* 8, 389–392.

Bocks, S. M., Brown, B. R. and Handley, W. R. C. (1963). The action of enzymes on plant polyphenols, Part I. *Rep. Forest Res., Lond.*, 1962, 93–96.

Bocks, S. M., Brown, B. R. and Handley, W. R. C. (1964). The action of enzymes on plant polyphenols, Part II. *Rep. Forest Res., Lond.*, 1963, 88–94.

Bonifay, E. (1962). Quaternaire et prehistoire des regions mediterraniennes françaises. *Quaternaria* 6, 343–370.

Bowen, H. C. (1961). "Ancient Fields." British Association for the Advancement of Science.

Bradley, R. (1972). Prehistorians and pastoralists in Neolithic and Bronze Age England. *World Archaeology* 4, 192–204.

Briner, G. P. (1963). Survey of clay minerals in some Victorian soils. *Proc. R. Soc. Vict.* 77, 191–195.

Brongers, J. A. (1962–1963). A chemical method for staining planes and profiles in an archaeological excavation. *Ber. Rijksdienst. oudheick. Bodemonderz.* 590–   .

Brown, B. R. and Love, C. W. (1961). Protein-fixing constituents of plants, Part II. *Rep. Forest Res., Lond.*, 1960, 102–106.

Brown, B. R., Love, C. W. and Handley, W. R. C. (1963). Protein-fixing constituents of plants, Part III. *Rep. Forest Res., Lond.*, 1962, 90–93.

Buckland, P. C., Greig, J. R. A. and Kenward, H. K. (1974). York: an early Medieval site. *Antiquity* 48, 25–33.

Burges, A. and Raw, F. (eds.) (1967). "Soil Biology." Academic Press, London.

Burges, N. A., Hurst, H. M. and Walkden, B. (1964). The phenolic constituents of humic acid and their relation to the lignin of plant cover. *Geochim. Cosmochim. Acta* 28, 1547–1554.

Buringh, P. (1960). "Soils and Soil Conditions in Iraq and Soil Conditions in Bagdad."

Butler, B. E. (1959). Periodic phenomena in landscape as a basis for soil studies. *C.S.I.R.O., Australia Soil Pub. No. 14.*

Butzer, K. W. (1963). Climatic-geomorphologic interpretation of Pleistocene sediments in the Eurafrican subtropics. *In* "African Ecology and Human Evolution" (F. C. Howell and F. Bourliere, eds.). Viking Fund Publ. in Anthrop. No. 36.

Butzer, K. W. and Cuerda, J. (1962). Coastal stratigraphy of southern Mallorca and its implications for the Pleistocene chronology of the Mediterranean Sea. *J. Geol.* **70**, 398–416.

Campbell, C. A., Paul, E. A., Rennie, D. A. and McCallum, K. J. (1967). Applicability of the carbon-dating method to soil humus studies. *Soil Sci.* **104**, 217–224.

Catt, J. A., Weir, A. H. and Bullock, P. (1970). Beta horizons in Hertfordshire soils. *Rothamsted Experimental Station Report for 1970.*

Catt, J. A., Corbett, W. M., Hodge, C. A. H., Madgett, P. A., Tatler, W. and Weir, A. H. (1971). Loess in the soils of north Norfolk. *J. Soil Sci.* **22**, 444–452.

Childs, E. C. (1969). "An Introduction to the Physical Basis of Soil Water Phenomena." Wiley, London.

Choni, L. I. (1971). Change in the structure of chernozems under the effect of forest plantations (as exemplified in the Maruipol Forest Experimental Station). *Pochvovdenie* No. 12, 37–41. (English abstract in *Soviet Soil Sci.,* 1971, p. 761).

Choubert, G. (1962). Reflexions sur les parallelisms probable des formations quaternaires atlantiques du Maroc avec celles de la Mediterranée. *Quaternaria* **6**, 137–175.

Choubert, G., Joly, F., Gigout, M., Marçais, J., Margat, J. and Raynal, R. (1956). Essai de classification du Quaternaire continental du Maroc. *C.r. hebd. Séanc. Acad. Sci., Paris,* **243**, 504–506.

Conry, M. J. (1971). Irish plaggen soils, their distribution, origin and properties. *J. Soil Sci.* **22**, 401–416.

Conway, V. M. (1947). Ringinglow Bog, near Sheffield. Part I: Historical. *J. Ecol.* **34**, 149–181.

Conway, V. M. (1954). Stratigraphy and pollen analysis of southern Pennine blanket peats. *J. Ecol.* **42**, 117–147.

Corbett, J. R. (1968). The genesis of some basaltic soils in New South Wales. *J. Soil Sci.* **19**, 174–85.

Cornwall, I. W. (1958). "Soils for the Archaeologist." Phoenix House Ltd., London.

Cornwall, I. W. (1968). Outline of a stratigraphic "bridge" between the Mexico and Puebla basins. *Bull. Inst. Archaeol., London,* No. 7, 89–140.

Coulson, C. B., Davies, R. I. and Lewis, D. A. (1960). Polyphenols in plant, humus and soil. I: Polyphenols of leaves, litter and humus from mull and mor sites. II: Reduction and transport by polyphenols of iron in model soil columns. *J. Soil Sci.* **11**, 20–44.

Crabtree, K. (1971). Overton Down experimental earthwork, Wiltshire 1968. *Proc. Univ. Bristol Spelaeol. Soc.* **12**, 237–244.

Dakiris, S. I., Higgs, E. S. and Hey, R. W. (1964). The climate, environment and industries of Stone Age Greece: Part I. *Proc. prehist. Soc.* **30**, 199–244.

Dan, J. and Yaalon, D. H. (1971). On the origin and nature of palaeopedological formations in the coastal desert fringe areas of Israel. *In* "Palaeopedology" (D. H. Yaalon, ed.), Int. Soc. Soil Sci. and Israel University Press, Jerusalem.

Darwin, C. (1881). "The Formation of Vegetable Mould Through the Action of Worms with Observations on their Habits" (republished by Faber and Faber Ltd., London, 1945, under the title "Darwin on Humus and the Earthworm," with an introduction by Sir Albert Howard).

Davies, R. I., Coulson, C. B. and Lewis, D. A. (1964). Polyphenols in plant, humus and soil. III: Stabilisation of gelatin by polyphenol tanning. IV: Factors leading to increase in biosynthesis of polyphenol in leaves and their relationship to mull and mor formation. *J. Soil Sci.* **15**, 299–318.

Dekker, L. W. and De Weerd, M. D. (1973). The value of soil survey for archaeology. *Geoderma* **10**, 169–178.

Dimbleby, G. W. (1952). Soil Regeneration on the north-east Yorkshire Moors. *J. Ecol.* **40**, 331–341.

Dimbleby, G. W. (1953). Natural regeneration of pine and birch on the heather moors of north-east Yorkshire. *Forestry* **26**, 41–52.

Dimbleby, G. W. (1962). "The Development of British Heathlands and their Soils." Oxford Forestry Memoir No. 23.

Dimbleby, G. W. (1965). Pollen Analysis. *In* P. A. M. Keefe, J. J. Wymer and G. W. Dimbleby. A Mesolithic Site on Iping Common, Sussex, England. *Proc. prehist. Soc.* **31**, 85–92.

Dimbleby, G. W. and Evans, J. G. (1974). Pollen and land-snail analysis of calcareous soils. *J. Archaeol. Sci.* **1**, 117–133.

Dimbleby, G. W. and Gill, J. M. (1955). The occurrence of podzols under deciduous woodland in the New Forest. *Forestry* **28**, 95–106.

Dobrovol'skiy, G. V., Afanas'yena, T. V., Vasilenko, V. I., Devirts, A. L. and Markova, N. G. (1970). Genesis and age of secondary podzolic soils in West Siberia. *Soviet Soil Sci.* No. 3, 291–293.

Duchaufour, P. (1970). "Précis de Pédologie" (3 ème ed.) Masson, Paris.

Dudal, R. (1963). Dark clay soils of tropical and sub-tropical regions. *Soil Sci.* **95**, 264–70.

Durand, J.-H. (1959). "Les Sols Rouges et les Croutes en Algerie." Service des Études Scientifiques, Alger.

Evans, J. G. (1972a). "Land Snails in Archaeology." Seminar Press, London.

Evans, J. G. (1972b). Ice-wedge casts at Broome Heath, Norfolk. *In* G. J. Wainwright. The Excavation of a Neolithic Settlement on Broome Heath, Ditchingham, Norfolk, England. *Proc. prehist. Soc.* **38**, 1–97.

Evans, J. G. and Limbrey, S. (1974). The Experimental Earthwork on Morden Bog, Wareham, Dorset, England: 1963–1972. Report of the Experimental Earthworks Committee of the British Association for the Advancement of Science. *Proc. Prehist. Soc.* **40**, 170–202.

F.A.O. (1966). "Soil Map of Europe, 1:2,500,000." Food and Agriculture Organisation of the United Nations.

Feeny, P. P. and Bostock, H. (1968). Seasonal changes in the tannin content of oak leaves. *Phytochem.* **7**, 871–880.

Ferguson, J. A. (1954). Transformation of clay minerals in black earths and red loams of basaltic origin. *Aust. J. agric. Res.* **5**, 98–108.

Filip, Z., Haider, K., Beutelspacher, H. and Martin, J. P. (1974). Comparison of I.R.-Spectra from melanins of microscopic soil fungi, humic acids and model phenol polymers. *Geoderma* **11**, 37–52.

Fitzpatrick, E. A. (1956). An indurated soil horizon formed by permafrost. *J. Soil Sci.* **7**, 248–254.

Fitzpatrick, E. A. (1971). "Pedology. A Systematic Approach to Soil Science." Oliver and Boyd, London.

Fleming, A. (1971). Bronze Age agriculture on the marginal lands of north-east Yorkshire. *Agric. Hist. Rev.* **19**, 1–24.

Fowler, P. J. and Evans, J. G. (1967). Plough-marks, lynchets and early fields. *Antiquity* **41**, 289–301.

Fox, C. (1959). "Life and Death in the Bronze Age." Routledge and Kegan Paul, London.

Gayel', A. G. and Malan'in, A. N. (1971). Soils of birch stands on sandy terraces of the Don Steppe. *Soviet Soil Sci.* No. 4, 402–413.

Gigout, M. (1951). "Études geologiques sur la meseta marocaine occidentale," Soc. de Sci. nat. du Maroc, Rabat.

Godwin, H. (1958). Pollen analysis in mineral soil. An interpretation of a podzol pollen analysis by Dr. G. W. Dimbleby. *Flora* 146, 321–327.

Godwin, H. (1962). Vegetational history of the Kentish chalk Downs as seen at Wingham and Frogholt. *Festschr. Franz Firbas., Veroff. geobot. Inst., Zurich* 37, 83–99.

Green, C. P. (1973). Pleistocene river gravels and the Stonehenge problem. *Nature, Lond.* 243, 214–216.

Greig, J. R. A. and Turner, J. (1974). Some pollen diagrams from Greece and their archaeological significance. *J. Archaeol. Sci.* 1, 177–194.

Haider, K. and Martin, J. P. (1970). Humic acid-type phenolic polymers from *Aspergillus sudowi* culture medium *Stachybotrys spp.* cells and auto-oxidised phenol mixtures. *Soil Biol. Biochem.* 2, 145–156.

Hallsworth, E. G. (1951). An interpretation of soils found on basalt in the Richmond–Tweed region of New South Wales. *Aust. J. agric. Res.* 2, 411–428.

Hallsworth, E. G. and Crawford, D. V. (eds) (1965). "Experimental Pedology. Proceedings of the Eleventh Easter School in Agricultural Science, University of Nottingham, 1964." Butterworth, London.

Hallsworth, E. G., Costin, A. B., Gibbons, F. R. and Robertson, G. K. (1952). Studies in pedogenesis in New South Wales. II: The Chocolate Soils. *J. Soil Sci.* 3, 103–124.

Handley, W. R. C. (1954). "Mull and Mor Formation in Relation to Forest Soils." *Bull. For. Commn., Lond.*, No. 23.

Hardan, A. (1971). Archaeological methods for dating soil salinity in the Mesopotamian Plain. *In* "Palaeopedology" (D. H. Yaalon, ed.), Int. Soc. Soil Sci. and Israel University Press, Jerusalem.

Harris, S. A. (1971). Podzol development on volcanic ash deposits in the Talamanea Range, Costa Rica. *In* "Palaeopedology" (D. H. Yaalon, ed.). Int. Soc. Soil Sci. and Israel University Press, Jerusalem.

Hay, R. L. (1968). Chert and its sodium silicate precursors in sodium carbonate lakes in East Africa. *Contr. Mineral. Petrol.* 17, 255–274.

Hay, R. L. (1971). Geological background of Beds I and II, Stratigraphic summary. *In* M. D. Leaky, "Olduvai Gorge", Volume 3. Cambridge University Press, Cambridge.

Hey, R. W. (1962). The Quaternary and Palaeolithic of northern Libya. *Quaternaria* 6, 435–449.

Hey, R. W. (1963). Pleistocene screes in Cyrenaica (Libya). *Eiszeitalter Gegenw.* 14, 77–84.

Hicks, S. P. (1971). Pollen analytical evidence for the effect of prehistoric agriculture on the vegetation of north Derbyshire. *New Phytol.* 70, 647–67.

Hicks, S. P. (1972). The impact of man on the East Moor of Derbyshire from Mesolithic times. *Archaeol. J.* 129, 1–21.

Higgs, E. S. and Vita-Finzi, C. (1966). The climate, environment and industries of Stone Age Greece: Part II. *Proc. prehist. Soc.* 32, 1–29.

Hodgson, J. M. (1967). "Soils of the West Sussex Coastal Plain." Soil Surv. Gt. Br., England and Wales, Bulletin No. 3.

Hodgson, J. M., Catt, J. A. and Weir, A. H. (1967). The origin and development of clay-with-flints and associated soil horizons on the South Downs. *J. Soil Sci.* 18, 85–102.

Int. Soc. Soil Sci. (1967). Proposal for a uniform system of soil horizon nomenclature. *Bull. int. Soc. Soil Sci.* 31, 4–7.

Ivanova, Ye. N., Lobova, Ye. V., Nogina, N. A., Rozov, N. N., Fridland, V. M. and Shuvalov, S. A. (1969). Present status of the doctrine of soil genesis in the U.S.S.R. *Soviet Soil Sci.* No. 3, 265–277.

Ivanova, Ye. N., Lobova, Ye. V., Nogina, N. A., Rozov, N. N., Fridland, V. M. and Shuvalov, S. A. (1970). Development and current status of research on the genesis of soils in the U.S.S.R. *Soviet Soil Sci.* No. 2, 129–144.

Iversen, J. (1964). Retrogressive vegetational succession in the post-glacial. *J. Ecol.* 52, 59–70.

Iversen, J. (1969). Retrogressive development of a forest ecosystem demonstrated by pollen diagrams from fossil mor. *Oikos* 12, 35–49.

Jacks, G. V. (1954). "Multilingual Vocabulary of Soil Science." Food and Agriculture Organisation of the United Nations.

Jackson, B. P. and Snowdon, D. (1974). "Powdered Vegetable Drugs." Thornes, London.

Jarman, M. R. (1972). European deer economies and the advent of the Neolithic. *In* "Papers in Economic Prehistory" (E. S. Higgs, ed.). Cambridge University Press, Cambridge.

Jewell, P. A. and Dimbleby, G. W. (1966). The experimental earthwork on Overton Down, Wiltshire, England: the first four years. *Proc. prehist. Soc.* 32, 313–342.

Jukes-Browne, A. J. (1906). The Clay-with-Flints; its origin and distribution. *Q.J. geol. Soc., Lond.* 62, 132–134.

Kellaway, G. A. (1971). Glaciation and the stones of Stonehenge. *Nature, Lond.* 233, 30–35.

Kerney, M. P., Brown, E. H. and Chandler, T. J. (1964). The Late-glacial and Post-glacial history of the chalk escarpment near Brook, Kent. *Phil. Trans. R. Soc. (B)* 248, 135–204.

Klinge, H. (1965). Podzol soils in the Amazon basin. *J. Soil Sci.* 16, 95–103.

Kononova, M. M. (1964). "Soil Organic Matter, its nature, its role in soil formation and in soil fertility" (2nd English edn). Pergamon Press, London.

Kononova, M. M. and Alexandrova, I. V. (1973). Formation of humic acids during plant residue humification and their nature. *Geoderma* 9, 157–164.

Kubiena, W. L. (1953). "The Soils of Europe." Murby, London.

Kuz'min, V. A. (1969). Organic matter of separates in the sod-podzolic soils with a second humus horizon in the Oka-Angara interfluve. *Soviet Soil Sci.*, No. 3, 278–285.

Lebedeva, I. I. (1969). Nature of the illuvial horizon of light grey forest soils on moraine loams. *Soviet Soil Sci.*, No. 1, 10–19.

Lewis, J. A. and Starkey, R. L. (1969). Decomposition of plant tannins by some soil micro-organisms. *Soil Sci.* 107, 235–241.

Limbrey, S. (1968). "The Basaltic Soils of Gran Canaria: Fossil, Relict and Recent Soils and Their Environmental Significance". Ph.D. Thesis, University of London.

Limbrey, S. (1975). Tlapacoya: problems of interpretation of lake margin sediments at an early occupation site in the Basin of Mexico. *In* "Geoarchaeology: Earth Science and the Past" (D. A. Davison and M. L. Shackley, eds). Duckworth, London.

Lloyd, P. S. (1971). Effects of fire on the chemical status of herbaceous communities of the Derbyshire dales. *J. Ecol.* **59**, 261–273.

Loughnan, F. C. (1969). "Chemical Weathering of Silicate Minerals." Elsevier, Amsterdam.

Loveday, J. (1958). "A Study of the Soils and Their Relation to Landscape Form in the Southern Chilterns." Ph.D. Thesis, University of London.

Loveday, J. (1962). Plateau deposits of the southern Chiltern Hills. *Proc. geol. Ass.* **73**, 83–102.

Low, A. J. (1973). Soil structure and crop yield. *J. Soil Sci.* **24**, 249–259.

Lukashev, K. I., Dromashko, S. G. and Dobrovol'skaya, I. A. (1970). Origin of carbonates in Belorussian loesses. *Soviet Soil Sci.* No. 6, 648–654.

Mackereth, F. J. H. (1965–6). Some chemical observations on Post-glacial lake sediments. *Phil. Trans. R. Soc. (B)* **250**, 165–213.

Manby, T. G. (1971). The Kilham long barrow excavations 1965–1969. *Antiquity* **45**, 50–53.

Marshall, C. E. (1964). "The Physical Chemistry and Mineralogy of Soils. Volume I. Soil Materials." Wiley, London.

Martin, J. P. and Haider, K. (1969). Phenolic substances of *Stachybotrys atra*, *S. chartarum* and *Epicoccum nigrum* in relation to humic acid formation. *Soil Sci.* **107**, 260–270.

McLaren, A. D. and Petersen, G. H. (eds) (1967). "Soil Biochemistry." Arnold, London.

McLaren, A. D. and Skujiņš, J. (eds) (1971). "Soil Biochemistry, Volume 2." Dekker, New York.

Milne, G. (1935). Some suggested units of classification particularly for East African soils. *Soil Res.* **4**, 183–198.

Oldfield, F. (1972). A paper presented to the Spring Conference of the Prehistoric Society, London.

Paepe, R. (1971). Dating and position of fossil soils in the Belgian Pleistocene stratigraphy. *In* "Palaeopedology" (J. Dan and D. H. Yaalon, eds), Int. Soc. Soil Sci. and Israel University Press, Jerusalem.

Paepe, R. and Van Hoorne, R. (1967). The stratigraphy and palaeobotany of the late Pleistocene in Belgium. *Mem. Cartes geol. et min. Belgiques.*

Pearl, I. A. (1967). "The Chemistry of Lignin." Arnold, London.

Pearsall, W. H. (1971). "Mountains and Moorlands" (revised edition, with revisions by Winifred Pennington). Collins/Fontana, London.

Pennington, W. (1969). "The History of British Vegetation." The English Universities Press, London.

Perrin, R. M. S., Davies, H. and Fysh, D. (1974). Distribution of late Pleistocene aeolian deposits in eastern and southern England. *Nature, Lond.* **248**, 320–324.

Pike, A. W. and Biddle, M. (1966). Parasite eggs in medieval Winchester. *Antiquity* **40**, 293–296.

Provan, D. M. J. (1971). Soil phosphate analysis as a tool in archaeology. *Norwegian Archaeol. Rev.* **4**, 37–50.

Rayner, J. H. (1966). Classification of soils by numerical methods. *J. Soil Sci.* **17**, 79–92.

Romans, J. C. C. (1962). The origin of the indurated B_3 horizon of podzolic soils in north-east Scotland. *J. Soil Sci.* **13**, 141–147.

Romans, J. C. C., Durno, S. E. and Robertson, L. (1973). A fossil brown forest soil from Angus. *J. Soil Sci.* **24**, 125–128.

Rudeforth, C. C. (1970). Soils of North Cardiganshire. *Mem. Soil Surv. Gt. Br.*, Agricultural Research Council, Harpenden.

Ruellan, A. (1965). Le rôle des climats et des roches sur le repartition des sols dans les plaines de la Basse Moulouya. *C. r. hebd. Séanc. Acad. Sci. Paris* **261**, 2379–2382.

Russell, E. J. (1957). "The World of the Soil." Collins/Fontana, London.

Russell, E. W. (1971). Soil structure: its maintenance and improvement. *J. Soil Sci.* **22**, 137–151.

Scharpenseel, H. W., Ronzani, C. and Pietig, F. (1968). Comparative age determinations on different humic-matter fractions. *In* "Isotopes and Radiation in Soil Organic-Matter Studies." Proceedings Series, Int. Atomic Energy Agency, Vienna.

Sieveking, G. de G., Longworth, I. H., Hughes, M. J., Clark, A. J. and Millett, A. (1973). A new survey of Grimes Graves—first report. *Proc. prehist. Soc.* **39**, 182–218.

Simmons, I. G. (1964). Pollen diagrams from Dartmoor. *New Phytol.* **63**, 165–180.

Simmons, I. G. (1969). Environment and early man on Dartmoor, Devon, England. *Proc. prehist. Soc.* **35**, 203–219.

Simonett, D. S. (1960). Soil genesis on basalt in north Queensland. *Trans. 7th Int. Congr. Soil Sci.* **4**, 238–243.

Simonett, D. S. and Bauleke, M. P. (1963). Mineralogy of soils on basalt in north Queensland. *Proc. Soil Sci. Soc. Am.* **27**, 205–212.

Singh, S. (1954). A study of the Black Cotton Soils with special reference to their coloration. *J. Soil Sci.* **5**, 289–299.

Singh, S. (1956). The formation of dark coloured clay-organic complexes in black soils. *J. Soil Sci.* **7**, 43–58.

Soil Survey Staff (1960). "Soil Classification, a Comprehensive System. 7th Approximation." U.S. Dept. Agric. Soil Conservation Service.

Soil Survey Staff (1967). "Supplement to Soil Classification System (7th Approximation)." U.S. Dept. Agric. Soil Conservation Service.

Soil Survey Staff (1968). "Supplement to Soil Classification System (7th Approximation)." U.S. Dept. Agric. Soil Conservation Service.

Solé-Sabarîs, L. (1962). La Quaternaire marin des Balearés et ses rapports avec les côtes méditerranéennes de la péninsule Ibérique. *Quaternaria* **6**, 309–342.

Stepanovits, P. (1971). "Brown Forest Soils of Hungary", Akadémiai Kiado, Budapest.

Stephens, C. G. (1971). Laterite and silcrete in Australia: a study of the genetic relationships of laterite and silcrete and their companion materials, and their collective significance in the formation of the weathered mantle, soils, relief and drainage of the Australian continent. *Geoderma* **5**, 5–52.

Stevenson, J. and Goh, K. M. (1971). Infrared spectra of human acids and related substances. *Geochim. Cosmochim. Acta* **35**, 471–483.

Teakle, L. J. H. (1952). An interpretation of the occurrence of diverse types of soils on basalt in northern New South Wales and Queensland. *Aust. J. Agric. Res.* **3**, 391–408.

Tolchel'nikov, Yn. S. (1970). Contribution to the description of the absolute age of the second horizon of sod-podzolic soils in West Siberia. *Soviet Soil Sci.* No. 3, 289–290.

Turner, C. (1970). The Middle Pleistocene deposits at Marks Tey, Essex. *Phil. Trans. R. Soc.* (B) **257**, 373–440.

Turner, J. (1964). The anthropogenic factor in vegetational history. *New Phytol.* **63**, 73–89.

Valentine, K. W. G. (1973). "The Identification, Lateral Variation and Chronology of three Buried Palaeocatenas in Lowland England." Ph.D. Thesis, University of Reading.

Veen, A. W. L., Slager, S. and Jongmans, A. G. (1971). A micromorphological study of four Pleistocene alluvial soils of Surinam, South America. *Geoderma* **6**, 81–100.

Veen, A. W. L. and Maaskant, P. (1971). Electron microprobe analysis of plasma in an impervious horizon of a tropical groundwater podzol. *Geoderma* **6**, 101–107.

Vinogradov, A. V., Memedov, E. D. and Stepanov, T. N. (1969). Ancient soils in the Kyzyl-Kum desert. *Soviet Soil Sci.* No. 5, 525–537.

Vita-Finzi, C. (1967). Late Quaternary alluvial chronology of northern Algeria. *Man* **2**, 205–215.

Vita-Finzi, C. (1969). "The Mediterranean Valleys. Geological Changes in Historical Times." Cambridge University Press, Cambridge.

Wainwright, G. J. (1972). The excavation of a Late Neolithic enclosure at Marden, Wiltshire. *Antiquaries J.* **LI**, 177–239.

Walker, D. (1966). The late Quaternary history of the Cumberland lowland. *Phil. Trans. R. Soc.* (B) **251**, 1–210.

Walker, T. W. (1965). The significance of phosphorus in pedogenesis. *In* "Experimental Pedology" (E. G. Hallsworth and D. V. Crawford, eds). Butterworths, London.

Waterbolk, H. T. (1957). Pollen analytisch onderzoek van twee nordbrabanstse tumuli. *In* G. Beex, Twee grafheuvels in Nord-Brabant. *Bijdr. Studie Brabantse Heem* **9**, 34–39.

Watson, J. P. (1962). The soil below a termite mound. *J. Soil Sci.* **13**, 46–51.

Watson, J. P. (1967). A termite mound in an Iron Age burial ground in Rhodesia. *J. Ecol.* **55**, 663–669.

Webster, R. (1965). An horizon of pea grit in gravel soils. *Nature, Lond.* **206**, 696–697.

Weir, A. H., Catt, J. A. and Madgett, P. A. (1971). Post-glacial soil formation in the loess of Pegwell Bay, Kent (England). *Geoderma* **5**, 131–149.

West, R. G. (1955–56). The Quaternary deposits at Hoxne, Suffolk. *Phil. Trans. R. Soc.* (B) **239**, 265–356.

West, R. G. (1961). Late and Post-glacial vegetational history in Wisconsin, particularly changes associated with the Valders readvance. *Am. J. Sci.* **259**, 766–783.

Williams, R. B. G. (1973). Frost and the works of Man. *Antiquity*, **47**, 19–31.

Yaalon, D. H. (ed.) (1971). "Paleopedology. Origin, Nature and Dating of Paleosols." Symposium on the Age of Parent Materials and Soils, Amsterdam, 1970. Int. Soc. Soil Sci. and Israel University Press, Jerusalem.

Index

Where a number of references are given, definitive or particularly important entries are given in heavy type. References to figures are given in italics

Climate (*cont.*)—
effect on leaching, 59
of Atlantic period, 159
of sub-Atlantic period, 168, 171
Climatic limits on tree growth, 109, 112, 113, 114
Close packing
in crystal structure, 13, 14, 20
of soil peds, 89, 125, 265, 297
Closure of clearings, of forest, 113, 114, 156
Coastal
deposits, 105
plains, lowland, 162, 204
resources of fertilizer, 170, 336, 337
Cobalt, 4
Collagen, 63
Collapse of buildings, 321–322
Collembola, 30, 31, 32, 137
Colloidal materials, 80, 92, 130, 134, 268 (*see also* soil colloid)
orientation of, 187
Colluvial deposits, 91, 154, 170, 182, 205–208, 211
Colluviation, 89–96, 100, 212, 236
Colonization
by micro-organisms, 30, 37
of clearings, 113, 149
Colour B horizon, 80
Colour
changes on drying, 258–259
charts, 254, 256–258
description of, 256–259
in gleyed soils, 82
of brickearth, 180
of burnt soils, 325
of clay-with-flints, 176
of loess, 107
of occupation deposits, 328, 329
of organic muds, 68
of plateau drift, 176
(*see also* descriptions of soil groups)
Columnar structure, 229, *266*, 268
Combustion, 118, 322, 323
Comminuted residues, 48, 78, 128, 138
Comminution of litter, 30, 77, 131, 137
Common land, 150
Compact layers, periglacial, 145, 253

Compaction, 269
of buried soils, 309
Completeness of soil profiles, 259, 264, 267
Composite crystals, 7
Compression of buried soils, 309, 315
Concentration
of artifacts by winnowing, 239
of nitrogen, 122
of phosphate, 72
of plant nutrients, 94, 121
Condensation reactions, 41–44
Condensed structure of humus substances, 44, 45
Condensed tannins, 44, 45
Concretions, 28, 58
Concretionary horizons, 214
Conifers, 43, 103, 109, 146, 337
Coniferous forest, 117, 130, **140–141**, 149, 199
Consistence, 256, **267–268**
Contained information, 233–235
Contamination
of dating samples, 46
of information, 239, 240
Continuity
of vegetation and soil
horizons, 88, 91, 151, 156
of settlement, 169
of thin iron pans, 144, 145, 312
Continuum,
soil properties, 84
particle sizes, 262
Control
of animals by hunters, 120, 156
of vegetation burning, 158, 173
Controlled exploitation of wild animals, 161
Convoluted
boundaries, 184, 270
layers, 103
Convolutions, 101, 151, 176, 178
Coombe deposits, 100, **178–179**
Coombe rock, 178–179, 180
Co-ordination, 5, **12–14**, 16, 22
Copper, 4
Cornish granite moors, 157
Cornwall, 252
Corpses, phosphatic, 72
Corrosion, 70, 71

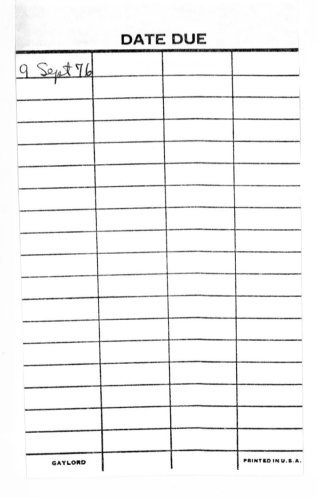

DATE DUE

9 Sept 76			